www.wadsworth.com

www.wadsworth.com is the World Wide Web site for Wadsworth and is your direct source to dozens of online resources.

At *www.wadsworth.com* you can find out about supplements, demonstration software, and student resources. You can also send email to many of our authors and preview new publications and exciting new technologies.

www.wadsworth.com
Changing the way the world learns®

Understanding Society

An Introductory Reader

Second Edition

MARGARET L. ANDERSEN
University of Delaware

KIM A. LOGIO
Saint Joseph's University

HOWARD F. TAYLOR
Princeton University

THOMSON
WADSWORTH

Australia • Canada • Mexico • Singapore • Spain
United Kingdom • United States

THOMSON

WADSWORTH

Sociology Editor: *Robert Jucha*
Assistant Editor: *Stephanie Monzon*
Editorial Assistant: *Melissa Walter*
Technology Project Manager: *Dee Dee Zobian*
Marketing Manager: *Matthew Wright*
Marketing Assistant: *Tara Pierson*
Advertising Project Manager: *Linda Yip*
Project Manager, Editorial Production:
   Cheri Palmer

Print/Media Buyer: *Doreen Suruki*
Permissions Editor: *Kiely Sexton*
Production Service: *Shepherd, Inc.*
Copy Editor: *Jeanne Patterson*
Cover Designer: *Yvo Riezebos*
Cover Image: *Hessam Abrishami*
Compositor: *Shepherd, Inc.*
Printer: *Webcom*

Printed in Canada
   2 3 4 5 6 7 08 07 06 05

For more information about our products,
contact us at:
Thomson Learning Academic Resource Center
1-800-423-0563
For permission to use material from this text
or product, submit a request online at
http://www.thomsonrights.com.
Any additional questions about permissions
can be submitted by email to
thomsonrights@thomson.com.

Library of Congress Control Number:
2004101007

ISBN 0-534-58815-8

Thomson Wadsworth
10 Davis Drive
Belmont, CA 94002-3098
USA

Asia
Thomson Learning
5 Shenton Way #01-01
UIC Building
Singapore 068808

Australia/New Zealand
Thomson Learning
102 Dodds Street
Southbank, Victoria 3006
Australia

Canada
Nelson
1120 Birchmount Road
Toronto, Ontario M1K 5G4
Canada

Europe/Middle East/Africa
Thomson Learning
High Holborn House
50/51 Bedford Row
London WC1R 4LR
United Kingdom

Latin America
Thomson Learning
Seneca, 53
Colonia Polanco
11560 Mexico D.F.
Mexico

Spain/Portugal
Paraninfo
Calle/Magallanes, 25
28015 Madrid, Spain

◆ Contents

PREFACE xv

ABOUT THE EDITORS xxi

PART I
Sociological Perspectives and Sociological Research

1   *The Sociological Imagination   1*
C. Wright Mills

2   *The Forest and the Trees   6*
Allan G. Johnson

3   *Not Our Kind of Girl   13*
Elaine Bell Kaplan

4   *Sense and Nonsense about Surveys   21*
Howard Schuman

   InfoTrac College Edition: Bonus Reading: "Promoting Bad Statistics"   27
Joel Best

   Search Terms   27

PART II

Culture

5 *Body Ritual Among the Nacirema   28*
   Horace Miner

6 *September 11, 2001: Mass Murder and Its Roots in the Symbolism
   of American Consumer Culture   33*
   George Ritzer

7 *Barbie Doll Culture and the American Waistland   39*
   Kamy Cunningham

8 *Buy This 24-Year-Old and Get All His Friends Absolutely Free   43*
   Jean Kilbourne

   *InfoTrac College Edition: Bonus Reading: "Why America
   Loves Reality TV"   50*
   Steven Reiss

   Search Terms   50

PART III

Socialization and the Life Course

9 *Leaving Home for College: Expectations for Selective Reconstruction
   of Self   51*
   David Karp, Lynda Lytle Holmstrom, and Paul S. Gray

10 *Navajo Women and the Politics of Identity   58*
   Amy J. Schulz

11 *Gender and Aging   70*
   Kathleen Slevin and Toni Calasanti

   *InfoTrac College Edition: Bonus Reading: "Footballs versus Barbies:
   Childhood Play Activities as Predictors of Sport Participation
   by Women"   74*
   Traci A. Giuliano, Kathryn E. Popp, and Jennifer L. Knight

   Search Terms   74

PART IV

Society and Social Interaction

12 *The Presentation of Self in Everyday Life   75*
   Erving Goffman

13   *Code of the Street*   *78*
Elijah Anderson

14   *Life Beyond the Screen: Embodiment and Identity
Through the Internet*   *84*
Michael Hardey

*InfoTrac College Edition: Bonus Reading: "Toward a Theory
of Disability and Gender"*   *94*
Thomas J. Gerschick

Search Terms   *94*

PART V

Groups and Organizations

15   *Clique Dynamics*   *95*
Patricia Adler and Peter Adler

16   *Fraternities and Collegiate Rape Culture: Why Are Some Fraternities More
Dangerous Places for Women?*   *102*
A. Ayres Boswell and Joan Z. Spade

17   *Social Networks: The Value of Variety*   *111*
Bonnie Erickson

*InfoTrac College Edition: Bonus Reading: "When
Agencies Sleep"*   *116*

Search Terms   *116*

PART VI

Deviance and Criminal Justice

18   *The Functions of Crime*   *117*
Emile Durkheim

19   *The Medicalization of Deviance*   *120*
Peter Conrad and Joseph W. Schneider

20   *The Rich Get Richer and the Poor Get Prison?*   *126*
Jeffrey H. Reiman

*InfoTrac College Edition: Bonus Reading: "The Social Cost of America's
Race to Incarcerate"*   *131*
Marc Mauer

Search Terms   *131*

PART VII

Social Class and Social Stratification

21   *The Communist Manifesto*   *132*
     Karl Marx and Frederich Engels

22   *Great Divides*   *136*
     Thomas M. Shapiro

23   *Wealth Matters*   *143*
     Dalton M. Conley

24   *Studying the Quagmire of Welfare Reform*   *148*
     Sharon Hays

     *InfoTrac College Edition: Bonus Reading: "Few Good Men: Why
     Poor Women Don't Remarry"*   *154*
     Kathyrn Edin

     Search Terms   *154*

PART VIII

Global Stratification

25   *The Garment Industry in the Restructuring Global Economy*   *155*
     Edna Bonacich, Lucie Cheng, Norma Chinchilla,
     Nora Hamilton, and Paul Ong

26   *The Nanny Chain*   *162*
     Arlie Russell Hochschild

27   *Michael Jordan and the New Global Capitalism*   *170*
     Walter LaFeber

     *InfoTrac College Edition: Bonus Reading: "How
     to Judge Globalism"*   *175*
     Amartya Sen

     Search Terms   *175*

PART IX

Race

28   *The Souls of Black Folk*   *176*
     W. E. B. Du Bois

29   *Seeing More than Black and White*   *178*
     Elizabeth Martinez

30   *Immigrant America: Who They Are and Why They Come*   *184*
     Alejandro Portes and Rubén Rumbaut

31   *Color-Blind Privilege: The Social and Political Functions of Erasing
     the Color Line in Post Race America*   *189*
     Charles A. Gallagher

     *InfoTrac College Edition: Bonus Reading: "How Does Racial/Ethnic
     Diversity Promote Education?"*   *196*
     Patricia Y. Gurin, Eric L. Dey, Gerald Gurin, and Sylvia Hurtado

     Search Terms   *196*

PART X
Gender

32   *The Social Construction of Gender*   *197*
     Margaret L. Andersen

33   *The Politics of Masculinities*   *202*
     Michael A. Messner

34   *Catching Sense: Learning from Our Mothers to Be Black
     and Female*   *208*
     Suzanne C. Carothers

35   *Challenges for Middle Eastern Women*   *218*
     Elizabeth Fernea

     *InfoTrac College Edition: Bonus Reading: "Sexuality in the Workplace:
     Organizational Control, Sexual Harassment, and the
     Pursuit of Pleasure"*   *224*
     Christine L. Williams, Patti A. Giuffre, and Kirsten Dellinger

     Search Terms   *224*

PART XI
Sexuality and Intimate Relationships

36   *Pluralistic Ignorance and Hooking Up*   *225*
     Tracy A. Lambert, Arnold S. Kahn, and Kevin J. Apple

37   *Masculinity as Homophobia*   *230*
     Michael S. Kimmel

38   *The Long Goodbye*   *234*
     Diane Vaughan

InfoTrac College Edition: Bonus Reading: "Perceptions of Homophobia and Heterosexism in Physical Education" 238
Ronald G. Morrow and Diane L. Gill

Search Terms 238

PART XII

Social Institutions
A. Family

39 *Weaving Work and Motherhood* 239
Anita Garey

40 *Family Rituals and the Construction of Reality* 245
Scott Coltrane

41 *Divorce and Remarriage* 253
Terry Arendell

42 *Caring for Our Young: Child Care in Europe and the United States* 263
Dan Clawson and Naomi Gerstel

InfoTrac College Edition: Bonus Reading: "Attitudes toward Violence against Women: A Cross-Nation Study" 271
Madhabika B. Nayak, Christina A. Byrne,
Mutsumi K. Martin, and Anna George Abraham

Search Terms 271

B. Religion

43 *The Protestant Ethic and the Spirit of Capitalism* 272
Max Weber

44 *Abiding Faith* 277
Mark Chaves and Dianne Hagaman

45 *Are American Jews Vanishing Again?* 281
Calvin Goldscheider

InfoTrac College Edition: Bonus Reading: "A Peaceful Faith, a Fanatic Few" 287
Kenneth L. Woodward

Search Terms 287

C. Education

46　*School Girls*　*288*
Peggy Orenstein

47　*Race in American Public Schools: Rapidly Resegregating*
School Districts　*292*
Erica Frankenberg and Chungmei Lee

48　*Racial Desegregation: Magnet Schools, Vouchers, Privatization, and Home*
Schooling　*297*
Loretta F. Meeks, Wendell A. Meeks, and Claudia A. Warren

InfoTrac College Edition: Bonus Reading: "Educational Reforms
and High-Stakes Testing: Are Public Schools Still for the Public?"　*304*
Jim Donlevy

Search Terms　*304*

D. Work

49　*The Service Society and the Changing Experience of Work*　*305*
Cameron Lynne Macdonald and Carmen Sirianni

50　*Toward a 24-Hour Economy*　*313*
Harriet B. Presser

51　*Nickel-and-Dimed: On (Not) Getting By in America*　*317*
Barbara Ehrenreich

InfoTrac College Edition: Bonus Reading: "Captive Labor: America's
Prisoners as Corporate Workforce"　*331*
Gordon Lafer

Search Terms　*331*

E. Government and Politics

52　*The Power Elite*　*332*
C. Wright Mills

53　*Diversity in the Power Elite*　*337*
Richard Zweigenhaft and G. William Domhoff

54　*Forever Seen as New: Latino Participation in American Elections*　*342*
Louis DeSipio and Rodolfo O. de la Garza

InfoTrac College Edition: Bonus Reading: "The Silent Voices: 2000
Presidential Election and the Minority Vote in Florida"　*347*
Revathi Hines

Search Terms　*347*

F. Health Care

55 *Latinos' Access to Employment-Based Health Insurance* *348*
E. Richard Brown and Hongjian Yu

56 *Beauty Myths and Realities and Their Impact
on Women's Health* *353*
Jane Sprague Zones

57 *Death and Social Structure* *361*
Robert Blauner

*Info Trac College Edition: Bonus Reading: "Identifying Male College
Students' Perceived Health Needs, Barriers to Seeking Help,
and Recommendations to Help Men Adopt Healthier Lifestyles"* *365*
Jon Davies et al.

Search Terms *365*

PART XIII

Population, Urbanization, and the Environment

58 *American Apartheid* *366*
Douglas S. Massey and Nancy A. Denton

59 *Black, Brown, Red, and Poisoned* *373*
Regina Austin and Michael Schill

60 *Mobilizing Minority Communities: Social Capital and Participation
in Urban Neighborhoods* *379*
Kent E. Portney and Jeffrey M. Berry

*Info Trac College Edition: Bonus Reading: "The Suburban Transformation
of the Globalizing American City"* *386*
Peter O. Muller

Search Terms *386*

PART XIV

Social Movements and Social Change

61 *Generations X, Y, and Z: Are They Changing America?* *387*
Duane F. Alwin

62 *Jihad vs. McWorld* *393*
Benjamin R. Barber

63   *The Genius of the Civil Rights Movement: Can It Happen Again?*   *397*
Aldon Morris

*InfoTrac College Edition: Bonus Reading: "How Biotechnology Is
Transforming What We Believe In and How We Live"*   *404*
Fred Edwords

Search Terms   *404*

GLOSSARY   405

INDEX   413

✦
Preface

"If you really acquire the sociological perspective, you can never be bored," writes June Jordan, a contemporary African American essayist. We agree, and we present these readings to help students see how fascinating the sociological perspective can be in interpreting human life. This anthology is intended for use in introductory sociology courses. Most of the students in these courses are first- or second-year students, many of whom are not majoring in sociology. We want to provide an anthology that will excite students about the sociological perspective and show what such a perspective can bring to understanding society.

This new edition keeps many of the same themes as in the first edition, but we have added many new articles (twenty-nine) that bring new material to the book and reflect some of the developments in society since the first edition was published. We have also reorganized and shortened the book to make it more "user-friendly." The outline is modeled on the companion text, *Sociology: The Essentials* by Andersen and Taylor, but it can easily be adapted for use with our more comprehensive text (*Sociology: Understanding a Diverse Society*) or other introductory books.

We have selected articles for this collection that will engage students and will show them what sociology can contribute to their understanding of the world. The readings include a variety of perspectives, research methodologies, and current topics. They have also been excerpted and kept short for student comprehension. The anthology has a strong focus on diversity, both in the sections on class, race, gender, and age and in various selections throughout the book. We have also included many articles that give students a global perspective on various sociological topics. And, the book features more current research than that found in competing anthologies. We have eliminated many of the classic readings from sociological theory, based on reviewers' suggestions and because faculty who want to include such readings in their introductory courses can use the Eve Howard anthology, *Classic Readings in Sociology* (also available from Wadsworth). We developed this book with several themes in mind:

- *Contemporary research:* We wanted students to see examples of strong contemporary research, presented in a fashion that would be accessible to beginning undergraduates. The articles included here feature different styles of sociological research. For example, Amy Schulz's article on identity among Navajo women uses a qualitative interview study to examine the construction and meaning of identity. In contrast, Dan Clawson and Naomi Gerstel use a comparative analysis to examine child-care policy in the United States and Western Europe. We have also included a new piece by Howard Schuman to help students understand more about survey research, but in a very accessible form.

- *Diversity:* In keeping with our knowledge that society is increasingly diverse, we have selected articles that show the range of experiences that people have by virtue of differences in race, gender, class, sexual orientation, disability, and other characteristics (like age and religion). These factors differentiate human experience in contemporary society. Numerous articles in the reader focus on African Americans, Native Americans, Latinos, Asian Americans, women, gays and lesbians, Jewish Americans, and people with disabilities, among others. Some of the selections bring a comprehensive analysis of race, class, and gender to the subject at hand, thus adding to students' understanding of how diverse groups experience the social structure of society. As an example, Suzanne Carothers' discussion of Black women and their mothers shows how race and gender are part of the socialization process.

- *Global perspective:* We have also incorporated a global perspective into the reader, with many sections including articles that broaden students' worldview beyond the borders of the United States. Articles like Arlie Hochschild's essay "The Nanny Chain" will help students see how patterns of domestic help and contemporary immigration link the experiences of those in U.S. families to women from other nations who are increasingly being employed to provide domestic help for professional workers in the United States. Similarly, Benjamin Barber's article "Jihad vs. McWorld" will help students understand some of the current global conflicts.

- *Applying sociological knowledge:* Our students commonly ask, "What can you do with a sociological perspective?" We think this is an important question and one with many different answers. Sociologists use their knowledge in a variety of ways: to influence social policy formation, to interpret current events, and to educate people about common misconceptions and stereotypes, to name a few. Because we want to show students how sociological knowledge can be used, we have included a number of readings that demonstrate how sociological analyses can be applied to specific issues. For example, Loretta Meeks et al. review current school "choice" options with an eye to improving social policies addressing class and race inequality in the schools. And the new article by Sharon Hays, "Studying the Quagmire of Welfare Reform," is an excellent and engaging discussion of the implications of current welfare policy.

- *Classical theory:* Although we have reduced the number of classical theory articles, we have maintained those that got high marks from reviewers. We think it is important that introductory students learn about the contributions of classical sociological theorists. Thus, we have kept articles by Weber, Marx and Engels, DuBois, and Goffman that showcase some of the most important classics. We have developed discussion questions for these readings to help students think about how such classic pieces are reflected in contemporary issues. For example, Max Weber's argument about the Protestant ethic and the spirit of capitalism is fascinating to think about in the contemporary context of increased consumerism and increased class inequality. Students might ask whether contemporary patterns of wealth and consumption no longer reflect the asceticism and moral calling about which Weber wrote. Likewise, W. E. B. Du Bois's reflections on double consciousness continue to be very important in discussions of race and group perceptions.

NEW TO THE SECOND EDITION

The second edition of *Understanding Society* is organized in fourteen parts, following the outline of most introductory courses. The part on "Social Institutions" is then subdivided into brief sections on six major social institutions (families, religion, education, work, government, and health care). This organization allows faculty to focus on different institutions in any order they choose.

We have added several new features. At the conclusion of each part, there is a Bonus Reading, using Wadsworth's excellent online library, InfoTrac® College Edition. We selected these readings to provide more depth in a given topic and to challenge students to think further about the subject matter in each section. Thus, we have also included a brief annotation to spark student curiosity; these annotations also include a critical think question to guide the students' interpretation of the article.

We include a brief introduction before each article to put the article in context and help frame the students' understanding of the selection. We have also included discussion questions at the end·of each reading to help students think further about the implications of what they have read.

Perhaps most important, we have added twenty-nine new readings (and shortened the overall length of the text) to analyze topics and current developments in society that students will find interesting. Thus, we have a piece on September 11 (see George Ritzer, "September 11, 2001: Mass Murder and Its Roots in the Symbolism of American Consumer Culture"); several other pieces also place this historic event in sociological context (see Benjamin Barber, "Jihad vs. McWorld"). We have included more articles in the Education section that examine racial segregation in the schools (see, for example, Erica

Frankenberg and Chungmei Lee, "Race in American Public Schools: Rapidly Resegregating School Districts"). We have added pieces on the media (Jean Kilbourne, "Buy This 24-Year-Old and Get All His Friends Absolutely Free"), cyberspace interaction (Michael Hardey, "Life Beyond the Screen: Embodiment and Identity through the Internet"), universal health care (E. Richard Brown and Hongjian Yu, "Latinos' Access to Employment-Based Health Insurance"), and youth and social change (Duane F. Alwin, "Generations X, Y, and Z: Are They Changing America?"), among other contemporary topics.

In sum, with this anthology we hope to capture student interest in sociology, provide interesting research and theory, incorporate the analysis of diversity into the core of the sociological perspective, analyze the increasingly global dimensions of society, and show students how what they learn about sociology can be applied to real issues and problems.

PEDAGOGICAL FEATURES

In addition to the sociological content of this reader, we have included a number of pedagogical features to enrich student learning and to help people teach with the book. Each essay has a *brief introductory paragraph* that identifies the major themes and questions being raised in the article. We follow each article with *discussion questions* that students can use to improve their critical thinking and to reinforce their understanding of the article's major points. Many of these questions could also be used as the basis for class discussion, student papers, or research exercises and projects.

In addition, students using the book will be given passwords to the online *InfoTrac College Edition* system. This allows students to use the *Bonus Reading* feature to have an additional article that we have selected for each section of the book. Some of these articles will challenge students further in thinking about the subject matter, and many will provide more detail about research methods than could reasonably be included in a short text. These articles and the *InfoTrac College Search Terms* that also end each section can also be the basis for students' papers and projects.

Unlike many anthologies, we have included a *glossary* at the end of the book that contains the definition of basic terms and concepts that students will encounter in the readings. Finally, we have also included a *subject/name index* to help students and faculty locate specific topics and authors in the book.

SUPPLEMENTS

InfoTrac College Edition

This fully searchable database offers 20 years' worth of full-text articles from almost 5,000 diverse sources, such as academic journals, newsletters, and up-to-the-minute periodicals including *Time, Newsweek, Science, Forbes,* and *USA Today.* This incredible depth and breadth of material—available 24 hours a day from any computer with Internet access—makes conducting research so easy that students will want to use it to enhance their work in every course! And, incorporating InfoTrac College Edition is easy as well, references to this virtual library are built into many of our texts in margins, exercises, and side-bars. New! Students now have instant access to critical thinking and paper writing tools through InfoWrite. In addition, special guides to InfoTrac College Edition are available on our Web sites and as handy booklets that can be packaged with our texts. Both adopters and their students receive unlimited access for 4 months.

Andersen/Logio/Taylor's Companion Web Site
at Wadsworth's Virtual Society:
http://sociology.wadsworth.com

Combine this text with Virtual Society's exciting range of Web resources, and you will have truly integrated technology into your learning system. And the best news of all: site access is FREE to adopters and their students. The Virtual Society resource center features a wealth of online study materials, such as two online chapters (on the environment and school violence), Virtual Explorations Web exercises, a student guide to Census 2000, Sociology in the News, a career center, and much more.

ACKNOWLEDGMENTS

Many people helped in a variety of ways as we revised this anthology. We especially thank Bethany Brown for her assistance in collecting articles. We also appreciate the support of Vicky Baynes, Linda Keen, and Judy Watson whose work makes a lot of things possible. And, this edition is substantially improved based on the recommendations of our reviewers. We give special thanks to: Jamee Wolfe, Roanoke College; Janet Hund, Long Beach City College; Linda Evans, Drake University; Catherine Marrone, State University of New York at Stony Brook; Rosalind Gottfried, San Joaquin Delta College; Cathy Coghlan, Texas Christian University; and Katheryn Dietrich, Texas A&M University.

We sincerely appreciate the enthusiasm and support provided by Wadsworth's outstanding editorial team. Thank you Bob Jucha, Julie Sakaue, Melissa Walter, and Eve Howard for all you do to support this work. And, most especially, we thank Richard, Jim, Nolan, Owen, and Pat for all the love, smiles, and support they give us every day.

◆ About the Editors

Margaret L. Andersen (Ph. D., University of Massachusetts, Amherst) is Professor of Sociology and Women's Studies at the University of Delaware. She is the author of *Race, Class and Gender* (with Patricia Hill Collins); *Thinking about Women: Sociological Perspectives on Sex and Gender; Sociology: Understanding a Diverse Society* (with Howard F. Taylor); and *Sociology: The Essentials* (with Howard Taylor). She recently won a national award, the SWS Feminist Lecturer Award, for her contributions to scholarship and social change on behalf of women. She is the former president of the Eastern Sociological Society. She is currently writing the life history of Paul R. Jones, an African American art collector. She is a recipient of the University of Delaware's Excellence in Teaching Award.

Kim A. Logio (Ph. D., University of Delaware) is Assistant Professor of Sociology at Saint Joseph's University in Philadelphia where she teaches courses in research methods, criminal justice, and women and health. Her article, "Gender, Race, Childhood Abuse, and Body Image among Adolescents" appeared in *Violence Against Women.* She also has an article on keeping victims informed during the juvenile justice process in the *International Review of Victimology.* She is the coauthor of *Adventures in Criminal Justice Research,* 3rd edition (with George Dowdall, Earl Babbie, and Fred Halley). She is currently working on a new project involving adolescent body image and overall health.

Howard F. Taylor (Ph. D., Yale University) is Professor of Sociology at Princeton University. He is the author of *Balance in Small Groups; The IQ Game; Sociology: Understanding a Diverse Society* (with Margaret L. Andersen);

and *Sociology: The Essentials* (with Margaret L. Andersen). He is the winner of the DuBois-Johnson-Frazier Award, given by the American Sociological Association for distinguished research in race and ethnic relations, and has received the Princeton University President's Award for Distinguished Teaching. He is past president of the Eastern Sociological Society and is currently writing a book on *Race, Class, and the Bell Curve in America*.

1

The Sociological Imagination

C. WRIGHT MILLS

First published in 1959, C. Wright Mills' essay, taken from his book, The Sociological Imagination, *is a classic statement about the sociological perspective. A man of his times, his sexist language intrudes on his argument, but the questions he posed about the connection between history, social structure, and people's biography (or lived experiences) still resonate today. His central theme is that the task of sociology is to understand how social and historical structures impinge on the lives of different people in society.*

Nowadays men often feel that their private lives are a series of traps. They sense that within their everyday worlds, they cannot overcome their troubles, and in this feeling, they are often quite correct: What ordinary men are directly aware of and what they try to do are bounded by the private orbits in which they live; their visions and their powers are limited to the close-up scenes of job, family, neighborhood; in other milieux, they move vicariously and remain spectators. And the more aware they become, however vaguely, of ambitions and of threats which transcend their immediate locales, the more trapped they seem to feel.

Underlying this sense of being trapped are seemingly impersonal changes in the very structure of continent-wide societies. The facts of contemporary history are also facts about the success and the failure of individual men and women. When a society is industrialized, a peasant becomes a worker; a feudal lord is liquidated or becomes a businessman. When classes rise or fall, a man is employed or unemployed; when the rate of investment goes up or down, a man takes new heart or goes broke. When wars happen, an insurance salesman becomes a rocket launcher; a store clerk, a radar man; a wife lives alone; a child grows up without a father. Neither the life of an individual nor the history of a society can be understood without understanding both.

Yet men do not usually define the troubles they endure in terms of historical change and institutional contradiction. The well-being they enjoy, they do not usually impute to the big ups and downs of the societies in which they live. Seldom aware of the intricate connection between the patterns of their own lives and the course of world history, ordinary men do not usually know what this connection means for the kinds of men they are becoming and for the kinds of history-making in which they might take part. They do not possess the quality of mind essential to grasp the interplay of man and society, of biography and history, of self and world. They cannot cope with their personal troubles in such ways as to control the structural transformations that usually lie behind them. . . .

The sociological imagination enables its possessor to understand the larger historical scene in terms of its meaning for the inner life and the external career of a variety of individuals. It enables him to take into account how individuals, in the welter of their daily experience, often become falsely conscious of their social positions. Within that welter, the framework of modern society is sought, and within that framework the psychologies of a variety of men and women are formulated. By such means the personal uneasiness of individuals is focused upon explicit troubles and the indifference of publics is transformed into involvement with public issues.

The first fruit of this imagination—and the first lesson of the social science that embodies it—is the idea that the individual can understand his own experience and gauge his own fate only by locating himself within his period, that he can know his own chances in life only by becoming aware of those of all individuals in his circumstances. In many ways it is a terrible lesson; in many ways a magnificent one. We do not know the limits of man's capacities for supreme effort or willing degradation, for agony or glee, for pleasurable brutality or the sweetness of reason. But in our time we have come to know that the limits of "human nature" are frighteningly broad. We have come to know that every individual lives, from one generation to the next, in some society; that he lives out a biography, and that he lives it out within some historical sequence. By the fact of his living he contributes, however minutely, to the shaping of this society and to the course of its history, even as he is made by society and by its historical push and shove.

The sociological imagination enables us to grasp history and biography and the relations between the two within society. That is its task and its promise. To recognize this task and this promise is the mark of the classic social analyst. . . .

No social study that does not come back to the problems of biography, of history and of their intersections within a society has completed its intellectual journey. Whatever the specific problems of the classic social analysts, however limited or however broad the features of social reality they have examined, those who have been imaginatively aware of the promise of their work have consistently asked three sorts of questions:

1. What is the structure of this particular society as a whole? What are its essential components, and how are they related to one another? How does it differ from other varieties of social order? Within it, what is the meaning of any particular feature for its continuance and for its change?

2. Where does this society stand in human history? What are the mechanics by which it is changing? What is its place within and its meaning for the development of humanity as a whole? How does any particular feature we are examining affect, and how is it affected by, the historical period in which it moves? And this period—what are its essential features? How does it differ from other periods? What are its characteristic ways of history-making?

3. What varieties of men and women now prevail in this society and in this period? And what varieties are coming to prevail? In what ways are they selected and formed, liberated and repressed, made sensitive and blunted? What kinds of "human nature" are revealed in the conduct and character we observe in this society in this period? And what is the meaning for "human nature" of each and every feature of the society we are examining?

Whether the point of interest is a great power state or a minor literary mood, a family, a prison, a creed—these are the kinds of questions the best social analysts have asked. They are the intellectual pivots of classic studies of man in society— and they are the questions inevitably raised by any mind possessing the sociological imagination. For that imagination is the capacity to shift from one perspective to another—from the political to the psychological; from examination of a single family to comparative assessment of the national budgets of the world; from the theological school to the military establishment; from considerations of an oil industry to studies of contemporary poetry. It is the capacity to range from the most impersonal and remote transformations to the most intimate features of the human self—and to see the relations between the two. Back of its use there is always the urge to know the social and historical meaning of the individual in the society and in the period in which he has his quality and his being.

That, in brief, is why it is by means of the sociological imagination that men now hope to grasp what is going on in the world, and to understand what is happening in themselves as minute points of the intersections of biography and history within society. In large part, contemporary man's self-conscious view of himself as at least an outsider, if not a permanent stranger, rests upon an absorbed realization of social relativity and of the transformative power of history. The sociological imagination is the most fruitful form of this self-consciousness. By its use men whose mentalities have swept only a series of limited orbits often come to feel as if suddenly awakened in a house with which they had only supposed themselves to be familiar. Correctly or incorrectly, they often come to feel that they can now provide themselves with adequate summations, cohesive assessments, comprehensive orientations. Older decisions that once appeared sound now seem to them products of a mind unaccountably dense. Their capacity for astonishment is made lively again. They acquire a new way of thinking, they experience a transvaluation of values: in a word, by their reflection and by their sensibility, they realize the cultural meaning of the social sciences.

Perhaps the most fruitful distinction with which the sociological imagination works is between "the personal troubles of milieu" and "the public issues of social structure." This distinction is an essential tool of the sociological imagination and a feature of all classic work in social science.

Troubles occur within the character of the individual and within the range of his immediate relations with others; they have to do with his self and with those limited areas of social life of which he is directly and personally aware. Accordingly, the statement and the resolution of troubles properly lie within the individual as a biographical entity and within the scope of his immediate milieu—the social setting that is directly open to his personal experience and to some extent his willful activity. A trouble is a private matter: values cherished by an individual are felt by him to be threatened.

Issues have to do with matters that transcend these local environments of the individual and the range of his inner life. They have to do with the organization of many such milieux into the institutions of an historical society as a whole, with the ways in which various milieux overlap and interpenetrate to form the larger structure of social and historical life. An issue is a public matter: some value cherished by publics is felt to be threatened. Often there is a debate about what that value really is and about what it is that really threatens it. This debate is often without focus if only because it is the very nature of an issue, unlike even widespread trouble, that it cannot very well be defined in terms of the immediate and everyday environments of ordinary men. An issue, in fact, often involves a crisis in institutional arrangements, and often too it involves what Marxists call "contradictions" or "antagonisms."

In these terms, consider unemployment. When, in a city of 100,000, only one man is unemployed, that is his personal trouble, and for its relief we properly look to the character of the man, his skills, and his immediate opportunities. But when in a nation of 50 million employees, 15 million men are unemployed, that is an issue, and we may not hope to find its solution within the range of opportunities open to any one individual. The very structure of opportunities has collapsed. Both the correct statement of the problem and the range of possible solutions require us to consider the economic and political institutions of the society, and not merely the personal situation and character of a scatter of individuals.

Consider war. The personal problem of war, when it occurs, may be how to survive it or how to die in it with honor; how to make money out of it; how to climb into the higher safety of the military apparatus; or how to contribute to the war's termination. In short, according to one's values, to find a set of milieux and within it to survive the war or make one's death in it meaningful. But the structural issues of war have to do with its causes; with what types of men it throws up into command; with its effects upon economic and political, family and religious institutions, with the unorganized irresponsibility of a world of nation-states.

Consider marriage. Inside a marriage a man and a woman may experience personal troubles, but when the divorce rate during the first four years of marriage is 250 out of every 1,000 attempts, this is an indication of a structural issue having to do with the institutions of marriage and the family and other institutions that bear upon them.

Or consider the metropolis—the horrible, beautiful, ugly, magnificent sprawl of the great city. For many upper-class people, the personal solution to "the problem of the city" is to have an apartment with private garage under it in the heart of the city, and forty miles out, a house by Henry Hill, garden by Garrett Eckbo, on a hundred acres of private land. In these two controlled environments—with a small staff at each end and a private helicopter connection—most people could solve many of the problems of personal milieux caused by the facts of the city. But all this, however splendid, does not solve the public issues that the structural fact of the city poses. What should be done with this wonderful monstrosity?

Break it all up into scattered units, combining residence and work? Refurbish it as it stands? Or, after evacuation, dynamite it and build new cities according to new plans in new places? What should those plans be? And who is to decide and to accomplish whatever choice is made? These are structural issues; to confront them and to solve them requires us to consider political and economic issues that affect innumerable milieux.

In so far as an economy is so arranged that slumps occur, the problem of unemployment becomes incapable of personal solution. In so far as war is inherent in the nation-state system and in the uneven industrialization of the world, the ordinary individual in his restricted milieu will be powerless—with or without psychiatric aid—to solve the troubles this system or lack of system imposes upon him. In so far as the family as an institution turns women into darling little slaves and men into their chief providers and unweaned dependents, the problem of a satisfactory marriage remains incapable of purely private solution. In so far as the overdeveloped megalopolis and the overdeveloped automobile are built-in features of the overdeveloped society, the issues of urban living will not be solved by personal ingenuity and private wealth.

What we experience in various and specific milieux, I have noted, is often caused by structural changes. Accordingly, to understand the changes of many personal milieux we are required to look beyond them. And the number and variety of such structural changes increase as the institutions within which we live become more embracing and more intricately connected with one another. To be aware of the idea of social structure and to use it with sensibility is to be capable of tracing such linkages among a great variety of milieux. To be able to do that is to possess the sociological imagination. . . .

AW

DISCUSSION QUESTIONS

1. Using either today's newspaper or some other source of news, identify one example of what C. Wright Mills would call an issue. How is this issue reflected in the personal troubles of people it affects? Why would Mills call it a social issue?

2. What are the major historical events that have influenced the biographies of people in your generation? In your parents' generation? What does this tell you about the influence of society and history on biography?

2

The Forest and the Trees

ALLAN G. JOHNSON

Allan Johnson uses the classic example of the forest and the trees as a metaphor to demonstrate that people in society are participating in something larger than themselves. He also argues that the strong cultural belief in individualism blunts the sociological imagination, because it makes you see only individuals, not the social structures that shape diverse group experiences.

As a form of sociological practice, I work with people in corporations, schools, and universities who are trying to deal with issues of diversity. In the simplest sense, diversity is about the variety of people in the world, the varied mix of gender, race, age, social class, ethnicity, religion, and other social characteristics. In the United States and Europe, for example, the workforce is changing as the percentages who are female or from non-European ethnic and racial backgrounds increase and the percentage who are white and male declines.

If the changing mix was all that diversity amounted to, there wouldn't be a problem since in many ways differences make life interesting and enhance creativity. Compared with homogeneous teams, for example, diverse work teams are usually better with problems that require creative solutions. To be sure, diversity brings with it difficulties to be dealt with such as language barriers and different ways of doing things that can confuse or irritate people. But we're the species with the "big brain," the adaptable ones who learn quickly, so learning to get along with people unlike ourselves shouldn't be a problem we can't handle. Like travelers in a strange land, we'd simply learn about one another and make room for differences and figure out how to make good use of them.

As most people know, however, in the world as it is, difference amounts to more than just variety. It's also used as a basis for including some and excluding others, for rewarding some more and others less, for treating some with respect and dignity and some as if they were less than fully human or not even there. Difference is used as a basis for privilege, from reserving for some the simple human dignities that everyone should have, to the extreme of deciding who lives and who dies. Since the workplace is part of the world, patterns of inequality and oppression that permeate the world also show up at work, even though people may like to think of themselves as "colleagues" or part of "the team." And just

From: Allan G. Johnson. 1997. *The Forest and the Trees: Sociology as Life, Practice, Promise.* Philadelphia: Temple University Press, pp. 7–27. Reprinted with permission.

as these patterns shape people's lives in often damaging ways, they can eat away at the core of a community or an organization, weakening it with internal division and resentment bred and fed by injustice and suffering. . . .

People tend to think of things only in terms of individuals, as if a society or a company or a university were nothing more than a collection of people living in a particular time and place. Many writers have pointed out how individualism affects social life. It isolates us from one another, promotes divisive competition, and makes it harder to sustain a sense of community, of all being "in this together." But individualism does more than affect how we participate in social life. It also affects how we *think* about social life and how we make sense of it. If we think everything begins and ends with individuals—their personalities, biographies, feelings, and behavior—then it's easy to think that social problems must come down to flaws in individual character. If we have a drug problem, it must be because individuals just can't or won't "say no." If there is racism, sexism, heterosexism, classism, and other forms of oppression, it must be because of people who for some reason have the personal "need" to behave in racist, sexist, and other oppressive ways. If evil consequences occur in social life, then it must be because of evil people and their evil ways and motives.

If we think about the world in this way—which is especially common in the United States—then it's easy to see why members of privileged groups get upset when they're asked to look at the benefits that go with belonging to that particular group and the price others pay for it. When women, for example, talk about how sexism affects them, individualistic thinking encourages men to hear this as a personal accusation: "If women are oppressed, then I'm an evil oppressor who wants to oppress them." Since no man wants to see himself as a bad person, and since most men probably don't *feel* oppressive toward women, men may feel unfairly attacked.

In the United States, individualism goes back to the nineteenth century and, beyond that, to the European Enlightenment and the certainties of modernist thinking. It was in this period that the rational mind of the individual person was recognized and elevated to a dominant position in the hierarchy of things, separated from and placed above even religion and God. The roots of individualistic thinking in the United States trace in part to the work of William James who helped pioneer the field of psychology. Later, it was deepened in Europe and the United States by Sigmund Freud's revolutionary insights into the existence of the subconscious and the inner world of individual existence. Over the course of the twentieth century, the individual life has emerged as a dominant framework for understanding the complexities and mysteries of human existence.

You can see this in bookstores and best-seller lists that abound with promises to change the world through "self-help" and individual growth and transformation. Even on the grand scale of societies—from war and politics to international economics—individualism reduces everything to the personalities and behavior of the people we perceive to be "in charge." If ordinary people in capitalist societies feel deprived and insecure, the individualistic answer is that the people who run corporations are "greedy" or the politicians are corrupt and incompetent and otherwise lacking in personal character. The same perspective argues that

poverty exists because of the habits, attitudes, and skills of individual poor people, who are blamed for what they supposedly lack as people and told to change
if they want anything better for themselves. To make a better world, we think we
have to put the "right people" in charge or make better people by liberating
human consciousness in a New Age or by changing how children are socialized
or by locking up or tossing out or killing people who won't or can't be better
than they are. Psychotherapy is increasingly offered as a model for changing not
only the inner life of individuals, but also the world they live in. If enough
people heal themselves through therapy, then the world will "heal" itself as well.
The solution to collective problems such as poverty or deteriorating cities then
becomes a matter not of collective solutions but of an accumulation of individual solutions. So, if we want to have less poverty in the world, the answer lies in
raising people out of poverty or keeping them from becoming poor, *one person at
a time.*

So, individualism is a way of thinking that encourages us to explain the world
in terms of what goes on inside individuals and nothing else. We've been able to
think this way because we've developed the human ability to be reflexive, which
is to say, we've learned to look at ourselves *as selves* with greater awareness and
insight than before. We can think about what kind of people we are and how we
live in the world, and we can imagine ourselves in new ways. To do this, however, we first have to be able to believe that we exist as distinct individuals apart
from the groups and communities and societies that make up our social environment. In other words, the *idea* of the "individual" has to exist before we think
about ourselves as individuals, and the idea of the individual has been around for
only a few centuries. Today, we've gone far beyond this by thinking of the social
environment itself as just a collection of individuals: Society *is* people and people
are society. To understand social life, all we have to do is understand what makes
the individual psyche tick.

If you grow up and live in a society that's dominated by individualism, the
idea that society is just people seems obvious. The problem is that this approach
ignores the difference between the individual people who participate in social
life and the relationships that connect them to one another and to groups and
societies. It's true that you can't have a social relationship without people to participate in it and make it happen, but the people and the relationship aren't the
same thing. That's why this book's title plays on the old saying about missing the
forest for the trees. In one sense, a forest is simply a collection of individual trees;
but it's more than that. It's also a collection of trees that exist in *a particular relation*
to one another, and you can't tell what that relation is by just looking at each
individual tree. Take a thousand trees and scatter them across the Great Plains of
North America, and all you have are a thousand trees. But take those same trees
and bring them close together and you have a forest. Same individual trees, but
in one case a forest and in another case just a lot of trees.

The "empty space" that separates individual trees from one another isn't a
characteristic of any one tree or the characteristics of all the individual trees
somehow added together. It's something more than that, and it's crucial to
understand the *relationships among* trees that make a forest what it is. Paying

attention to that "something more"—whether it's a family or a corporation or an entire society—and how people are related to it is at the heart of sociological practice.

THE ONE THING

If sociology could teach everyone just one thing with the best chance to lead toward everything else we could know about social life, it would, I believe, be this: *We are always participating in something larger than ourselves, and if we want to understand social life and what happens to people in it, we have to understand what it is that we're participating in* and *how we participate in it.* In other words, the key to understanding social life isn't just the forest and it isn't just the trees. It's the forest *and* the trees and how they're related to one another. Sociology is the study of how all this happens.

The "larger" things we participate in are called social systems, and they come in all shapes and sizes. In general, the concept of a system refers to any collection of parts or elements that are connected in ways that cohere into some kind of whole. We can think of the engine in a car as a system, for example, a collection of parts arranged in ways that make the car "go." Or we could think of a language as a system, with words and punctuation and rules for how to combine them into sentences that mean something. We can also think of a family as a system—a collection of elements related to one another in a way that leads us to think of it as a unit. These include things such as the positions of mother, father, wife, husband, parent, child, daughter, son, sister, and brother. Elements also include shared ideas that tie those positions together to make relationships, such as how "good mothers" are supposed to act in relation to children or what a "family" is and what makes family members "related" to one another as kin. If we take the positions and the ideas and other elements, then we can think of what results as a whole and call it a social system.

In similar ways, we can think of corporations or societies as social systems. They differ from one another—and from families—in the kinds of elements they include and how those are arranged in relation to one another. Corporations have positions such as CEOs and stockholders, for example; but the position of "mother" isn't part of the corporate system. People who work in corporations can certainly be mothers in families, but that isn't a position that connects them to a corporation. Such differences are a key to seeing how systems work and produce different kinds of consequences. Corporations are sometimes referred to as "families," for example, but if you look at how families and corporations are actually put together as systems, it's easy to see how unrealistic such notions are. Families don't usually "lay off" their members when times are tough or to boost the bottom line, and they usually don't divide the food on the dinner table according to who's the strongest and best able to grab the lion's share for themselves. But corporations dispense with workers all the time as a way to raise dividends and the value of stock, and top managers routinely take a huge share of each year's profits even while putting other members of the corporate "family" out of work.

What social life comes down to, then, is social systems and how people participate in and relate to them. Note that people *participate* in systems without being *parts* of the systems themselves. In this sense, "father" is a position in my family, and I, Allan, am a person who actually occupies that position. It's a crucial distinction that's easy to lose sight of. It's easy to lose sight of because we're so used to thinking solely in terms of individuals. It's crucial because it means that people aren't systems, and systems aren't people, and if we forget that, we're likely to focus on the wrong thing in trying to solve our problems. . . .

INDIVIDUALISTIC MODELS DON'T WORK

Probably the most important basis for sociological practice is to realize that *the individualistic perspective that dominates current thinking about social life doesn't work.* Nothing we do or experience takes place in a vacuum; everything is always related to a context of some kind. When a wife and husband argue about who'll clean the bathroom, for example, or who'll take care of a sick child when they both work outside the home, the issue is never simply about the two of them even though it may seem that way at the time. We have to ask about the larger context in which this takes place. We might ask how this instance is related to living in a society organized in ways that privilege men over women, in part by not making men feel obliged to share equally in domestic work except when they choose to "help out." On an individual level, he may think she's being a nag; she may think he's being a jerk; but it's never as simple as that. What both may miss is that in a different kind of society, they might not be having this argument in the first place because both might feel obliged to take care of the home and children. In similar ways, when we see ourselves as a unique result of the family we came from, we overlook how each family is connected to larger patterns. The emotional problems we struggle with as individuals aren't due simply to what kind of parents we had, for their participation in social systems— at work, in the community, in society as a whole—shaped them as people, including their roles as mothers and fathers.

An individualistic model is misleading because it encourages us to explain human behavior and experience from a perspective that's so narrow it misses most of what's going on. A related problem is that *we can't understand what goes on in social systems simply by looking at individuals.* In one sense, for example, suicide is a solitary act done by an individual, typically while alone. If we ask why people kill themselves, we're likely to think first of how people feel when they do it— hopeless, depressed, guilty, lonely, or perhaps obliged by honor or duty to sacrifice themselves for someone else or some greater social good. That might explain suicides taken one at a time, but what do we have when we add up all the suicides that happen in a society for a given year? What does that number tell us, and, more importantly, about what? The suicide rate for the entire

U.S. population in 1994, for example, was twelve suicides per 100,000 people. If we look inside that number, we find that the rate for males was twenty per 100,000, but the rate for females was only five per 100,000. The rate also differs dramatically by race and country and varies over time. The suicide rate for white males, for example, was 71 percent higher than for black males, and the rate for white females was more than twice that for black females. While the rate in the United States was twelve per 100,000, it was thirty-four per 100,000 in Hungary and only seven per 100,000 in Italy. So, in the United States, males and whites are far more likely than females and blacks to kill themselves; and people in the United States are almost twice as likely as Italians to commit suicide but only one third as likely as Hungarians.

If we use an individualistic model to explain such differences, we'll tend to see them as nothing more than a sum of individual suicides. If males are more likely to kill themselves, then it must be because males are more likely to feel suicidally depressed, lonely, worthless, and hopeless. In other words, the psychological factors that cause individuals to kill themselves must be more common among U.S. males than they are among U.S. females, or more common among people in the United States than among Italians. There's nothing wrong with such reasoning; it may be exactly right *as far as it goes.* But that's just the problem: It doesn't go very far because it doesn't answer the question of *why* these differences exist in the first place. Why, for example, would males be more likely to feel suicidally hopeless and depressed than females, or Hungarians more likely than Italians? Or why would Hungarians who feel suicidally depressed be more likely to go ahead and kill themselves than Italians who feel the same way? To answer such questions, we need more than an understanding of individual psychology. Among other things, we need to pay attention to the fact that words like "female," "white," and "Italian" name positions that people occupy in social systems. This draws attention to how those systems work and what it means to occupy those positions in them.

Sociologically, a suicide rate is a number that describes something about a group or a society, not the individuals who belong to it. A suicide rate of twelve per 100,000 tells us nothing about you or me or anyone else. Each of us either commits suicide during a given year or we don't, and the rate can't tell us who does what. In the same way, how individuals feel before they kill themselves isn't by itself enough to explain why some groups or societies have higher suicide rates than others. Individuals can feel depressed or lonely, but groups and societies can't feel a thing. We could consider that Italians might tend to be less depressed than people in the United States, for example, or that in the United States, people might tend to deal with feelings of depression more effectively than Hungarians. It makes no sense at all, however, to say that the United States is more depressed or lonely than Italy.

While it might work to look at what goes on in individuals as a way to explain why one person commits suicide, this can't explain *patterns* of suicide found in social systems. To do this, we have to look at how people feel and behave *in relation* to systems and how these systems work. We need to ask, for example, how societies are organized in ways that encourage people who participate in them to feel more or less depressed or to respond to such feelings in suicidal or nonsuicidal ways. We need to see how belonging to particular groups shapes people's experience as they participate in social life, and how this limits the alternatives they think they can choose from. What is it about being male or being white that can make suicide a path of least resistance? How, in other words, can we go to the heart of sociological practice to ask how people participate in something larger than themselves and see how this affects the choices they make? How can we see the relationship between people and systems that produces variations in suicide rates or, for that matter, just about everything else that we do and experience, from having sex to going to school to working to dying?

Just as we can't tell what's going on in a system just by looking at individuals, we also can't tell what's going on in individuals just by looking at systems. Something may look like one thing in the system as a whole, but something else entirely when we look at the people who participate in it. If we look at the kind of mass destruction and suffering that war typically causes, for example, an individualistic model suggests a direct link with the "kinds" of people who participate in it. If war produces cruelty, bloodshed, aggression, and conquest, then it must be that the people who participate in it are cruel, bloodthirsty, aggressive people who want to conquer and dominate others. Viewing the carnage and destruction that war typically leaves in its wake, we're likely to ask, "What kind of people could do such a thing?" Sociologically, however, this question misleads us by reducing a social phenomenon to a simple matter of "kinds of people" without looking at the systems those people participate in. Since we're always participating in one system or another, when someone drops a bomb that incinerates thousands of people, we can't explain what happened simply by figuring out "what kind of person would do such a thing." In fact, if we look at what's known about people who fight in wars, they appear fairly normal by most standards and anything but bloodthirsty and cruel. Most accounts portray men in combat, for example, as alternating between boredom and feeling scared out of their wits. They worry much less about glory than they do about not being hurt or killed and getting themselves and their friends home in one piece. For most soldiers, killing and the almost constant danger of being killed are traumatic experiences that leave them forever changed as people. They go to war not in response to some inner need to be aggressive and kill, but because they think it's their duty to go, because they'll go to prison if they dodge the draft, because they've seen war portrayed in books and movies as an adventurous way to prove they're "real men," or because they don't want to risk family and friends rejecting them for not measuring up as true patriots.

People aren't systems, and systems aren't people, which means that social life can produce horrible or wonderful consequences without necessarily meaning that the people who participate in them are horrible or wonderful. Good people participate in systems that produce bad consequences all the time. I'm often aware of this in the simplest situations, such as when I go to buy clothes or food. Many of the clothes sold in the United States are made in sweatshops in cities like Los Angeles and New York and in Third World countries, where people work under conditions that resemble slavery in many respects, and for wages that are so low they can barely live on them. A great deal of the fruit and vegetables in stores are harvested by migrant farm workers who work under conditions that aren't much better. If these workers were provided with decent working conditions and paid a living wage, the price of clothing and food would probably be a lot higher than it is. This means that I benefit directly from the daily mistreatment and exploitation of thousands of people. The fact that I benefit doesn't make me a bad person; but my participation in that system does involve me in what happens to them. . . .

DISCUSSION QUESTIONS

1. Johnson argues that there is a tendency in the United States for people to explain everything in individual terms. Using the example of suicide, why are individualistic explanations inadequate? What would sociological explanations emphasize instead?

2. Johnson opens his discussion by noting the diversity that characterizes U.S. society. How does he apply the sociological perspective to his understanding of diversity and its significance?

3

Not Our Kind of Girl

ELAINE BELL KAPLAN

Elaine Bell Kaplan's research on African American teen mothers debunks a number of myths about teenage pregnancy and about African American mothers. She conducted her research using the method of participant observation. Here she describes how she did her research, including how her position as both an insider and an outsider influenced her research project. Her project and its results are presented in her book, Not Our Kind of Girl: Unraveling the Myths of Black Teenage Motherhood *(1995).*

From: Elaine Bell Kaplan. 1995. *Not Our Kind of Girl: Unraveling the Myths of Black Teenage Motherhood*. Berkeley: University of California Press, pp. xviii–xxiii, 10–26.

"If we want to solve the problems of the Black community, we have to do something about illegitimate babies born to teenage mothers." The caller, who identified himself as Black, was responding to a radio talk show discussion about the social and economic problems of the Black community. According to this caller's view, Black teen mothers' children grow up in fatherless households with mothers who have few moral values and little control over their offspring. The boys join gangs; the girls stand a good chance of becoming teen mothers themselves. The caller's perspective captures the popular view of many Americans: that marital status and age-appropriate sexual behavior ensure the well-being of the family and the community. . . .

According to mainstream ideology, men who through hard work have moved up the career ladder and provide their families with decent food on the dinner table, clothes on their backs, and an occasional family vacation have achieved the American Dream. Women's achievements are measured by their marriage and child rearing, done in proper order and at an appropriate age. Teenage girls are expected to replicate these values by refraining from sexual relations before adulthood and marriage.

Certainly, such traditional ideas held sway over the Black community I knew. Two decades ago unmarried teenagers with babies were a rare and unwelcome presence in my Harlem community. These few girls would be subjected to gossip about their lack of morals and stigmatized if they were on welfare. But by the 1980s so many young Black girls were pushing strollers around inner-city neighborhoods that they became an integral part of both the reality and the myth concerning the sexuality of Black underclass culture and Black family values. These Black teenage mothers did not fit in with the American ethic of hard work and strong moral character. . . .

If this conservative ideology is extended to teen mothers, their situation can be explained only as a result of aberrant moral character. If Black adolescent girls fail to achieve, something in their nature prevents them from doing so. As president, Ronald Reagan often urged teenage mothers to "just say no" so that taxpayers would no longer be forced to pay for their sexual behavior. The "Just Say No" slogan invoked by the Reagan and Bush administrations in the 1980s was utilized in the 1990s by both Black and White conservatives in the attempt "to change welfare as we know it." If these politicians have their way, teenage mothers will be shunned, hidden, and ignored.

As I made my way through East Oakland and downtown Richmond to interview teen mothers,* I witnessed a different scenario from the one devised by politicians. Teenage mothers are housed in threatening, drug-infested environments, schooled in jail-like institutions, and obstructed from achieving the American Dream. In our ostensibly open society, teenage mothers are disqualified from full participation and are marked as deviant. Black teenage girls aged

---

* Except when necessary for clarification, all teenage mothers and older women who were previously teenage mothers will for the sake of brevity be referred to as teen mothers—a term they use. When appropriate the teen mothers' own mothers will be referred to as adult mothers. All names and places have been changed to protect confidentiality.

fourteen, fifteen, and sixteen—many of them just beginning to show an adolescent interest in wearing makeup, dressing in the latest fashions, and reading teen magazines—are stigmatized. These teen mothers attempt to cope as best they can by redefining their situation in terms that involve the least damage to their self-respect.

Are Black teenage mothers responsible for the socioeconomic problems besetting the Black community, as the radio show caller would have us believe? Do Black teenage mothers have different moral values than most Americans? Do they have babies in order to collect welfare, as politicians suggest? Do the families of Black teenage mothers condone their deviant behavior, as the popular view contends? Or, as William J. Wilson's economic theory suggests, is Black teenage motherhood simply a response to the economic problems of the Black community? Black teenage girls confront a world in which gender norms, poverty, and racism are intertwined. Accordingly, to answer questions about these young mothers, we must sort out a host of complex economic and social problems that pervade their lives. I hope the questions I have asked and the answers given will provide portraits of real teenage mothers involved in real experiences.

The reality of these teenage mothers is that they have had to adopt strategies for survival that seem to them to make sense within their social environment but are as inadequate for them as they were for teenage mothers in the past.

These ethnographic pictures illuminate the way structural contradictions act on psychological well-being and the way people construct and reconstruct their lives in order to cope on a daily basis. One issue that comes through quite clearly in this study, and one that is often overlooked by politicians and various studies on Black teenage mothers, is that these teenagers know what constitutes a successful life. Black teenage mothers . . . struggle against being considered morally deviant, underclass, and unworthy.

If we are to understand the stories of these teenage mothers and generalize from their experiences in any significant way, we must place them within the current theoretical and political discussions concerning Black teenage mothers. As T. S. Eliot noted long ago, reality is often more troubling than myth.

What is begging for our attention is the fact that adolescence is a time when Black girls, striving for maturity, lose the support of others in three significant ways. First, they are abandoned by the educational system; second, they become mere sexual accompanists for boys and men; third, these problems create a split between the girls and their families and significant others. What is needed to understand the losses, the stresses, and the large and small violences that render such teenage girls incapable of successfully completing their adolescent tasks is a gender, race, and class analysis. When early motherhood is added to these challenges, they become insurmountable. The adolescent mothers I saw were deprived of every resource needed for any human being to function well in our society: education, jobs, food, medical care, a secure place to live, love and respect, the ability to securely connect with others. In addition, these girls were silenced by the insidious and insistent stereotyping of them as promiscuous and aberrant teenage girls. . . .

TALKING TO TEEN MOTHERS

I began my search to understand the rise in Black motherhood by interviewing two teen mothers referred to me by friends. They came to my house early one Saturday morning and stayed for three hours. Although I had prepared a series of general questions, the young women had so much more to say that I was compelled to create a more extensive set. Next, the director of a local family planning center let me attend a teen parent meeting, where I left a letter of introduction inviting those who were interested in my project to contact me. These teen mothers referred me to others. Eventually, I created a list of fifteen teen mothers.

The director of the family planning center also introduced me to Mary Higgins, the director of the Alternative Center in East Oakland. The Center operated with a grant from a large charity organization that allowed it to develop outreach programs geared to the needs of the local teenage population. These programs included an alternative school, day care, self-esteem development, parenting skills training, and personal counseling. Mary in turn introduced me to Ann Getty, a counselor at the center. Through Ann I met Claudia Wilson, a counselor for the Richmond Youth Counseling program. A short time after that meeting, I began to work as a volunteer consultant for the Alternative Center and to attend meetings with counselors and others who visited the center.

Through my contacts at the center, in the autumn of 1985 I met De Vonya Smalls and twenty of the sample of thirty-two teen mothers who participated in this study. The rest of my sample was drawn from other contacts I made in a network of community workers at the Richmond Youth Service Agency and through my work as a volunteer consultant there. The youth agency's counselors introduced me to teenage mothers who lived in the downtown Richmond area. As a consultant, I was able to talk extensively with the adolescents who took part in teen mother programs.

After several months of making contacts, losing some, and making new ones, I was able to pull together the sample of thirty-two teenage mothers. Of this sample, I "hung out" with a core group of seven teen mothers for a period of seven months, including sixteen-year-old De Vonya Smalls. The other six teen mothers who participated were sixteen-year-old Susan Carter, a mother of a two-month-old baby, who was living with her mother and sister in East Oakland; seventeen-year-old Shana Leeds, a mother with a nine-month-old baby, who was living with a family friend in downtown Richmond; and eighteen-year-old Terry Parks, a mother of a two-year-old, who was sharing her East Oakland apartment with twenty-year-old Dana Little and her five-year-old son. The group also included twenty-year-old Diane Harris, who had become pregnant at seventeen and within months had exchanged a middle-class lifestyle for that of a welfare mother and was now living in a run-down apartment in East Oakland; Lois Patterson, a twenty-seven-year-old mother of two and long-term welfare recipient, who was living with her extended family in a small, crowded house in East Oakland; and Evie Jenkins, a forty-three-year-old mother of two, who was living on monthly disability insurance in a housing project near downtown Richmond. Like Diane Harris, Evie lost her middle-class status when she became a teenage welfare mother at age seventeen.

I accompanied these women to the Alternative Center, to the welfare office, and to visits with their mothers. Some of the teen mothers could not find private places to talk, so we talked in the back seat of my car, over lunch or dinner in coffee shops, in a shopping mall, at teenage program meetings, or while moving boxes to a new apartment—in other words, anywhere they would let me join them.

Interviewing the teen mothers on a regular basis was difficult: they frequently moved, appointments were missed, telephones were disconnected. One day I tried to call five mothers about planned participant observation sessions only to find all their telephones disconnected. A few mothers were willing to be interviewed because they thought they would benefit in some way. One mother let me interview her because she thought I had access to housing and could get her an apartment. Another thought I would be able to get her into a teen parent program. A few mothers did not bother returning my telephone calls once they discovered I could not pay them.

I did not pay the teen mothers or the others for taking part in these interviews. In exchange for their information, I told the teen mothers about my own family, gave out information about welfare assistance and teen parent programs, and drove them to various stores. I helped De Vonya Smalls move into her first apartment. I went out with the teen mothers to eat Chinese food, shared takeout dinners, and bought potato chips and sodas for, so it seemed, everyone's sisters, brothers, and cousins. I was in some homes so often that the families began to treat me like a friend.

I found myself caught up in the teen mothers' lives more than I had planned. I was able to capture changes in their lives. I watched a teen mother break up with her baby's father. I witnessed DeVonya Smalls and Shana Leeds move in and out of three different homes. I saw Shana Leeds go through the process of applying for AFDC. I sat through long afternoons with Diane Harris discussing her baby's "womanizing" father, only to attend their wedding a few months later.

I also talked to everyone else I could, including the teen mothers' mothers, Black and White teenage girls who were not mothers, teachers, counselors, directors of teen programs, social workers, and Planned Parenthood counselors. Many have definite views about teenage mothers, some representing a more conservative voice than we usually hear in the Black community.

Sadly, most of the teen mothers' fathers and their babies' fathers were not involved in their lives in any significant way. The teen mothers' lack of knowledge about the babies' fathers' whereabouts made it impossible for me to interview the men. The few men who were still involved with the teen mothers refused to be interviewed. The best I could do was to observe some of the dynamics between two teen fathers and mothers.

PERSONAL HISTORIES

The teen mothers' ages ranged from fourteen to forty-three. Seventeen of them were currently teen mothers (aged fourteen to nineteen), and fifteen were older women who had previously been teen mothers (aged twenty to forty-three). The presence of the two age groups enabled me to appreciate the dynamic quality and long-term effects of teenage pregnancy on the mothers. The current teen mothers brought to the study a "here and now" aspect: I witnessed some of the family drama as it unfolded. The older women brought a sense of history and their reflective skills; the problems of being a teenage mother did not disappear when the teenage mothers became adults. The older women's stories served two goals for this book: to show that the black community has a history of not condoning teenage motherhood, and to locate emerging problems within the structural changes of our society that have affected everyone in recent years. . . .

As a group, the teen mothers' personal histories reveal both common and not so common patterns among teenage mothers. The youngest teen mother was fourteen and the oldest was eighteen at the time of their first pregnancies. Seventeen teen mothers were currently receiving welfare aid. But contrary to the commonly held assumption that welfare mothers beget welfare mothers, only five teen mothers reported that their families had been on welfare for longer than five years. Twenty-four of the teen mothers had grown up in families headed by a single mother—a common pattern among teenage mothers. Thirteen reported that their mothers had been teenage mothers. Unlike other studies that focus on poor teenage mothers, this study also included five middle-class and three working-class teenage mothers whose parents were teachers, civil service managers, or nursing assistants. Nine of the teen mothers were attending high school (of whom six were attending alternative high school). Several had taken college courses, and two had managed to obtain a college degree.

Along with capturing an ethnographic snapshot of the seven teen mothers, I conducted semistructured interviews in which I asked all the teen mothers specific questions about their experiences before, during, and after their pregnancies. I asked questions about various common perceptions: the idea of passive and promiscuous teenage girls, the role of men in their lives, the notion of strong cultural support for their pregnancies, the concept of extended family support networks, and the idea that teenage mothers have babies in order to receive welfare aid. Each teenage mother was interviewed for two to two and one-half hours. I audiotaped and transcribed all of the interviews.

I transcribed the material verbatim except for names and other identifying markers, which were changed during the transcription. I coded each teen mother on background variables and patterns. I read and reread my fieldnotes, supporting documents, and relevant literature. For this book I chose those quotations that would best represent typical responses, overall categories, and major themes. I used quotations from the core sample of seven as well as from the larger sample of thirty-two to include a wide range of responses.

Whenever possible I have tried to capture the teen mothers' emotional responses to the questions or issues. Often a teen mother would express through a sigh or a laugh feelings about some issue that contradicted her verbal response. For instance, when Terry Parks laughed as she described her feelings about being on welfare, I added a note about her laughter because it indicated to me that she was embarrassed about the subject. Without that notation, I would not have been able to communicate the emotional intensity with which she said the word "welfare" as she talked about her welfare experiences.

THROUGH THE ETHNOGRAPHIC LENS

I use an ethnographic approach to provide an intricate picture of how gender and poverty dictate the lives of these young teenage mothers and how societal gender, race, and class struggles are played out at the personal level. An ethnographic approach can bridge the gap between the sociological discussion of field research and the actual field experience. Studying these women through the lens of ethnography helped me move the teen mothers' personal stories to an objective level of analysis. The ethnographic method allowed the teen mothers to express to me personal information that was close to the heart. The method also allowed me to bring these Black teenage mothers into sociology's purview, to better understand them as persons, to make their voices heard, and to make their lives important to the larger society. The interviews and observations show that Black teenage girls' experiences are structural and troublesome. At all times I have attempted to make these teen mothers' stories real and visible by presenting the teen mothers' own words with as little editing as possible and by revealing their own insights into the interlocking structures of gender, race, and class.

THE INSIDER INTERVIEWER

I could not walk easily into some teen mothers' lives. Being close to the people being interviewed made me both pleased and tense. Being an insider—someone sharing the culture, community, ethnicity, or gender background of the study participants—has its advantages and disadvantages. When the interviewer can identify with the class and ethnic background of the person being interviewed, there is a greater chance of establishing rapport. The person will express a greater range of attitudes and opinions, especially when the opinions to be expressed are somewhat opposed to general public opinion. The situation is more complex when interviewees are asked to reveal information that may serve the researcher's interest but not that of the group involved. "Don't wash dirty linen in public," they remind the researcher.

The most difficult questions I faced, as do most insider interviewers, had to do with the politics of doing interviews in my own community. As an insider I had to decide whether making certain issues public would benefit the group at the same time that it served my research goals. I imagine that these interviews will raise questions. How will the White community perceive Black families if I discuss the conflicts between teen mothers and their mothers, or fathers who refuse to support their children, or the heavy negative sanctioning of these teen mothers by some in the Black community? My work would be taken out of context, several people warned me.

Every Black researcher who works on issues pertaining to her or his community grapples with these questions. We think about the possibility that our findings may contradict what the Black community wants outsiders to know. Some researchers select nonthreatening topics. Others romanticize Black life despite the evidence that life is hard for those on the bottom. And others simply adopt a code of silence, taking a position similar to that of the Black college teacher who in another context made the point to me, "I'm socialized to bear my pain in silence and not go blabbing about my problems to White folks, let alone strangers."

Being an insider did not help me gain the confidence of the teen mothers and others immediately. Most were suspicious of researchers. I lost a chance to interview one group of teen mothers involved in a special school project because the counselors who worked with them did not like the way a White male researcher had treated the teen mothers previously. Indeed, these teen mothers had the right to be suspicious. What these girls and women say about their lives can be used against them by public policy makers, since the Black community is often blamed for its own social and economic situations.

But overall, being a Black woman was helpful, because eventually the teen mothers, realizing we had much in common, stopped being suspicious of me and began to talk candidly of their lives. Occasionally I could not find a babysitter and had to bring my little boy along. I found my son's presence helped reduce the aloofness of my role as researcher and the powerlessness of the teens' position as interview subjects. I was surprised at how helpful my son was in breaking through the first awkward moments. We made him the topic of discussion—mothers can always compare child-care problems. His presence also helped me counter some of the teenagers' tendencies to deny problems. When I talked to De Vonya Smalls about my son's effects on my own schedule, like having to get up at five in the morning instead of at seven, she relaxed and told me about her efforts to study for a test while her baby cried for attention. She also admitted to doing poorly in school.

I decided to study these teenage mothers because Black teenage mothers are not going away, no matter how much we ignore, romanticize, or remain silent about their lives. I strongly disagree with approaches that let the group's code of silence supersede the need to understand the problems and issues of Black teenage mothers. That kind of false ideology only perpetuates the myths about Black teenage motherhood and causes researchers to neglect larger sociological issues or fail to ask pertinent questions about the lives of these mothers. In the name of racial pride, then, we essentially overlook how the larger society shares a great deal of responsibility for these problems. The only way to reduce the number of teenage pregnancies or to improve the lives of teenage mothers is to understand the societal causes by examining the realities of these girls' lives. The time had arrived, as Nate Hare put it, for an end to the unrealistic view of Black lives.

DISCUSSION QUESTIONS

1. How did Elaine Bell Kaplan's status as a Black woman give her an insider's view of Black teenage mothers? In what ways did she remain an outsider and how does this affect her research?
2. Why is participant observation a particularly good research method for investigating the questions that Kaplan was asking about Black teen mothers?

4

Sense and Nonsense about Surveys

HOWARD SCHUMAN

Survey research in social science research continues to be the most useful and popular method of data collection. Schuman points out that many media summaries of polls and other survey results lack a sophisticated understanding of how the results were obtained. This article provides some guidance on how to interpret and evaluate the results of polls and surveys.

Surveys draw on two human propensities that have served us well from ancient times. One is to gather information by asking questions. The first use of language around 100,000 years ago may have been to utter commands such as "Come here!" or "Wait!" Questions must have followed soon after: "Why?" or "What for?" From that point, it would have been only a short step to the use of interrogatives to learn where a fellow hominid had seen potential food, a dangerous animal, or something else of importance. Asking questions continues to be an effective way of acquiring information of all kinds, assuming of course that the person answering is able and willing to respond accurately.

The other inclination, learning about one's environment by examining a small part of it, is the sampling aspect of surveys. A taste of something may or may not point to appetizing food. A first inquiry to a stranger, a first glance around a room,

From: Howard Schuman. 2002. "Sense and Nonsense About Surveys." Contexts 1
Summer: 40–47. Reprinted with permission.

a first date—each is a sample of sorts, often used to decide whether it is wise to proceed further. As with questions, however, one must always be aware of the possibility that the sample may not prove adequate to the task.

SAMPLING: HOW GALLUP ACHIEVED FAME

Only within the past century—and especially in the 1930s and 1940s—were major improvements made in the sampling process that allowed the modern survey to develop and flourish. A crucial change involved recognition that the value of a sample comes not simply from its size but also from the way it is obtained. Every serious pursuit likes to have a morality tale that supports its basic beliefs: witness Eve and the apple in the Bible or Newton and his apple in legends about scientific discovery. Representative sampling has a marvelous morality tale also, with the additional advantage of its being true.

The story concerns the infamous *Literary Digest* poll prediction—based on 10 million questionnaires sent out and more than 2 million received back—that Roosevelt would lose decisively in the 1936 presidential election. At the same time, George Gallup, using many fewer cases but a much better method, made the more accurate prediction that FDR would win. Gallup used quotas in choosing respondents in order to represent different economic strata, whereas the *Literary Digest* had worked mainly from telephone and automobile ownership lists, which in 1936 were biased toward wealthy people apt to be opposed to Roosevelt. (There were other sources of bias as well.) As a result, the *Literary Digest* poll disappeared from the scene, and Gallup was on his way to becoming a household name.

Yet despite their intuitive grasp of the importance of representing the electorate accurately, Gallup and other commercial pollsters did not use the probability sampling methods that were being developed in the same decades and that are fundamental to social science surveys today. Probability sampling in its simplest form calls for each person in the population to have an equal chance of being selected. It can also be used in more complex applications where the chances are deliberately made to be unequal, for example, when oversampling a minority group in order to study it more closely; however, the chances of being selected must still be known so that they can later be equalized when considering the entire population.

INTUITIONS AND COUNTERINTUITIONS
ABOUT SAMPLE SIZE

Probability sampling theory reveals a crucial but counterintuitive point about sample size: the size of a sample needed to accurately estimate a value for a population depends very little on the size of the population. For example, almost the same size sample is needed to estimate, with a given degree of precision, the proportion of left-handed people in the United States as is needed to make the same estimate for, say, Peoria, Illinois. In both cases a reasonably accurate estimate can be obtained with a sample size of around 1,000. (More cases are needed when extraordinary precision is called for, for example, in calculating unemployment rates, where even a tenth of a percent change may be regarded as important.)

The link between population size and sample size cuts both ways. Although huge samples are not needed for huge populations like those of the United States or China, a handful of cases is not sufficient simply because one's interest is limited to Peoria. This implication is often missed by those trying to save time and money when sampling a small community.

Moreover, all of these statements depend on restricting your interest to over-all population values. If you are concerned about, say, left-handedness among African Americans, then African Americans become your population, and you need much the same sample size as for Peoria or the United States.

WHO IS MISSING?

A good sample depends on more than probability sampling theory. Surveys vary greatly in their quality of implementation, and this variation is not captured by the "margin of error" plus/minus percentage figures that accompany most media reports of polls. Such percentages reflect the size of the final sample, but they do not reveal the sampling method or the extent to which the targeted individuals or households were actually included in the final sample. These details are at least as important as the sample size.

When targeted members of a population are not interviewed or do not respond to particular questions, the omissions are a serious problem if they are numerous and if those missed differ from those who are interviewed on the matters being studied. The latter difference can seldom be known with great confidence, so it is usually desirable to keep omissions to a minimum. For example, sampling from telephone directories is undesirable because it leaves out those with unlisted telephones, as well as those with no telephones at all. Many survey reports are based on such poor sampling procedures that they may not deserve to be taken seriously. This is especially true of reports based on "focus groups," which offer lots of human interest but are subject to vast amounts of error. Internet surveys also cannot represent the general population adequately at present, though this is an area where some serious attempts are being made to compensate for the inherent difficulties.

The percentage of people who refuse to take part in a survey is particularly important. In some federal surveys, the percentage is small, within the range of 5 to 10 percent. For even the best non-government surveys, the refusal rate can reach 25 percent or more, and it can be far larger in the case of poorly executed surveys. Refusals have risen substantially from earlier days, becoming a major cause for concern among serious survey practitioners. Fortunately, in recent years research has shown that moderate amounts of nonresponse in an otherwise careful survey seem in most cases not to have a major effect on results. Indeed, even the *Literary Digest*, with its abysmal sampling and massive nonresponse rate, did well predicting elections before the dramatic realignment of the electorate in 1936. The problem is that one can never be certain as to the effects of refusals and other forms of nonresponse, so obtaining a high response rate remains an important goal.

QUESTIONS ABOUT QUESTIONS

Since survey questions resemble the questions we ask in ordinary social interaction, they may seem less problematic than the counterintuitive and technical aspects of sampling. Yet survey results are every bit as dependent on the form, wording and context of the questions asked as they are on the sample of people who answer them.

No classic morality tale like the *Literary Digest* fiasco highlights the question-answer process, but an example from the early days of surveys illustrates both the potential challenges of question writing and the practical solutions.

In 1940 Donald Rugg asked two slightly different questions to equivalent national samples about the general issue of freedom of speech:

- Do you think the United States should forbid public speeches against democracy?
- Do you think the United States should allow public speeches against democracy?

Taken literally, forbidding something and not allowing something have the same effect, but clearly the public did not view the questions as identical. Whereas 75 percent of the public would not allow such speeches, only 54 percent would forbid them, a difference of 21 percentage points. This finding was replicated several times in later years, not only in the United States but also (with appropriate translations) in Germany and the Netherlands. Such "survey-based experiments" call for administering different versions of a question to random subsamples of a larger sample. If the results between the subsamples differ by more than can be easily explained by chance, we infer that the difference is due to the variation in wording. . . .

SOLUTIONS TO THE QUESTION WORDING PROBLEM

. . . [O]ther difficulties (including the order in which questions are asked) suggest that responses to single survey questions on complex issues should be viewed with considerable skepticism. What to do then, other than to reject all survey data as unusable for serious purposes? One answer can be found from the replications of the forbid/allow experiment above: Although there was a 21 percentage points difference based on question wording in 1940 and a slightly larger difference (24 percentage points) when the experiment was repeated some 35 years later, both the forbid and the allow wordings registered similar declines in Americans' intolerance of speeches against democracy. . . . No matter which question was used—as long as it was the same one at both times—the conclusion about the increase in civil libertarian sentiments was the same.

More generally, what has been called the "principle of form-resistant correlations" holds in most cases: if question wording (and meaning) is kept constant, differences over time, differences across educational levels, and most other careful comparisons are not seriously affected by specific question wording. Indeed, the distinction between results for single questions and results based on comparisons or associations holds even for simple factual inquiries. Consider, for example, a study of the number of rooms in American houses. No God-given rule states what to include when counting the rooms in a house (bathrooms? basements? hallways?); hence the average number reported for a particular place and time should not be treated as an absolute truth. What we can do, however, is try to apply the same definitions over time, across social divisions, even across nations. That way, we gain confidence in the comparisons we make—who has more rooms than who, for example. . . .

Survey researchers should also ask several different questions about any important issue. In addition to combining questions to increase reliability, the different answers can be synthesized rather than depending on the angle of vision provided by any single question. A further safeguard is to carry out frequent experiments like that on the forbid/allow wordings. By varying the form, wording, and context of questions, researchers can gain insight into both the questions and the relevant issues. Sometimes variations turn out to make no difference, and that is also useful to learn. For example, I once expected support for legalized abortion to increase when a question substituted *end pregnancy* for the word *abortion* in the phrasing. Yet no difference was found. Today, more and more researchers include survey-based experiments as part of their investigations, and readers should look for these sorts of safeguards when evaluating survey results.

THE NEED FOR COMPARISONS

To interpret surveys accurately, it's important to use a framework of comparative data in evaluating the results. For example, teachers know that course evaluations can be interpreted best against the backdrop of evaluations from other similar courses: a 75 percent rating of lectures as "excellent" takes on a quite different meaning depending on whether the average for other lecture courses is 50 percent or 90 percent. Such comparisons are fundamental for all survey results, yet they are easily overlooked when one feels the urge to speak definitively about public reactions to a unique event.

Comparative analysis over time, along with survey-based experiments, can also help us understand responses to questions about socially sensitive subjects. Experiments have shown that expressions of racial attitudes can change substantially for both black and white Americans depending on the interviewer's race. White respondents, for instance, are more likely to support racial intermarriage when speaking to a black than to a white interviewer. Such self-censoring mirrors variations in cross-race conversations outside of surveys, reflecting not a methodological artifact of surveys but rather a fact of life about race relations in America. Still, if we consider time trends, with the race of interviewer kept constant, we can also see that white responses supporting intermarriage have clearly increased over the past half century . . . , that actual intermarriage rates have also risen (though from a much lower level) over recent years, and that the public visibility of cross-race marriage and dating has also increased. It would be foolish to assume that the survey data on racial attitudes reflect actions in any literal sense, but they do capture important *trends* in both norms and behavior.

Surveys remain our best tool for learning about large populations. One remarkable advantage surveys have over some other methods is the ability to identify their own limitations, as illustrated by the development of both probability theory in sampling and experiments in questioning. In the end, however, with surveys as with all research methods, there is no substitute for both care and intelligence in the way evidence is gathered and interpreted. What we learn about society is always mediated by the instruments we use, including our own eyes and ears. As Isaac Newton wrote long ago, error is not in the art but in the artificers. . . .

DISCUSSION QUESTIONS

1. Do you want large samples or small samples for survey research? Why? Is there any circumstance when a small sample would be sufficient?

2. Look at a recent newspaper article that provides some basic statistical survey results. Does the article explain the sampling procedures and the sample size? Does the article give sufficient information about who was included and who was excluded in the study? Do you know how the questions were worded?

http://infotrac.thomsonlearning.com

InfoTrac College Edition

BONUS READING

Best, Joel. "Promoting Bad Statistics." *Society* 38 (March 2001): 10–15.

This article talks about how the public relies on statistics as scientific fact. But, this reliance is often naïve, failing to examine statistics critically, with special attention to how they were created or calculated. The author calls for greater attention to measurement and definitions in the use of statistics to determine the extent of social problems. Find an example of the use of statistics to describe a social problem in the media and ask yourself the kinds of critical questions that the author here suggests. How does this change your view of the social problem you have identified?

SEARCH TERMS

sociological imagination
diversity and suicide
sociological methods
uses of statistics
race and teen mothers

5

Body Ritual Among the Nacirema

HORACE MINER

Miner's classic article shows us that sociologists and anthropologists study other cultures with as objective a view as possible. The body rituals of the Nacirema strike most of us as very peculiar; yet, they give us insight into the culture of the Nacirema people. We can read this and think of our own cultural practices and what they say about our society.

The anthropologist has become so familiar with the diversity of ways in which different peoples behave in similar situations that he is not apt to be surprised by even the most exotic customs. In fact, if all of the logically possible combinations of behavior have not been found somewhere in the world, he is apt to suspect that they must be present in some yet underscribed tribe. This point has, in fact, been expressed with respect to clan organization by Murdock (1949:71). In this light, the magical beliefs and practices of the Nacirema present such unusual aspects that it seems desirable to describe them as an example of the extremes to which human behavior can go.

Professor Linton first brought the ritual of the Nacirema to the attention of anthropologists twenty years ago (1936:326), but the culture of this people is still very poorly understood. They are a North American group living in the territory between the Canadian Cree, the Yaqui and Tarahumare of Mexico, and the Carib and Ara-wak of the Antilles. Little is known of their origin, although tradition states that they came from the east. According to Nacirema mythology, their nation was originated by a culture hero, Notgnihsaw, who is otherwise known for two great feats of strengths—the throwing of a piece of wampum across the river Pa-To-Mac and the chopping down of a cherry tree in which the Spirit of Truth resided.

Nacirema culture is characterized by a highly developed market economy which has evolved in a rich natural habitat. While much of the people's time is devoted to economic pursuits, a large part of the fruits of these labors and a considerable portion of the day are spent in ritual activity. The focus of this activity is the human body, the appearance and health of which loom as a dominant concern in the ethos of the people. While such a concern is certainly not unusual, its ceremonial aspects and associated philosophy are unique.

The fundamental belief underlying the whole system appears to be that the human body is ugly and that its natural tendency is to debility and disease.

From: "Body Ritual Among the Nacirema," by Horace Miner from *American Anthropologist* 58(3), 503–507, 1956.

Incarcerated in such a body, man's only hope is to avert these characteristics through the use of the powerful influences of ritual and ceremony. Every household has one or more shrines devoted to this purpose. The more powerful individuals in this society have several shrines in their houses and, in fact, the opulence of a house is often referred to in terms of the number of such ritual centers it possesses. Most houses are of wattle and daub construction, but the shrine rooms of the more wealthy are walled with stone. Poorer families imitate the rich by applying pottery plaques to their shrine walls.

While each family has at least one such shrine, the rituals associated with it are not family ceremonies but are private and secret. The rites are normally only discussed with children, and then only during the period when they are being initiated into these mysteries. I was able, however, to establish sufficient rapport with the natives to examine these shrines and to have the rituals described to me.

The focal point of the shrine is a box or chest which is built into the wall. In this chest are kept the many charms and magical potions without which no native believes he could live. These preparations are secured from a variety of specialized practitioners. The most powerful of these are the medicine men, whose assistance must be rewarded with substantial gifts. However, the medicine men do not provide the curative potions for their clients, but decide what the ingredients should be and then write them down in an ancient and secret language. This writing is understood only by the medicine men and by the herbalists who, for another gift, provide the required charm.

The charm is not disposed of after it has served its purpose, but is placed in the charm-box of the household shrine. As these magical materials are specific for certain ills, and the real or imagined maladies of the people are many, the charm-box is usually full to overflowing. The magical packets are so numerous that people forget what their purposes were and fear to use them again. While the natives are very vague on this point, we can only assume that the idea in retaining all the old magical materials is that their presence in the charm-box, before which the body rituals are conducted, will in some way protect the worshipper.

Beneath the charm-box is a small font. Each day every member of the family, in succession, enters the shrine room, bows his head before the charm-box, mingles different sorts of holy water in the font, and proceeds with a brief rite of ablution. The holy waters are secured from the Water Temple of the community, where the priests conduct elaborate ceremonies to make the liquid ritually pure.

In the hierarchy of magical practitioners, and below the medicine men in prestige, are specialists whose designation is best translated "holy-mouth-men." The Nacirema have an almost pathological horror of and fascination with the mouth, the condition of which is believed to have a supernatural influence on all social relationships. Were it not for the rituals of the mouth, they believe that their teeth would fall out, their gums bleed, their jaws shrink, their friends desert them, and their lovers reject them. They also believe that a strong relationship exists between oral and moral characteristics. For example, there is a ritual ablution of the mouth for children which is supposed to improve their moral fiber.

The daily body ritual performed by everyone includes a mouth-rite. Despite the fact that these people are so punctilious about care of the mouth, this rite involves a practice which strikes the uninitiated stranger as revolting. It was reported to me that the ritual consists of inserting a small bundle of hog hairs into the mouth, along with certain magical powders, and then moving the bundle in a highly formalized series of gestures.

In addition to the private mouth-rite, the people seek out a holy-mouth-man once or twice a year. These practitioners have an impressive set of paraphernalia, consisting of a variety of augers, awls, probes, and prods. The use of these objects in the exorcism of the evils of the mouth involves almost unbelievable ritual torture of the client. The holy-mouth-man opens the client's mouth and, using the above-mentioned tools, enlarges any holes which decay may have created in the teeth. Magical materials are put into these holes. If there are no naturally occurring holes in the teeth, large sections of one or more teeth are gouged out so that the super-natural substance can be applied. In the client's view, the purpose of these ministrations is to arrest decay and to draw friends. The extremely sacred and traditional character of the rite is evident in the fact that the natives return to the holy-mouth-man year after year, despite the fact that their teeth continue to decay.

It is to be hoped that, when a thorough study of the Nacirema is made, there will be careful inquiry into the personality structure of these people. One has but to watch the gleam in the eye of a holy-mouth-man, as he jabs an awl into an exposed nerve, to suspect that a certain amount of sadism is involved. If this can be established, a very interesting pattern emerges, for most of the population shows definite masochistic tendencies. It was to these that Professor Linton referred in discussing a distinctive part of the daily body ritual which is performed only by men. This part of the rite involves scraping and lacerating the surface of the face with a sharp instrument. Special women's rites are performed only four times during each lunar month, but what they lack in frequency is made up in barbarity. As part of this ceremony, women bake their heads in small ovens for about an hour. The theoretically interesting point is that what seems to be a preponderantly masochistic people have developed sadistic specialists.

The medicine men have an imposing temple, or *latipso,* in every community of any size. The more elaborate ceremonies required to treat very sick patients can only be performed at this temple. These ceremonies involve not only the thaumaturge but a permanent group of vestal maidens who move sedately about the temple chambers in distinctive costume and headdress.

The *latipso* ceremonies are so harsh that it is phenomenal that a fair proportion of the really sick natives who enter the temple ever recover. Small children whose indoctrination is still incomplete have been known to resist attempts to take them to the temple because "that is where you go to die." Despite this fact, sick adults are not only willing but eager to undergo the protracted ritual purification, if they can afford to do so. No matter how ill the supplicant or how grave the emergency, the guardians of many temples will not admit a client if he cannot give a rich gift to the custodian. Even after one has gained admission and survived the ceremonies, the guardians will not permit the neophyte to leave until he makes still another gift.

The supplicant entering the temple is first stripped of all his or her clothes. In everyday life the Nacirema avoids exposure of his body and its natural functions.

Bathing and excretory acts are performed only in the secrecy of the household shrine, where they are ritualized as part of the body-rites. Psychological shock results from the fact that body secrecy is suddenly lost upon entry into the *latipso*. A man, whose own wife has never seen him in an excretory act, suddenly finds himself naked and assisted by a vestal maiden while he performs his natural functions into a sacred vessel. This sort of ceremonial treatment is necessitated by the fact that the excreta are used by a diviner to ascertain the course and nature of the client's sickness. Female clients, on the other hand, find their naked bodies are subjected to the scrutiny, manipulation, and prodding of the medicine men.

Few supplicants in the temple are well enough to do anything but lie on their hard beds. The daily ceremonies, like the rites of the holy-mouth-men, involve discomfort and torture. With ritual precision, the vestals awaken their miserable charges each dawn and roll them about on their beds of pain while performing ablutions, in the formal movements of which the maidens are highly trained. At other times they insert magic wands in the supplicant's mouth or force him to eat substances which are supposed to be healing. From time to time the medicine men come to their clients and jab magically treated needles into their flesh. The fact that these temple ceremonies may not cure, and may even kill the neophyte, in no way decreases the people's faith in the medicine men.

There remains one other kind of practitioner, known as a "listener." This witch-doctor has the power to exorcise the devils that lodge in the heads of people who have been bewitched. The Nacirema believe that parents bewitch their own children. Mothers are particularly suspected of putting a curse on children while teaching them the secret body rituals. The counter-magic of the witch-doctor is unusual in its lack of ritual. The patient simply tells the "listener" all his troubles and fears, beginning with the earliest difficulties he can remember. The memory displayed by the Nacirema in these exorcism sessions is truly remarkable. It is not uncommon for the patient to bemoan the rejection he felt upon being weaned as a babe, and a few individuals even see their troubles going back to the traumatic effects of their own birth.

In conclusion, mention must be made of certain practices which have their base in native esthetics but which depend upon the pervasive aversion to the natural body and its functions. There are ritual fasts to make fat people thin and ceremonial feasts to make thin people fat. Still other rites are used to make women's breasts larger if they are small, and smaller if they are large. General dissatisfaction with breast shape is symbolized in the fact that the ideal form is virtually outside the range of human variation. A few women afflicted with almost inhuman hypermammary development are so idolized that they make a handsome living by simply going from village to village and permitting the natives to stare at them for a fee.

Reference has already been made to the fact that excretory functions are ritualized, routinized, and relegated to secrecy. Natural reproductive functions are similarly distorted. Intercourse is taboo as a topic and scheduled as an act. Efforts are made to avoid pregnancy by the use of magical materials or by limiting intercourse to certain phases of the moon. Conception is actually very infrequent. When pregnant, women dress so as to hide their condition. Parturition takes place in secret, without friends or relatives to assist, and the majority of women do not nurse their infants.

Our review of the ritual life of the Nacirema has certainly shown them to be a magic-ridden people. It is hard to understand how they have managed to exist so long under the burdens which they have imposed upon themselves. But even such exotic customs as these take on real meaning when they are viewed with the insight provided by Malinowski when he wrote (1948:70):

> Looking from far and above, from our high places of safety in the developed civilization, it is easy to see all the crudity and irrelevance of magic. But without its power and guidance early man could not have mastered his practical difficulties as he has done, nor could man have advanced to the higher stages of civilization.

REFERENCES

Linton, Ralph, 1936. *The Study of Man*. New York: Appleton-Century.

Malinowski, Bronislaw, 1948. *Magic, Science, and Religion*. Glencoe, IL: Free Press.

Murdock, George P., 1949. *Social Structure*. New York: Macmillan.

DISCUSSION QUESTIONS

1. What cultural practice among Americans might seem peculiar to someone from outside this society? What is the cultural meaning of that practice?

2. What do the body rituals of the Nacirema tell us about what is important to them as a society?

6

September 11, 2001

Mass Murder and Its Roots in the Symbolism of American Consumer Culture

GEORGE RITZER

Ritzer discusses the terrorist acts of September 11, 2001, as attacks against American culture. He explains how McDonald's fast-food chain, credit cards, large shopping malls, and discount stores are symbols of America's consumption culture that affront the rest of the world regularly. This discussion offers insight into why other nations feel such animosity toward the United States.

. . . On September 11, 2001, the terrorists not only killed thousands of innocent people and destroyed buildings of various sorts, they also sought to destroy (and in one case succeeded) major symbols of America's preeminent position in the globalization process: The World Trade Center was a symbol of America's global hegemony in the economic realm, and the Pentagon is obviously the icon of its military preeminence around the world. In addition, there is a widespread belief that the fourth plane, the one that crashed in Pennsylvania, was headed for the symbol of American political power—the White House. Obviously, the common element in all these targets is that they are, among other things, cultural icons, with the result that the terrorist attacks can be seen as assaults on American culture. (This is not, of course, to deny the very material effects on people, buildings, the economy, and so on.) Furthermore, although symbols, jobs, businesses, and lives were crippled or destroyed, the main objective was symbolic—the demonstration that the most important symbols of American culture were not only vulnerable but could be, and were, badly hurt or destroyed. The goal was to show the world that the United States was not an invulnerable superpower but that it could be assaulted successfully by a small number of terrorists. One implication was that if such important symbols could be attacked successfully, nothing in the United States (as well as in U.S. interests around the world) was safe from

From: George Ritzer. "September 11, 2001: Mass Murder and Its Roots in the Symbolism of American Consumer Culture." In *McDonaldization: The Reader*, edited by George Ritzer, 199–212. Thousand Oaks, CA: Pine Forge Press, 2002.

the wrath of terrorists. Thus, we are talking about an assault on, among other things, culture—an assault designed to have a wide-ranging impact throughout the United States and the world.

In emphasizing culture, I am not implying that economic, political, and military issues (to say nothing of the loss of life) are less important. Indeed, these domains are encompassed, at least in part, under the broad heading of culture and attacks on cultural icons. Clearly, many throughout the world are angered by a variety of things about the United States, especially its enormous economic, political, and military influence and power. In fact, this essay will focus on one aspect of the economy—consumption—and its role in producing hostility to the United States. Again, this is just one factor in the creation of this hostility, but it is certainly worthy of further discussion. Others, with greater expertise in those areas, will certainly be analyzing the military, political, and other economic dimensions and implications of the events of September 11, 2001.

By focusing on America's role in consumption and its impact around the world, I am not condoning the terrorist attacks (they are among the most heinous of acts in human history) or blaming the United States for those attacks. Rather, my objective is to discuss one set of reasons that people in many different countries loathe (while a far larger number of people love) the United States. Indeed, it is a truism that, often, love and hate coexist in the same people. However, needless to say, those involved in these terrorist acts had nothing but hatred for the United States.

CONSUMPTION

American hegemony throughout the world is most visible and, arguably, of greatest significance, in economic and cultural terms, in the realm of consumption. On a day-to-day basis in much of the world, people are far more likely to be confronted by American imperialism in the realm of consumption than they are in other economic domains. . . .

Rather than focus on consumption in general, I will discuss three of its aspects of greatest concern to me: fast-food restaurants, credit cards, and "cathedrals of consumption" (for example, discounters such as Wal-Mart). Before getting to these, it is important to point out that they, and many other components of our consumer culture, are not only physical presences throughout the world, they are media presences by way of television, movies, the Internet, and so on. Furthermore, even in those countries where these phenomena are not yet material realities, they are already media presences. As a result, their impact is felt even though they have not yet entered a particular country, and in those countries where they already exist, their impact is increased because they are also media presences.

I want to focus on the ways in which, from the perspective of those in other nations and cultures, fast-food restaurants, credit cards, and cathedrals of consumption bring with them (a) an American way of doing business, (b) an American way of consuming, and (c) American cultural icons . . .

FAST-FOOD RESTAURANTS

. . . McDonald's is today a global corporation with restaurants in nearly 150 nations throughout the world, and we can expect expansion into other nations in the coming years.

As it moves into each new nation, it brings with it a variety of American ways of doing business. In fact, in more recent years, the impact of its ways of doing business were surely felt long before the restaurant chain itself became a physical presence. McDonald's has been such a resounding success, and has offered so many important business innovations, that business leaders in other nations were undoubtedly incorporating many of its ideas almost from the inception of the chain. Of course, the major business innovation here is the franchise system. Although the franchise system predated McDonald's by many years . . . , the franchise system came of age with the development of the McDonald's chain. Kroc made a number of innovations in franchising . . . that served to make it a far more successful system. The central point, given the interest of this essay, is that this system has been adopted, adapted, and modified by all sorts of businesses not only in the United States but throughout the world. In the case of franchise systems in other nations, they are doing business in a way that is similar to, if not identical with, comparable American franchises.

The fact that indigenous businesses (e.g., Russkoye Bistro in Russia, Nirulas in India) are conducting their business based, at least to a larger degree, on an American business model is not visible to most people. However, they are affected in innumerable ways by the ways in which these franchises operate. Thus, day-to-day behaviors are influenced by all this, even if consumers are unaware of these effects.

What is far more obvious, even to consumers, is that people are increasingly consuming like Americans. This is clear not only in American chains in other countries but in indigenous clones of those chains. In terms of the former, in Japan, to take one example, McDonald's has altered long-standing traditions about how people are expected to eat. Thus, although eating while standing has long been taboo, in McDonald's restaurants, many Japanese eat just that way. Similarly, long expected not to touch food with their hands or drink directly from containers, many Japanese are doing just that in McDonald's and elsewhere. Much the same kind of thing is happening in indigenous clones of American fast-food restaurants in Japan such as Mos Burger.

These and many other changes in the way people consume are obvious, and they affect the way people live their lives on a daily basis. Just as many Japanese may resent these incursions into, and changes in, the ways in which they have traditionally conducted their everyday lives, those in many other cultures are likely to have their own wide-ranging set of resentments. However, these changes involve much more than transformations in the way people eat. The "McDonaldization thesis" involves far more than restaurants; universities . . . , churches . . . , and museums, among many other settings, can be seen as becoming McDonaldized. Almost no sector of society is immune from McDonaldization, and this means that innumerable aspects of people's everyday lives are transformed by it.

. . . McDonald's itself has become such an icon, as has its "golden arches," Ronald McDonald, and many of its products—Big Mac, Egg McMuffin, and so on. Other fast-food chains have brought with them their own icons—Burger King's Whopper, Colonel Sanders of Kentucky Fried Chicken, and so on. These icons are accepted, even embraced, by most, but others are likely to be angered by them. For example, traditional Japanese foods such as sushi and rice are being replaced, at least for some, by Big Macs and "supersized" French fried potatoes. Because food is such a central part of any culture, such a transformation is likely to enrage some. More important, perhaps, is the ubiquity of the McDonald's restaurant, especially its golden arches, throughout so many nations of the world. To many in other societies, these are not only important symbols in themselves but have become symbols of the United States and, in some cases, even more important than more traditional symbols (such as the American embassy and the flag). In fact, there have been a number of incidents in recent years in which protests against the United States and its actions have taken the form of actions against the local McDonald's restaurants. To some, a McDonald's restaurant, especially when it is placed in some traditionally important locale, represents an affront, a "thumb in the eye," to the society and its culture. It is also perceived as a kind of "Trojan Horse," and the view is that hidden within its bright and attractive wrappings and trappings are all manner of potential threats to local culture. Insulted by, and fearful of, such "foreign" entities, a few react by striking out at them and the American culture and business world that stands behind them.

CREDIT CARDS

. . . The credit card represents an American way of doing business, especially a reliance on the extension of credit to maintain and to increase sales. Many nations have been dominated, and some still are, by "cash-and-carry" business. Businesses have typically been loathe to grant credit, and when they do, it has usually not been for large amounts of money. When credit was granted, strong collateral was required. This is in great contrast to the credit card industry, which has granted billions of dollars in credit with little or no collateral. In these and many other ways, traditional methods of doing business are being threatened and eroded by the incursion of the credit card. . . .

. . . [C]redit cards are perceived as playing a key role in the development and expansion of consumer culture, a role characterized by hyper consumption. Although many are overjoyed to be deeply immersed in consumer culture and others would dearly love to be so involved, still others are deeply worried by it on various grounds. One of the concerns, felt not only in the United States but perhaps even more elsewhere in the world, is the degree to which immersion in the seeming superficialities of consumption and fashion represents a threat, if not an affront, to deep-seated cultural and religious values. For example, many have viewed modern consumption as a kind of religion, and I have described malls and other consumption settings as cathedrals of consumption. As such, they can be seen as alternatives, and threats, to conventional religions in many parts of the world. At the minimum, the myriad attractions of consumption and a day at the mall serve as powerful alternatives to visiting one's church, mosque, or synagogue.

Finally, credit cards in general, to say nothing of the major brands—Visa and MasterCard (as well as the "charge card" and its dominant brand, *American Express*)—are seen as major icons of American culture. While these icons are similar to, say, McDonald's and its golden arches, there is something quite unique and powerful about credit cards. Although one who lives outside the United States may encounter a McDonald's and its arches every day, or maybe every few days, a Visa credit card, for example, is *always* with those who have one. It is always there in one's wallet, and it is probably a constant subconscious reality. Furthermore, one is continually reminded of it every time one passes a consumption site, especially one that has the logo of the credit card on its door or display window. Even without the latter, the mere presence of a shop and its goods is a reminder that one possesses a credit card and that the shop can be entered and goods can be purchased. The credit card is a uniquely powerful cultural icon because it is with cardholders all the time and they are likely to be reminded of it continually.

CATHEDRALS OF CONSUMPTION

Cathedrals of consumption, many of which are also American innovations, are increasing presences elsewhere in the world. There is a long list of these cathedrals of consumption . . . , but let us focus on two—shopping malls and discounters, especially Wal-Mart. American-style fully enclosed shopping malls are springing up all over the world . . . . Most of these are indigenous developments, but the model is the American mall. . . .

These, of course, represent American ways of doing business. In the case of the mall, this involves the concentration of businesses in a single setting devoted to them. In the case of discounters, it represents the much greater propensity of American businesses (in comparison with their peers around the world) to compete on a price basis and to offer consumers deep discounts. Although appealing to many people, resentment may develop not only because these represent American rather than indigenous business practices but because they pose threats to local businesses. As in the United States, still more resentment is likely to be generated because small local shops are likely to be driven out of business by the development of a mall or the opening of a Wal-Mart on the outskirts of town.

Again, more obvious is the way consumers are led to alter their behaviors as a result of these developments. For example, instead of walking or bicycling to local shops, increasing numbers are more likely to drive to the new and very attractive malls and discounters. This can also lead to movement toward the increasing American reality that such trips are not just about shopping; such settings have become *destinations* where people spend many hours wandering from shop to shop, having lunch, and even seeing a movie or having a drink. Consumption sites have become places to while away days, and as such, they pose threats to alternative public sites, such as parks, zoos, and museums. In the end, malls and discounters are additional and very important contributors to the development of hyperconsumption and all the advantages and problems associated with it. Settings such as a massive shopping mall with a huge adjacent parking lot and a large Wal-Mart with its parking lot are abundantly obvious to people, as are the changes they help to create in the way natives consume.

Settings such as these are perceived as American cultural icons. Wal-Mart may be second only to McDonald's in terms of the association of consumption sites with things American, and the suburban mall is certainly broadly perceived in a similar way. Furthermore, malls are likely to house a number of other cultural icons, such as McDonald's, the Gap, and so on. And still further, the latter are selling yet other icons in the form of products such as Big Macs, blue jeans, Nike shoes, and so on. Many of those icons will be taken from the malls and eaten, worn, and otherwise displayed in public. Their impact is amplified because their well-known logos and names are likely to be plastered all over these products. Again, there is an "in-your-face" quality to all of this, and although many will be led to want these things, others will react negatively to the ubiquity of these emblems of America and its consumer culture and that these emblems tend to supplant indigenous symbols.

The argument here is that the recent terrorist attacks can be seen as assaults on American cultural icons—specifically the World Trade Center (business and consumption), the Pentagon (military), and potentially, the White House (political). . . .

Although I have focused on three American cultural icons in this essay and their worldwide proliferation, in vast portions of the world they are of minimal importance or completely nonexistent. Even where they are not physical presences, however, they are known through movies, television, magazines, and newspapers, and even by word of mouth. Thus, their influence throughout the world far exceeds their material presence in the world. . . .

CONCLUSION

Wars are always about culture, at least to some degree, but the one we have embarked on seems to reek with cultural symbolism. We live in an era—the era of globalization—in which not only cultural products and the businesses that sell them but also the symbols that go to their essence are known throughout the world. In fact, some—Nike and Tommy Hilfiger come to mind—are *nothing but symbols*; they manufacture nothing (except symbols). Although we deeply mourn the loss of life, the terrorists were after more. Surely, they wanted to kill people, but mainly because their deaths represented symbolically the fact that America could be made to bleed. But they also wanted to destroy some of the American symbols best known throughout the world; in destroying them, they were, they thought, symbolically destroying the United States. Interestingly, the initial response from the United States was largely symbolic—American flags were displayed everywhere; the sounds of the national anthem wafted through the air on a regular basis; red, white, and blue ribbons were wrapped around trees and tele-phone poles; and so on. Of course, the response soon went beyond symbols: Missiles have been launched and bombs dropped, special forces are in action, and people are dying. However, should the accused mastermind of the terrorist acts—Osama Bin Laden—be caught or killed, he will quickly become an even greater cultural icon than he is at the moment (his likeness already adorns T-shirts in Pakistan and elsewhere). Destroying his body may be satisfying to some, but it may well create a greater cultural problem for the United States, one that might translate into still more American citizens killed and structures destroyed. . . .

DISCUSSION QUESTIONS

1. Think back to the days immediately following September 11, 2001. What cultural symbols did Americans display as a sign of unity against terrorists? What are some other examples of American culture that some nations may take issue with and find offensive?

2. What are the issues to consider when the United States attempts to bring democracy to other nations? How do we balance bringing mostly welcomed American products and practices to a foreign culture and possibly destroying native cultural practices in the process?

7

Barbie Doll Culture and the American Waistland

KAMY CUNNINGHAM

Cultural norms establish expectations about beauty and body image that, as Kamy Cunningham points out, can be unrealistic to attain. Yet, many women (young and old) try to meet these standards, often harming their bodies and their self-concepts. The profits generated from cultural icons like Barbie bring enormous benefits to those who produce these cultural artifacts.

From: Kamy Cunningham. "Barbie Doll Culture and the American Waistland" *Symbolic Interaction* 16, no. 1 (1993). Copyright © 1993 by JAI Press.

AWaistland is a land where, if you're a woman, you have to have a tiny waist in order to not feel like something the cat drug out of the garbage bin. I remember at age ten gazing at my first Barbie, the blue-eyed version with painted toenails in its zebra-striped suit, and deciding, well, I guess this pneumatic creature (I already had a pretty sophisticated vocabulary back then) with the long, horsey legs and Scarlett O'Hara waist was what I was supposed to grow up to look like.

Doesn't every little, and big, American girl want to look like Barbie? And doesn't she want to *be* Barbie, wholesome and popular and perky? In short, a plastic doll.

Barbie beckons us little, and big, girls, but toward what? And if it's toward beauty, what sort of beauty is this, with its tiny waist?

During a moment of epiphany in the middle of a television commercial the other night (most people just go to the bathroom), I speculated that it might be the beauty of the Heartland of America, cholesterol free and patriotically waving tubs of margarine called Promise.

The show between the commercials was the Ms. Teenage America Pageant followed, a couple of days later, by a grown-up beauty contest, the Supermodel of the Year. The teenage hopefuls were all ruffles and tans and soufflés of clichés, each determined to be herself and not succumb to peer pressure and to work with the handicapped, the learning disabled, the old, and the terminally ill because, of course, the most wonderful thing in life is to help others and be the best you can be.

Their supermodel counterparts were slinky, and slid along the stage like skinny eels, in that funny model posture, pelvis jutting forward, small bosoms receding onto the terrace of the breastbone, that makes a woman look like a limp piece of spaghetti about to fall over—backwards.

Barbie combines the prototypes of the two pageants—she's all ruffles and cuteness *and* all experienced slinkiness. A recent version, the Fashion Play Barbie, is a good illustration of what I mean. Clad in a Frederick's of Hollywood wisp of lingerie and topped by luxuriant platinum tresses, the doll has lavender eyes, both willing and innocent, that look out of a face cutely dimpled and empty of feminine guile, yet somehow eerily seductive.

In her own plastic person, Barbie carries the Virgin/Whore paradox to an even more tensile extreme than does, say, a Marilyn Monroe, or a Madonna. Marilyn combined helpless, yielding child with voluptuous, knowing woman in a caricature of the two that was almost obscene. Madonna—shrewd, ruthless, experienced, slightly perverse—is a walking, strutting contradiction to the name of the Virgin she has appropriated.

Slip off that wisp of lingerie, barely clinging to those fulsome curves, and a naked Barbie doll is a sexy thing. Pouty bosom, that tiny waist so oft spoke of, flared hips, lissome legs. Squeeze her and knead her and she has a rubbery life of her own. Cup her, King Kong fashion, and feel the points of the breasts press into your palm. Run your hand down the full 11½" and experience cool, clean silk feel of plastic. Wholesome and seductive.

Is this beauty? Egyptian women painted their eyelids a heavy charcoal black. Medieval paintings show that women with small tulip breasts and big, oven-rounded bellies were desirable. Rubens and Titian and Ingres thought that women layered like lily-white, hothouse marshmallows were best. In some cultures the male is the heavily painted and artificial one. Like an obscene, opalescent peacock or an aquamarine bird of paradise with blue dragons curling around his arms, in sinuous indigo, he gyrates in front of the womenfolk, hoping (and hopping) to be "pretty" enough to be picked.

The question is not really one of beauty, of course, but of the oppressive equation of beauty, however we define it, with worth. Surface so dominates essence in America that the equation has gotten out of hand. The reason is obvious. We are bombarded by images of Barbie doll women. On a recent *Smithsonian World*, a popular culture critic called advertising "one of the predominate art forms of our time." Advertising is so dominate, the show goes on to say, that "its messages are the only ones being heard." "America is about selling" and "we accept the marketplace as the arbiter of values."

Ads create the symbols of our culture; they suggest that the Johnson's make-me-your-baby-powder woman is the only acceptable version of the feminine.

When I was ten years old, I didn't know that I was longing for a Rubenesque or a Titianesque, rather than a Barbiesque, visual model. I didn't know at the time that she was influencing and reinforcing impossible cultural norms of physical beauty, norms that I would never be able to even approximate. I didn't know that most men want Barbie doll women, the ones with long blonde hair, innocent baby-blue (or baby-lavender) eyes, substantial Cosmo cover melons, tiny waists, flat tummies, taut bottoms, and long graceful legs. (And absurdly small feet: Barbie doesn't even have to wear heels in order to be "hobbled"—it's built into those ridiculous concubine feet.) I didn't know that to be considered desirable, I would have to be a centerfold, zipped into my nakedness like a shrimp in its casing.

If I had known all of this, I would probably have thrown myself off Hoover Dam and never reached age eleven.

There are some other things wrong with Barbie too. She's a simulacrum of a human being, a sad grotesquerie: her creators gave her breasts but no nipples, flared hips but no womb, seductively spread legs but no vagina. No milk, no sucklings, no procreation. A twilight zone creature, as strange as her life-sized counterpart—the department store mannequin with the sterility of a lavender sheen on its cadaverous, blue-grey cheeks—she is an emblem of frustration and unfulfillment.

In Las Vegas, at Caesar's Palace, an enormous figurehead of Cleopatra juts out over the casino. With her huge bronze breasts that dangle above your head and her ample but shapely girth, she looks as if she could have mothered the whole human race. Instead, taut in every disappointed muscle, she strains out into nothing, gazing at this sterile indoor cosmos of star-spangled chandeliers.

Las Vegas showgirls look manufactured—identical lanky clones carrying ten pounds of feathers above eyes so mascaraed no eyes are there. Ads across the country misrepresent them as voluptuous; actually, by some ironic twist, they're all tiny-bosomed because big breasts bounce around on the stage with the least step or jiggle. All the girls would look like cumbersome, milkheavy cows. Go to a Las Vegas show and the sensation is eerie: two-hundred identical breasts with tiny peppermint nipples point your way, like the pink noses of puppies. Beneath the nipples, identical Rockettes' legs. Large breasts might be an improvement: the hilarity would relieve the manufactured look of the women.

The "simply irresistible" clone women, of Pepsi commercial and MTV fame, produce a similar shuddering sensation. Painted over with that lavender sheen of the mannequin, with big, hard, dark eyes, and starved cheeks, their faces look like those of boxed dolls—identical and inexpressive. Only their bodies are alive, in a mechanical way, as they move. Their eyes are dead. Some even wear goggles, blinders that make them look like horses in harness. They have been zapped of all their vitality, by being turned into mindless doll-like clones. Dead dolls, vampire women. Plastic. Manufactured. Artificial. Unreal. No room for the appealing flaws and living warmth of "real" women, those whose Rubenesque curves might spread a little and whose Titianesque arms might have a bit of the soft sway of the basset hound. Warm arms, motherly arms.

Barbie, with all of her accessories (thousands of little outfits, and dozens of pieces of pink plastic furniture, and Hollywood hot tubs and sleek racing cars) brings in three quarters of a billion dollars a year for Mattel. Every two seconds someone somewhere in the world buys a Barbie. Numerically, there are 2.5 Barbies for every household in America.

She is obviously a powerful cultural icon, but what is her iconography? What text is she illustrating? The text of woman as manufactured cadaver? Woman robbed of any insides because she has to be all outside?

Barbie's living clone, Vanna White, Goddess of the Empty Woman, seems to be illustrating a depressing blankness (note the name *White*). Vanna's message is that if you look like her and dress beautifully and smile warmly and turn letters with great skill and remain forever, mentally and emotionally, on the level of an untroubled child, then you will be valued and given lots of money. Turning letters counts for far more, apparently, than turning a phrase. Rarely, in the history of womankind mankind, has so little been so richly rewarded.

One night, on *Wheel of Fortune,* she gushed, in a see-Spot-run vocabulary, over an "island paradise" vacation she'd just taken that was "simply wonderful" and "so great." (Her narrative, childishly adjectival and without a story line, had not quite reached the level of sophistication of the "cow jumped over the moon.") I feel resentful that Vanna and her ilk (all those manufactured mannequins and cadaverous clones) can pile up fortunes by selling their bodies and that I can't make anything by selling my mind.

But, to temper my tirade (a bit), I feel a little sorry for her (and them). And happy for me, a little. In a world where you have to sell something to survive, maybe it's better to have a mind to market than a transient body.

Perhaps, decades from now, I may be able to entertain myself with books and thoughts after all the centerfolds have sagged and the Vannas have died away from their own untroubled boredom.

DISCUSSION QUESTIONS

1. When you were growing up, what toys did you play with? How do you think these might have influenced your concepts of good-looking women and men? Has this established any norms for your own appearance now?

2. What beauty images do you see as most frequently represented in various forms of popular culture now? What cultural expectations do these images establish based on gender? age? race? class?

8

Buy This 24-Year-Old and Get All His Friends Absolutely Free

JEAN KILBOURNE

Kilbourne's article uncovers the cultural importance of advertising in America. Her work reveals how the true product in advertising is each of us, the individual consumer. The strategies employed by advertisers single out groups of people, categorize them, and market products specifically to them. An analysis of advertisements reveals information about our culture, specifically the importance of consumerism and mass buying power.

If you're like most people. you think that advertising has no influence on you. This is what advertisers want you to believe. But, if that were true, why would companies spend over $200 billion a year on advertising? Why would they be willing to spend over $250,000 to produce an average television commercial and another $250,000 to air it? If they want to broadcast their commercial during the Super Bowl, they will gladly spend over a million dollars to produce it and over one and a half million to air it. After all, they might have the kind of success that Victoria's Secret did during the 1999 Super Bowl. When they paraded bra-and-panty-clad models across TV screens for a mere thirty seconds, one million people turned away from the game to log on to the Website promoted in the ad. No influence?

Ad agency Arnold Communications of Boston kicked off an ad campaign for a financial services group during the 1999 Super Bowl that represented eleven

From: Jean Kilbourne. 1999. In *Can't Buy My Love: How Advertising Changes the Way We Think and Feel*, 33–56. New York: Simon & Schuster.

months of planning and twelve thousand "man-hours" of work. Thirty hours of footage were edited into a thirty-second spot. An employee flew to Los Angeles with the ad in a lead-lined bag, like a diplomat carrying state secrets or a courier with crown jewels. Why? Because the Super Bowl is one of the few sure sources of big audiences—especially male audiences, the most precious commodity for advertisers. Indeed, the Super Bowl is more about advertising than football: The four hours it takes include only about twelve minutes of actually moving the ball.

Three of the four television programs that draw the largest audiences every year are football games. And these games have coattails: twelve prime-time shows that attracted bigger male audiences in 1999 than those in the same time slots the previous year were heavily pushed during football games. No wonder the networks can sell this prized Super Bowl audience to advertisers for almost any price they want. The Oscar ceremony, known as the Super Bowl for women, is able to command one million dollars for a thirty-second spot because it can deliver over 60 percent of the nation's women to advertisers. Make no mistake: The primary purpose of the mass media is to sell audiences to advertisers. *We* are the product. Although people are much more sophisticated about advertising now than even a few years ago, most are still shocked to learn this.

Magazines, newspapers, and radio and television programs round us up, rather like cattle, and producers and publishers then sell us to advertisers, usually through ads placed in advertising and industry publications. "The people you want, we've got all wrapped up for you," declares *The Chicago Tribune* in an ad placed in *Advertising Age,* major publication of the advertising industry, which pictures several people, all neatly boxed according to income level.

Although we like to think of advertising as unimportant, it is in fact the most important aspect of the mass media. It *is* the point. Advertising supports more than 60 percent of magazine and newspaper production and almost 100 percent of the electronic media. Over $40 billion a year in ad revenue is generated ,for television and radio and over $30 billion for magazines and newspapers. . . .

Once we begin to count, we see that magazines are essentially catalogs of goods, with less than half of their pages devoted to editorial content (and much of that in the service of the advertisers). . . . And, in fact, there are magazines for everyone from dirt-bike riders to knitters to mercenary soldiers, from *Beer Connoisseur* to *Cigar Aficionado.* There are plenty of magazines for the wealthy such as *Coastal Living* "for people who live or vacation on the coast." *Barron's* advertises itself as a way to "reach faster cars, bigger houses and longer prenuptial agreements" and promises a readership with an average household net worth of over a million.

The Internet advertisers target the wealthy too, of course. "They give you Dick," says an ad in *Advertising Age* for an Internet news network. "We give you Richard." The ad continues, "That's the Senior V. P. Richard who lives in L.A., drives a BMW and wants to buy a DVD player and a kayak." Not surprisingly, there are no magazines or Internet sites or television programs for the poor or for people on welfare. They might not be able to afford the magazines or computers but, more important, they are of no use to advertisers.

This emphasis on the affluent surely has something to do with the invisibility of the poor in our society. Since advertisers have no interest in them, they are not reflected in the media. We know so much about the rich and famous that it becomes a problem for many who seek to emulate them, but we know very little about the lifestyles of the poor and desperate. It is difficult to feel compassion for people we don't know.

Publications and programs that target minorities are also only interested in the affluent. "At $446 billion, African American buying power is more than the GNP of Switzerland," says an ad in *Advertising Age*. Another, for a "Black-owned agency," implies it can "get you inside the soul of the African American consumer." "Are You Skirting a Major Market?" asks an ad for a local Florida television station picturing a Latina in a very short skirt. It concludes, "Channel 23. Because South Florida spends a lot of dinero!" An ad for *Latina* magazine says, "She's Latina, She spends more." And the Hispanic Network tells advertisers that "Hispanic families are more responsive to advertising. And ads in Spanish are 5 times more persuasive." . . .

Ethnic minorities will soon account for 30 percent of all consumer purchases. No wonder they are increasingly important to advertisers. Nearly half of all Fortune 1000 companies have some kind of ethnic marketing campaign. Nonetheless, minorities are still underrepresented in advertising agencies. African-Americans, who are over 10 percent of the total workforce, are only 5 percent of the advertising industry. Minorities are underrepresented in ads as well—about 87 percent of people in mainstream magazine ads are white, about 3 percent are African-American (most likely appearing as athletes or musicians), and less than 1 percent are Hispanic or Asian. As the spending power of minorities increases, so does marketing segmentation. Mass marketing aimed at a universal audience doesn't work so well in a multicultural society, but cable television, the Internet, custom publishing, and direct marketing lend themselves very well to this segmentation. The multiculturalism that we see in advertising is about money, of course, not about social justice.

The same is true for the increased visibility of gay men and lesbians in advertising. The gay media have provided the lucrative market of gay men to advertisers, such as IBM, Benetton, Johnson & Johnson, and United Airlines, for years. Hartford Financial Services Group launched a 1998 campaign aimed at gays that not only appeared in the gay media but crossed over to mainstream media. Ads picturing pairings of pink and blue cars promoted Hartford's discounts to gay and lesbian couples with the tagline, "Commitment. Bring it on."

Lesbians are still on the fringes of the gay marketing movement, probably because most women still have lower incomes than men. A male couple would be likely to have a higher joint income than a heterosexual couple, but this is not usually the case for a female couple. Nonetheless, Olivia Cruises ran an overtly lesbian commercial during the coming-out episode of *Ellen,* and Molson beer launched a commercial featuring a lesbian kiss in Canada in 1997. A Subaru print ad featured two women with the headline, "It loves camping, dogs and long-term commitment. Too bad it's only a car." And American Express ran an ad featuring a real-life lesbian couple with copy that included, "When you're ready to plan a future together, who can you trust to understand the financial challenges that gay men and lesbians face?"

Advertisers don't casually decide to target gays. They spend significant amounts of money to conduct research on the market and to find out how their products are faring. American Express spent $250,000 for research in 1997 before committing additional funds to the market. More advertisers are seeking the gay audience on the Internet, ever since a major gay-market study in 1997 found that gays are large online subscribers (51.5 percent use for gays compared with 15.8 percent for the general population).

However, many advertisers who target gay consumers still prefer to remain closeted about it, for fear of offending their heterosexual customers. More than two-thirds of gay-market advertisers contacted in 1997 by *Advertising Age* chose not to comment for an article on the topic. . . .

So, the media round us up—gay and straight, male and female, African-American, white, Latino, young, and middle-aged (advertisers are not interested in old people, who usually already have brand loyalty and often have limited incomes). Then they spend a fortune on research to learn a lot about us, using techniques like polls, trends analysis, focus groups, and PRIZM, a marketing program that garners information about consumers from their ZIP codes—and that is advertised in *Advertising Age* as "the targeting tool that turns birds of a feather into sitting ducks."

Many companies these days are hiring anthropologists and psychologists to examine consumers' product choices, verbal responses, even body language for deeper meanings. They spend time in consumers' homes, listening to their conversations and exploring their closets and bathroom cabinets. . . .

Through focus groups and depth interviews, psychological researchers can zero in on very specific target audiences—and their leaders. "Buy this 24-year-old and get all his friends absolutely free," proclaims an ad for MTV directed to advertisers. MTV presents itself publicly as a place for rebels and nonconformists. Behind the scenes, however, it tells potential advertisers that its viewers are lemmings who will buy whatever they are told to buy. . . .

Home pages on the World Wide Web hawk everything from potato chips to cereal to fast food—to drugs. Alcohol and tobacco companies, chafing under advertising restrictions in other media, have discovered they can find and woo young people without any problem on the Web. Indeed, children are especially vulnerable on the Internet, where advertising manipulates them, invades their privacy, and transforms them into customers without their knowledge. Although there are various initiatives pending, there are as yet no regulations against targeting children online. Marketers attract children to Websites with games and contests and then extract from them information that can be used in future sales pitches to the child and the child's family. . . .

Some sites offer prizes to lure children into giving up the e-mail addresses of their friends too. Online advertising targets children as young as four in an attempt to develop "brand loyalty" as early as possible. . . .

The United States is one of the few industrialized nations in the world that thinks that children are legitimate targets for advertisers. Belgium, Denmark, Norway, and the Canadian province of Quebec ban all advertising to children on television and radio, and Sweden and Greece are pushing for an end to all advertising aimed at children throughout the European Union. An effort to pass similar legislation in the United States in the 1970s was squelched by a coalition of food and toy companies, broadcasters, and ad agencies. Children in America appear to have value primarily as new consumers. . . .

Just as children are sold to the toy industry and junk food industry by programs, video games, and films, women are sold to the diet industry by the magazines we read and the television programs we watch, almost all of which make us feel anxious about our weight. "Hey, Coke," proclaims an ad placed by *The Ladies' Home Journal,* "want 17-1/2 million very interested women to think Diet?" It goes on to promise executives of Coca-Cola a "very healthy environment for your ads." What's being sold here isn't Diet Coke—or even *The Ladies' Home Journal.* What's really being sold are the readers of *The Ladies' Home Journal,* first made to feel anxious about their weight and then delivered to the diet industry. Once there, they can be sold again—*Weight Watchers Magazine* sells its readers to the advertisers by promising that they "reward themselves with $4 billion in beauty and fashion expenditures annually."

In the same way, female drinkers are sold to the alcohol industry. As alcohol consumption has been falling in recent years, the alcohol industry has been directly targeting groups that traditionally have been lighter drinkers. One important target is women. Women's magazines are happy to cooperate. *Cosmopolitan* sells its readers to the alcohol industry in a trade publication ad proclaiming, "Cosmopolitan Readers Drank 21,794,000 Glasses of Beer in the Last Week." . . .

Young people are also an important market for alcohol. *Sport* magazine reminds the alcohol industry that "what young money spends on drinks is a real eye-opener," and, through its ad space, *Sport* is more than willing to help. "Black people drink too much," says an ad for the Black Newspaper Network. "Too much, that is," the copy continues, "for you to ignore." "Diario Las Américas readers Pour It On," echoes an ad in *Advertising Age* for a Spanish-language newspaper sold in Florida. The truth is that African-Americans and Latinos don't drink nearly as much as Caucasians, but they represent desirable new territory for the alcohol industry. And so the African-American and Latino media hand them over.

Perhaps this wouldn't matter very much if it didn't affect the content of the media. But it does. "Uncork the black market," says an ad for *Ebony* magazine in *Advertising Age,* which promises alcohol advertisers that "nothing sells black consumers better." A few years later *Ebony* did a story on the ten most serious health problems affecting blacks—but did not include the fact that alcohol is related to nine out of the ten health problems. There were eleven alcohol ads in this same issue of *Ebony.*

Magazines, television programs, newspapers, all in the business of attracting advertisers, certainly can't afford to offend them. On the contrary, they promise their advertisers an editorial climate in which their ads will be favorably received. . . .

Advertising's influence on media content is exerted in two major ways: via the suppression of information that would harm or "offend the sponsor" and via the inclusion of editorial content that is advertiser-friendly, that creates an environment in which the ads look good. The line between advertising and editorial content is blurred by "advertorials" (advertising disguised as editorial copy) "product placement" in television programs and feature films, and the widespread use of "video news releases," corporate public-relations puff pieces aired by local television stations as genuine news. Up to 85 percent of the news we get is bought and paid for by corporations eager to gain positive publicity. . . .

Nowhere is this more obvious than in most women's and girls' magazines, where there is a very fine line, if any, between advertising and editorial content. Most of these magazines gladly provide a climate in which ads for diet and beauty products will be looked at with interest, even with desperation. And they suffer consequences from advertisers if they fail to provide such a climate. . . .

An informal survey of popular women's magazines in 1996 found cover stories on some of the following health issues: skin cancer, Pap smears, leukemia, how breast cancer can be fought with a positive attitude, how breast cancer can be held off with aspirin, and the possibility that dry-cleaned clothes can cause cancer. There were cigarette ads on the back covers of all these magazines—and not a single mention inside of lung cancer and heart disease caused by smoking. In spite of increasing coverage of tobacco issues in the late 1990s, the silence in women's magazines has continued, in America and throughout the world. In my own research, I continue to find scanty coverage of smoking dangers, no feature stories on lung cancer or on smoking's role in causing many other cancers and heart disease . . . and hundreds of cigarette ads. . . .

Americans rely on the media for our health information. But this information is altered, distorted, even censored on behalf of the advertisers—advertisers for alcohol, cigarettes, junk food, diet products. We get most of our information from people who are likely to be thinking. "Is this going to cause Philip Morris or Anheuser-Busch big problems?" Of course, in recent years there has been front-page coverage of the liability suits against the tobacco industry and much discussion about antismoking legislation. However, there is still very little information about the health consequences of smoking, especially in women's magazines. The Partnership for a Drug-Free America, made up primarily of media companies dependent on advertising, basically refuses to warn children against the dangers of alcohol and tobacco. The government is spending $195 million in 1999 on a national media campaign to dissuade adolescents from using illicit drugs, but not a penny of the appropriated tax dollars is going to warn about the dangers of smoking or drinking. . . .

Although the conglomerates are transnational, the culture they sell is American. Not the American culture of the past, which exported writers like Ernest Hemingway and Edgar Allan Poe, musical greats like Louis Armstrong and Marian Anderson, plays by Eugene O'Neill and Tennessee Williams, and Broadway musicals like *West Side Story.* These exports celebrated democracy, freedom, and vitality as the American way of life.

Today we export a popular culture that promotes escapism, consumerism, violence, and greed. Half the planet lusts for Cindy Crawford, lines up for blockbuster films like *Die Hard 1 & 2* with a minimum of dialogue and a maximum of violence (which travels well, needing no translation), and dances to the monotonous beat of the Backstreet Boys. *Baywatch,* a moronic television series starring Ken and Barbie, has been seen by more people in the world than any other television show in history. And at the heart of all this "entertainment" is advertising. As Simon Anholt, an English consultant specializing in global brand development, said, "The world's most powerful brand is the U.S. This is because it has Hollywood, the world's best advertising agency. For nearly a century, Hollywood has been pumping out two-hour cinema ads for Brand U.S.A., which audiences around the world flock to see." When a group of German advertising agencies placed an ad in *Advertising Age* that said, "Let's make America great again," they left no doubt about what they had in mind. The ad featured cola, jeans, burgers, cigarettes, and alcohol—an advertiser's idea of what makes America great.

Some people might wonder what's wrong with this. On the most obvious level, as multinational chains replace local stores, local products, and local character, we end up in a world in which everything looks the same and everyone is Gapped and Starbucked. Shopping malls kill vibrant downtown centers locally and create a universe of uniformity internationally. Worse, we end up in a world ruled by, in John Maynard Keynes's phrase, the values of the casino. On this deeper level, rampant commercialism undermines our physical and psychological health, our environment, and our civic life and creates a toxic, society. Advertising corrupts us and, I will argue, promotes a dissociative state that exploits trauma and can lead to addiction. To add insult to injury, it then co-opts our attempts at resistance and rebellion.

Although it is virtually impossible to measure the influence of advertising on a culture, we can learn something by looking at cultures only recently exposed to it. In 1980 the Gwich'in tribe of Alaska got television, and therefore massive advertising, for the first time. Satellite dishes, video games, and VCRs were not far behind. Before this, the Gwich'in lived much the way their ancestors had for a thousand generations. Within ten years, the young members of the tribe were so drawn by television they no longer had time to learn ancient hunting methods, their parents language, or their oral history. Legends told around campfires could not compete with *Beverly Hills 90210*. Beaded moccasins gave way to Nike sneakers, sled dogs to gas-powered skimobiles, and "tundra tea" to Folger's instant coffee.

Human beings used to be influenced primarily by the stories of our particular tribe or community, not by stories that are mass-produced and market-driven. As George Gerbner, one of the world's most respected researchers on the influence of the media, said, "For the first time in human history, most of the stories about people, life, and values are told not by parents, schools, churches, or others in the community who have something to tell, but by a group of distant conglomerates that have something to sell." The stories that most influence our children these days are the stories told by advertisers.

DISCUSSION QUESTIONS

1. Look for two different advertisements of the same product. How does the place where the ad is located alter the content of the ad? What cultural stereotypes does this difference reveal?

2. How are advertising strategies that focus only on selling the product dangerous for society? Can you think of examples when the information in the ad is counter to what we know about health and well-being?

http://infotrac.thomsonlearning.com

InfoTrac College Edition

BONUS READING

Reiss, Steven. "Why America Loves Reality TV." *Psychology Today* 34 (September–October 2001): 52–53.

This brief reading examines the reasons why people like to watch reality television programs. The pervasive presence of reality television in our culture today makes these shows hard to avoid. People may feel the need to watch in order to engage in conversations with their peers about current happenings on these shows. Why do you believe people are drawn to these programs? Given what you know from Jean Kilbourne's article, why do the networks continue to develop new ideas for reality programming?

SEARCH TERMS

culture and body image
gender and eating disorders
gender roles and advertising
body modification
terrorism and culture

9

Leaving Home for College: Expectations for Selective Reconstruction of Self

DAVID KARP, LYNDA LYTLE HOLMSTROM, AND PAUL S. GRAY

Many young adults leave home for the first time when they go away to college. This article addresses the changes young students go through when they leave high school and their family home. The authors discuss how personal changes in identity and perceptions of self are more significant than the geographical move to college.

In their important and much discussed critique of American culture, *Habits of the Heart* (1984). Robert Bellah and his colleagues remark that American parents are of two minds about the prospect of their children leaving home. The thought that their children will leave is difficult, but perhaps more troublesome is the thought that they might not. In contrast to many cultures, American parents place great emphasis on their children establishing independence at a relatively early age. Still, as Bellah's wry comment suggests, they are deeply ambivalent about their children leaving home. The data presented in this paper, part of a larger project on family dynamics during the year that a child applies for admission to college, show that such ambivalence is shared by the children. Our goal here is to document some of the social psychological complexities of achieving independence in America by analyzing the perspectives of 23 primarily upper-middle-class high school seniors as they moved through the college application process and contemplated leaving home.[1]

Of course, a great deal has been written about the internal conflict that surrounds any significant personal change (most obviously, Erik Erikson 1963, 1968, 1974, 1980; see also Manaster 1977; O'Mally 1995). Although researchers have attended to the phenomenon of "incompletely launched young adults" (Heer, Hodge, and Felson 1985; Grigsby and McGowan 1986; Schnaiberg and Goldenberg 1989), little has been written about how relatively sheltered, middle- to upper-middle-class children think about "leaving the nest." Leaving home for college is perhaps among the greatest changes that the economically comfortable students we interviewed have thus far encountered in their lives. For them, going to college carries great significance as a coming-of-age moment, in part because it has been long anticipated and not to do so would be unacceptable from a normative stand point. Literature on students who "beat the odds" by going to college suggests that this is also an important transition for them, but one carrying fundamentally different meanings. Unlike the middle- or upper-middle-class students we interviewed, who are trying, at the least, to retain their class position, students arriving at college from less privileged backgrounds must confront wholly new cultural worlds (Rodriguez 1982; Smith 1993; Hooks 1993).

The 23 students with whom we were able to complete interviews simply assumed, as did their parents, that they would go to college.[2] Among the 30 sets of parents, all but four individuals had attended college (and two received some different training beyond high school). All of the adults, however, felt strongly about the necessity of college attendance for their children. One of the four who did not go to college, a self-made and extraordinarily successful entrepreneur, did offer some reservation about the utility of an education in the rough and tumble "real world." Even so, both he and his wife were highly invested in getting their son into a prestigious college. While all of the children knew their parents' expectations and fully expected to meet them, we did speak with two students who had some misgivings about whether they really wanted to go to college. Like their counterparts in our sample, these students knew they would go, but still entertained private doubts about their interest in and motivation for college work.[3]

What does it mean to become independent of one's parents, family, and high school friendship groups? As Anna Freud noted, "few situations in life are more difficult to cope with than the attempts of adolescent children to liberate themselves" (Bassoff 1988, p. xi). Young people are ambivalent regarding independence; it is hard to break away. Their ambivalence embodies both symbolic and pragmatic dimensions. Symbolically, independence is the desired outcome of a necessary process of differentiation (Blos 1962). The task for adolescents is "to find their own way in the world and develop confidence that they are strong enough to survive outside the protective family circle" (Bassoff 1988, p. 3). To establish their own identity and sense of purpose, ". . . they need to wrench themselves away from those who threaten their developing selfhood" (Bassoff 1988, p. 3; see also Campbell, Adams, and Dobson 1984; Katchadourian and Boli 1994). However, independence also has a pragmatic side. In college, young people can "start over"; they can make new friends, establish intimate relationships, and develop the skills and knowledge to help them become self-supporting adults. "But the truth is that they are not sure they can take care of themselves or that they want to be left alone" (Bassoff 1988, p. 3). . . .

IDENTITY AFFIRMATION, IDENTITY RECONSTRUCTION, AND IDENTITY DISCOVERY

While the students in this study anticipated college as a time during which they would maintain, refine, build upon, and elaborate certain of their identities, they also anticipated negotiating some fundamental identity changes. The students saw college as the time for discovering who they *really* were. They anticipated finding wholly new and permanent life identities during the college years. In addition, they believed that going to college provides a unique opportunity to consciously establish some new identities. Repeatedly, students described the importance of going away to college in terms of an opportunity to discard disliked identities while making a variety of "fresh starts." Their words suggest that college-bound students look forward to re-creating themselves in a context far removed (often geographically, but always symbolically) from their family, high school, and community. The immediately following sections attend, in turn, to how upper-middle-class high school seniors (1) anticipate change, (2) strategize about solidifying certain identities, [and] (3) evaluate identities they wish to escape. . . .

Anticipating Change

Along with such turning points as marriage, having children, and making an occupational commitment, it is plain that leaving for college is self-consciously understood as a dramatic moment of personal transformation. The students with whom we spoke all saw leaving home as a critical juncture in their lives. One measure of consensus in the way our 23 respondents interpreted the meaning of leaving home is the similarity of their words. Students used nearly identical phrases in describing the transition to college as the time to "move on," "discover who I really am," to "start over," to "become an adult," to "become independent," to "begin a new life." The students, moreover, explicitly saw going to college as the "next stage" of their lives. . . .

While all the students interviewed recognized the need for change and were looking forward to it, their certainty about the appropriateness of moving on did not prevent them from feeling anxiety and ambivalence about the transition to college. Theirs is an anticipation composed of optimism, excitement, anxiety, and sometimes fear.

> [I'm] starting the rest of my life. I mean, deciding what I'm going to do and figuring out my future. I mean, that's one thing I'm looking forward to, but it's also one thing I'm not looking forward to. I have mixed feelings about that. It's exciting to figure out your future. In another sense it's scary to have all of the responsibility. (White male attending a public school)

These comments suggest that the prospect of leaving home generates an anticipatory socialization process characterized by multiple and sometimes contradictory feelings and emotions. Students long for independence, anticipate the excitement that accompanies all fresh starts, but worry about their ability to fully meet the challenge. . . .

Affirming Who I Really Am

The one concrete and critical choice that college-bound students must make is which school, in fact, to attend. This decision is often an agonizing one for both students and their parents and involves very high levels of "emotion work" (Hochschild 1983). The significance of making the college choice and the anxiety that it occasions go well beyond questions of money, course curricula, or the physical amenities of the institutions themselves. What makes the decision so difficult is that the students know they are choosing the context in which their new identities will be established, . . . The fateful issue in the minds of the students is whether people with their identity characteristics and aspirations will be able to flourish. Consequently, it is not surprising that the most consistent and universal pattern in our data is the effort expended by students to find a school where "a person like me" will feel comfortable. . . .

In the most global way, prospective students were searching for a place where the students seemed friendly. On several occasions, students remarked that they were turned on or off to a school because their "tour guide" was either really nice or not friendly enough. . . .

In contrast to the students-like-me theme, an interesting sub-set of seniors expressed a strong interest in diversity. These students not only wanted to meet new people, but different kinds of new people. Students who wanted diversity were excited at the prospect of meeting people different from themselves as a critical learning experience. It is important to note that it was primarily the minority students we interviewed who looked for diversity as they contemplated colleges. An Asian student put it this way:

> The more mixed the better. I think interaction with other ethnic and racial groups is very healthy. If possible, I would not mind having, you know, like an Afro-American roommate. I'd love to. (Asian male attending a public school) . . .

While the statements immediately above illustrate that students make careful assessments about the goodness of fit between certain aspects of themselves and the character of different colleges, a dominant theme in the interviews concerned change. Students repeatedly commented that, during their college years, they expected their identities to shift in two fundamental ways. First, they anticipated discovering "who I am" in the broadest sense. Second, they saw college as providing a fresh start because they could discard some of their disliked, sticky identities, often acquired as early as grade school.

Creating the Person I Want to Be

. . . Seen in terms of Erving Goffman's (1959) dramaturgical model of interaction, going to college provides a new stage and audience, together allowing for new identity performances. Goffman notes (1959, p. 6) that "When an individual appears before others his actions will influence the definition of the situation which they come to have. Sometimes the individual will act in a thoroughly calculating manner, expressing himself in a given way solely in order to give the kind of impression to others that is likely to evoke from them a specific response he is concerned to obtain." To the extent that such impression-management is most centrally dependent upon information control, leaving home provides an unparalleled opportunity to abandon labels that have most contributed to disliked and unshakable identities. When students speak of college as providing a fresh start, they have in mind the possibility of fashioning new roles and identities. Going to college promises the chance to edit, to revise, to re-write certain parts of their biographies.

It's sort of like starting a new life. I'll have connections to the past, but I'm obviously starting with a clean slate. . . . Because no one cares how you did in your high school after you're in college. So everyone's equal now. (White male attending a public high school) . . .

As students described their hopes about college, the theme of "fresh starts" was almost universally voiced. . . . Leaving home, friends, and community offers students the possibility to jettison identities which are the product of others' consistent definitions of them over many years. Going to college provides a unique opportunity to display new identities consistent with the person they wish to become.

The data presented thus far are meant to convey the symbolic weightiness of the transition from high school to college. Every student with whom we spoke saw leaving home as a critical biographical moment. They see it as a definitive life stage when their capacity for independence will be fully tested for the first time. Some have had a taste of independence at summer camps and the like, but the transition to college is viewed as the "real thing." Their words, we have been suggesting, indicate that they see strong connections among leaving home, gaining independence, achieving adult status, and transforming their identities. Students carefully attempt to pick a college where they will fit in, thus indicating the importance of retaining and consolidating certain parts of their identities (see Shreier 1991). In addition, they believe that they will discover, in a holistic sense, who they "really" are during the college years.

WILL THEY MISS ME?

. . . The family is a social system in which roles are interconnected and interdependent. When a child goes off to college, the system is disturbed and the family will try to adapt to the new circumstances. College-bound seniors worry about this process of adaptation. They speculate that their remaining siblings will miss them, or will be left to face the unremitting attentiveness and concern of parents. They also wonder about prospective changes in their parents' marital relationship. In particular, they are concerned for their mothers, whom they identify as being more invested than their fathers in keeping the family system *status quo ante*. Finally, and most significantly, these late adolescents manifest insecurity about their place in the family, especially now that they are leaving. Several of them remarked ruefully, "I should hope they feel some grief [laughter]" "I think they'll be lonelier. I hope they will." "They'll miss me, I hope. . . . I hope they feel my presence being gone. . . . They don't have to be, like, mourning my departure, but just a little bit would be nice." It's not that they actually want their parents and siblings to suffer, but missing them would be proof positive that their membership in the family was valued, and that their future place in the family system is assured, in spite of their changing addresses. . . .

In many of our conversations, it appeared that the worst thing about going away to college was that the young people would no longer be able to participate in many aspects of family life. However, perhaps no issue symbolizes the worry associated with leaving home as powerfully as pending decisions over space in the household. How quickly one's bedroom is claimed by other members of the family is, for many of these students, a commentary on the fragility of their position. Although Silver (1996) points out that both the home room and college dorm room are used to symbolically affirm family relations, our conversations with students were more focused on the meanings they attached to their bedrooms at home. One senior said, "They always joke around and they say, 'Oh, we're going to make your room into a den.'". . .

Some of the seniors are beginning to understand that the nature of relations with their parents will be altered forever. They will have much more discretion concerning what to reveal about themselves, and therefore much more control over the impression they choose to give their parents. As one young woman put it, "I will experience a lot of things without them there, so that they won't know that they've happened . . . [unless] I tell them or if they can see a difference in me." Others expressed shared anxieties about personal transformations and the consequent stability of their place in the family constellation. . . .

What are we to make of these worries, speculations, and musings? College-bound young adults genuinely want to remain attached to their families, even as they are yearning for true independence. Getting into college is understood as a point of departure which has the potential to alter fundamentally their relationship with their family. However, in spite of their worries, most students see the transition to college as a good thing—a positive transformation with life-long consequences. They cannot predict precisely how their relations with parents and siblings will change, but they know for sure that they have initiated a process that will alter the character of these primary relationships. Such knowledge is plainly implicated in the calculus of ambivalence they feel about leaving home:

> It's like, if you want to be treated like an adult, you have to act like an adult. If you want to be treated like a child, act like a child. If you want to be treated like an adult the rest of your life, you've got to start sometime. (White male attending a public high school)

"You've got to start sometime." That, of course, is exactly what they are doing as they embark on their great adventure of self-discovery, into college first and hopefully, thereby, toward full adulthood.

NOTES

1. We used father's occupation as a proxy for social class. We characterized our sample as predominantly upper-middle class. A sampling of the types of father's occupations that warrant this description includes: physician, lawyer, professor, administrator, and architect. A few occupations were either higher or lower in status.

2. Either because we could not reach them or because they declined to be interviewed, we did not speak to eight of the 31 students originally included in our sample. The number is 31 because one of the 30 families had twins.

3. One student, who declined to be interviewed, did not complete the college application process during his senior year in high school. He was the only student in our sample who did not anticipate attending college in the year following high school graduation.

REFERENCES

Basoff, Evelyn. 1988. *Mothers and Daughters: Loving and Letting Go.* New York: Penguin Books.

Bellah, Robert, Richard Madsen, William Sullivan, Ann Swidler, and Steven Tipton. 1985. *Habits of the Heart: Individualism and Commitment in American Life.* Berkeley: University of California Press.

Blos, Peter. 1962. *On Adolescence: A Psychoanalytic Interpretation.* New York: Free Press.

Campbell, Eugene, Gerald Adams, and William Dobson. 1984. "Familial Correlates of Identity Formation in Late Adolescence: A Study of the Predictive Utility of Connectedness and Individuality in Family Relations." *Journal of Youth and Adolescence* 13: 509–525.

Erikson, Erik. 1963. *Childhood and Society,* 2nd ed. New York: W. W. Norton.

———. 1968. *Identity: Youth and Crisis.* New York: W. W. Norton.

———. 1974. *Dimensions of a New Identity.* New York: W. W. Norton.

———. 1980. *Identity and the Life Cycle.* New York: W. W. Norton.

Goffman, Erving. 1959. *The Presentation of Self in Everyday Life.* Garden City, NY: Doubleday Anchor.

Grigsby, Jill, and Jill McGowan. 1986. "Still in the Nest: Adult Children Living with Their Parents." *Sociology and Social Research* 70: 146–148.

Heer, David, Robert Hodge, and Marcus Felson. 1985. "The Cluttered Nest: Evidence that Young Adults Are More Likely to Live at Home Now Than in the Recent Past." *Sociology and Social Research* (69): 436–441.

Hochschild, Arlie. 1983. *The Managed Heart: Commercialization of Human Feeling.* Berkeley: University of California Press.

Hooks, Bell. 1993. "Keeping Close to Home: Class and Education." Pp. 99–111 in *Working-Class Women in the Academy,* edited by Michelle Tokarczyk and Elizabeth Fay. Amherst, MA: The University of Massachusetts Press.

Katchadourian, Herant, and John Boli. 1994. *Cream of the Crop: The Impact of Elite Education in the Decade After College.* New York: Basic Books.

Manaster, Guy. 1977. *Adolescent Development and the Life Tasks.* Boston: Allyn and Bacon.

O'Mally, Dawn. 1995. *Adolescent Development: Striking a Balance Between Attachment and Autonomy.* Ph.D. dissertation, Department of Psychology, Harvard University, Cambridge, MA.

Rodriguez, Richard. 1982. *Hunger of Memory: The Education of Richard Rodriguez.* Boston: David R. Godine.

Schnaiberg, Allan, and Sheldon Goldenberg. 1998. "From Empty Nest to Crowded Nest: The Dynamics of Incompletely-Launched Young Adults." *Social Problems* 36: 251–269.

Schreier, Barbara. 1991. *Fitting In: Four Generations of College Life.* Chicago: Chicago Historical Society.

Silver, Ira. 1996. "Role Transitions, Objects, and Identity." *Symbolic Interaction* 19: 1–20.

Smith, Patricia. 1993. "Grandma Went to Smith, All Right, But She Went from Nine to Five: A Memoir." Pp. 126–139 in *Working-Class Women in the Academy,* edited by Michelle Tokarczyk and Elizabeth Fay, Amherst, MA: The University of Massachusetts Press.

DISCUSSION QUESTIONS

1. What changes did you go through (or are you going through) during your first year of college? How do these changes influence your self-identity and how others perceive you?

2. When you go home for vacations and visits, how does home feel different now that you have lived away? What feels the same?

10

Navajo Women and the Politics of Identity

AMY J. SCHULZ

In this original research, Schulz examines the construction of American Indian identity among three generations of Navajo women. Her interviews reveal how these women balance multiple identities in response to historical and social situations. Against the backdrop of American Indian history, Schulz concludes that Navajo women construct and negotiate their identities as both Indian and Navajo.

The resurgence—or continuation—of ethnic identification and conflict in the United States belies the American myth of the "melting pot" and calls into question the inevitability of the melting process. The continued salience of group identities invites attention to the processes and structures that influence their construction, deconstruction, and reconstruction over time (Anderson 1983; Barth 1969; Nagel 1996; Waters 1990). How are collective identities created and transformed? How do changes in social, political, economic, and cultural contexts influence the problematics as well as the possibilities of identities? Identities both reflect and potentially disrupt or re-create social and political relationships within and between groups. Examining the construction and reconstruction of identities over time contributes to our understanding of social and political processes through which individuals and groups locate themselves in relation to others, understand themselves, and define their possibilities.

From: *Social Problems* 45 (August 1997): 336–53. Reprinted with permission.

This analysis addresses these questions by examining intergenerational changes in the construction of collective identities and the meanings associated with those identities. The analysis focuses on two of many dimensions of identity available and salient to the women who participated in the study: "Indian" as a supratribal identity; and "Navajo" or tribal identity. I examine differences between younger and older Navajo women in the social, economic, political, and cultural contexts they experienced, the problems associated with Indian and Navajo identities, and the resources available to them to negotiate those identities. Drawing on in-depth interviews conducted between 1990 and 1992, I explore these dimensions of identity as they are constructed in the women's narratives and in their interactions with me, an Anglo researcher. The approach used in the analysis emphasizes the situated nature of identities, and the creative—but also socially structured—actions of individuals in constructing collective and personal identities. Furthermore, as collective identities are inextricably connected to efforts to construct a coherent sense of self, my analysis explores the use of multiple, sometimes fragmented and conflicting identities as women define themselves in relationship to others (Calhoun 1994; Lemert 1994; Woodward 1997). . . .

CONCEPTUAL FRAMEWORK

Identities are not unilateral or constant. Rather, there are many different dimensions or layers of identity, including nationality, ethnicity, gender, family, social class, and sexuality (Barth 1969; Hall 1990; Taylor 1989; Woodward 1997). The salience of these identities may vary with situational and political factors (Cohen 1985; Cornell 1988; Nagel 1996). Furthermore, the identities themselves may be contradictory, conflicting, or fragmented, creating tension within the individual as well as within and between groups as identities and their meanings are negotiated (Calhoun 1994; Waters 1990; Wiley 1994; Woodward 1997). The politics of identity involve the disruption or reconstruction of identities, or the meanings that have been associated with them, and are associated with efforts to shift power relations within or between groups (Calhoun 1994).

The identities examined in the following pages are arenas in which social and political relationships between Anglos and American Indians are negotiated and contested. Before contact with Europeans, those indigenous to North America organized their political, social, and economic relationships in a wide variety of ways, from hierarchical to loosely-knit networks (Cornell 1988). Kinship and clan networks, bands, and broader social groups came together to create tribes, as they searched for ways to respond to the social and political disruptions created by contact with European-Americans (Cornell 1988). The multiple levels of group identity available to contemporary American Indians, including supratribal or pan-Indian, tribal, and subtribal (clan, kinship), were shaped by these processes and reflect historical and contemporary relationships between Anglos and Indians. In this study I focus on self and collective identities as key conceptual frames through which individuals define or locate themselves, and through which I examine political and social relationships. Excerpts from women's narratives emphasize the creative actions of individuals as they negotiate identities within particular social, economic, political, and cultural contexts (Nagel 1996; Waters 1990; Woodward 1997). Furthermore, these narrative accounts speak to the important role played by women's day-to-day actions in resisting assimilation into dominant social and cultural systems (Ward 1993).

METHODS

The analysis presented here used primary and secondary sources to examine United States Indian policies as records of the political context, and of Anglo representations of—and beliefs about—American Indians between the late 1800s and the late 1900s. In addition, I present findings from an inductive analysis of thirty-one in-depth interviews conducted between 1990 and 1992 with women living on the Navajo Nation, ranging in age from 15 to 76 at the time of the interviews, to discuss the implications of these policies and beliefs for Navajo women. My purpose was to understand women's experience of, and their strategies for managing, pressures to be incorporated into Western political, economic, and cultural systems.

The interviews consisted of accounts of personal life histories, guided by questions to encourage elaboration about experiences with the schools, families, and work. In addition, participants were asked to describe important challenges or conflicts they had experienced and their responses to those challenges. Finally, the interviews included questions about the salience and meanings of their identities as Indian, Navajo, and women[1]: for example, "what has it meant to you to be Navajo?"; and "what does it mean to be 'Indian'?"[2] Each participant received a transcript of her interview and was invited to comment or to discuss it with me. In addition, I shared preliminary analysis with several respondents as a means of member validation and to encourage further discussion of the material (Erikson 1976; Mbilinyi 1989; Oakley 1981). This process led to many rich discussions with participants that were subsequently incorporated into the analysis.

I located participants through snowball sampling, beginning with interviews with women who were friends or acquaintances, and who then helped identify other potential study participants. Criteria for participation in the study were broadly defined to include women who currently lived on the Navajo Nation; spoke English; and were interested in and willing to talk about their experiences of social and cultural change. This selection process favored women with relatively more exposure to United States educational and labor systems than was the norm for women living on the Navajo Nation in 1990. For example, 68 percent of the participants in this study had completed high school and 12 percent had completed college, as compared to 42 percent and 3 percent, respectively, of all women living on the Navajo Nation in 1990.[3] Similarly, the women who participated in this study were more likely than the average woman on the Navajo Nation to participate in wage work, with 57 percent of participants engaged in the labor force as compared to 39 percent of all women on the Navajo Nation in 1990.[4] Most of my respondents live within extended family networks on the Navajo Nation; however, they have had relatively more exposure to, and day-to-day experience with, United States educational and labor systems than the average woman living on the Nation. Thus, the women who participated in this study live on the boundaries between the Navajo and Anglo worlds.

All of the women interviewed for this study spoke Navajo more or less fluently, in addition to English. Women in the oldest cohort and most of those in the middle cohort spoke Navajo as their first language, and had learned English in school. For most of the youngest cohort, English was their first language, and most had struggled to learn Navajo in school with varying degrees of success. Nearly all of the respondents had lived for some period of their life away from the reservation: at boarding school; doing migrant farm work; at college; or living or working in an urban community. The timing of these experiences varied, as some had left during childhood, some during adolescence, and others during early adulthood. Most had returned to the reservation by middle adulthood, although a few of the older women had retired from wage work in nearby communities and then returned to their home communities on the Navajo Nation.

The analysis presented here draws on the thirty-one interviews described above, divided into three cohorts: thirteen women born prior to 1946 (aged 46 to 76 at the time of the interviews); nine women born between 1946 and 1960 (aged 32–44 at the time of the interviews); and nine women born between 1961 and 1976 (aged 15–27 at the time of the interviews). These three cohorts were chosen because of their relationship to changes in United States Indian policies, these policies influenced the social conditions and formal educational institutions that women experienced as they came to adulthood.

I argue in the following pages that Navajo women's experiences are socially and historically patterned, and their responses to social and historical circumstances are shaped by the resources available to them at various historical moments. On the average, the women in this study have more exposure to educational institutions and labor force participation than is the norm for women living on the Navajo Nation and more day-to-day exposure to extended kinship networks, spiritual practices, and Navajo language than many Navajo who live in urban communities. Their particular social location, as members of extended family networks on the Navajo Nation and as participants in United States educational and labor force systems, mean that they experience tensions related to the construction of "Indian" and "Navajo" identities in their daily lives. This social location, along with their historical location as members of the cohorts described above, shaped the issues they confronted and the resources at their disposal as they negotiated these identities. The women in this study, in the language of grounded theory methodology, represent "particularly rich cases" (Strauss and Corbin 1990) whose experiences offer insights into the construction and reconstruction of identity group boundaries and meanings precisely because of the salience of these negotiations in their lives. . . .

NEGOTIATING INDIANNESS

. . . The different historical periods in which women came of age influenced their personal histories and the specific issues, concerns, resources, and constraints that they encountered as they negotiated their identities. Despite their varied experiences, as women talked about what it meant to "be Indian" a central theme that cut across cohorts was the idea of difference. For women of all ages to be Indian meant to be set apart, to be distinguished from others (Schulz 1994). For example, Nora, born in 1937, left her Tewa family of origin to live with an Anglo foster family in Oklahoma to attend junior and senior high school in the 1950s. She described what it had meant to her to be identified as Indian at that time:

> Sometimes, like when I went away to school, I felt among all the Anglo that I was different. Then inside, I knew that I wasn't all that different from them. I'm alive just like them, my body is just like them, only the skin is different. But I have a mind just like them and think like them. I guess the only thing is the Indian culture is different from yours. You have a culture too, but being an Indian—I'm proud that I am an Indian. . . . It was not really how they . . . they treated me well. It's just . . . I always thought that Anglos knew more than I did. Like when you watch movies, in the wild west kind, they make the Indians lower than themselves and I guess I had that feeling in myself.
>
> (Nora, age 54)

In this excerpt, Nora struggled with an externally imposed identity that defined her as not only "different," but "lower than" Anglos. The salience of Indianness was highlighted as she lived among, and interacted with, non-Indians. Living in a predominantly White community in post World War II America, she encountered systems of beliefs that constructed her as a member of a group that was not only different, but devalued. The "antipodal categories" articulated by Cornell (1988) . . . were salient in the 1950s, juxtaposing Anglos as good and right and powerful against Indians as "pagan, savage, nearly dumb." These dichotomies served to devalue Indians and to legitimize their political, social, and economic marginalization (Berkhofer 1979; Burt 1986).

As Nora pulled apart the strands of her experience, she noted that her White foster family "treated me well" but they were unable to protect her from the effects of these deeply rooted ideologies of racial hierarchy. Looking back on this experience some 40 years later, Nora noted that while she *knew* intellectually that she was not different or inferior to Anglos—that both had skin, minds, and cultures—she still *felt* that she was different and believed that she was inferior to Anglos.

As she described her experience as a child, Nora was unable to construct an alternative identity that would disrupt or challenge the representations of Indianness she encountered in her everyday life. As an adult, she addressed these representations by pointing out to me that I too had a culture even though it differed from her own. Emphasizing our sameness, with the exception of the color of our skins and our cultures, she challenged the differentiation underlying the devaluation of Indians as a group and of herself as a member of that group.

By the later 1940s, more elementary schools began to be located on the Navajo Nation and children increasingly lived with family members while they attended elementary school. Adele, born in 1945, attended day school through her elementary years, where she encountered strictly enforced policies that forbade the use of Navajo language or other "Indian" practices within the school. As she spoke of her school experiences, Adele also described her parents' efforts to confront school personnel and bring the attention of other tribal members to bear on the harsh treatment of children in the schools. In addition to modeling active resistance, her parents also provided her with an alternative sense of herself, grounded in her Navajo and Apache heritage. Adele's connection to those tribal identities was apparent as she spoke about what it meant to be Indian:

> I'm considered as an Indian but within Indian society there are different tribes that have different languages and cultures. There are differences among the Indian people. I think it's unique that I'm from two tribes—from Navajo and Apache. I'm able to speak and understand two languages. I was raised up by two cultures and . . . I understand two languages and two cultures too.
>
> (Adele, age 46)

Adele's choice of words in this excerpt accentuated the externally imposed nature of Indian identity—her use of passive voice emphasized that she was considered *by others* (non-Indians) to be Indian, but that she did not claim that identity. Adele's response interrupted stereotypic and homogeneous images of Indians by highlighting tribal differences. Finally, she used herself as an example to drive home the point that "Indians" are actually many different communities with different languages, cultures, and identities. In so doing, she actively negotiated the meanings associated with Indian identity, resisting externally imposed and reductionist ideas and promoting instead a more complex notion of what it meant to be Indian.

Born in 1952, Ursula also lived on the Navajo Nation within an extended family as she attended school. Like Adele, she spoke of incidents in school in which her Navajo spiritual beliefs and practices were explicitly devalued by school personnel, describing the impact of those experiences as she said ". . . and the more and more messages I got like that, I had really bad feelings about myself." However, Ursula also described support for the development of a positive sense of herself, noting that her father "used to encourage us to speak up, he'd teach us that it was important for our voice to be heard and our opinion was important." Thus, like Adele, Ursula had support from family members that provided alternatives to the sense of self derived from representations of Indians constructed by outsiders. When I asked what it meant to her to be Indian, Ursula replied:

> I guess when I think about myself as an Indian I get in touch with my humanity. Because to me, the way I was taught, being Indian is a label. And who you are is *ashdlaa'o*[5]—man with five fingers. You exist as part of the human race. As an Indian you have skin of color—*bitsj' yishtlizh*—they call it—the color of the earth. Your skin is like the color of the earth, so therefore you are an earth person. So this is the Navajo coming in, for the way they teach us is—first you're a man with five fingers, that puts you in touch with the human race. When you get in touch with your humanness, then you have respect for others that are like you . . . whether they have different colors or not, they're still humans. They teach you to have respect for that.
>
> (Ursula, age 44)

Ursula's response highlighted multiple dimensions of identity and the ways in which they may be intertwined or in conflict. She began by defining Indian as a label imposed by outsiders, not an identity that she would claim. Like Nora, she emphasized similarities among human beings, challenging the construction of an Indian "other" and representations of Indians as less than human, or as entities to be civilized, assimilated, or annihilated. Her words challenged the meanings that outsiders have associated with the term Indian. . . .

Each of these women constructed her response to questions about identity that I—an Anglo/outsider—had asked them. Their responses cannot be separated from the context in which they were elicited, shaped by the salience of my outsider status and what that represented in our interactions (Cooley 1902). Their responses were also shaped by a history in which Indians have been defined as "different" by people like me, and that difference has been used as justification for their political and economic marginalization as well as the destruction of their unique identities as Navajo. The women drew upon more complex identities as Navajo, human beings, and in Adele's case, Apache. They used the Navajo language and knowledge of their own and others' histories, to disrupt and challenge stereotypic images and to construct alternative images of what it meant to "be Indian."

The responses of women born after 1960 were qualitatively different from those of the older cohorts when asked what it meant to be Indian. Most women in this youngest cohort embraced an Indian identity at the same time that they recognized its stigmatizing and marginalizing potential. For example, Reyna, born in 1974, said:

> For me, being an Indian is the greatest thing that ever happened to me. Even though . . . some [in] White society put us down. But we're still there, and we have all these cultures and traditions, different values that we carry are really sacred and spiritual. There's no way no one can take that away from us. Because we've come through that and it's helped us along the way . . . it's our own values and our traditions and we should keep it. It makes us stay together—it holds us together with a lot of strength.
>
> (Reyna, age 17)

Unlike the women in the previous cohort, Reyna did not deny or deconstruct the meanings associated with the term Indian. However, as she embraced it, she also recognized conflicts associated with that identity—in particular, the denigration and marginalization of those identified as Indian. She noted that culture, tradition, and values have been central to the ability of the group to stay together as a coherent entity, and to overcome struggles for survival. She did not, however, explicitly recognize *different* cultures, traditions, and values that are specific to different groups included within the term "Indian"—a distinction made repeatedly by the women in the previous cohort.

Reyna's response reflects the general tendency of women in the youngest cohort not to deny an Indian identity. In contrast to the women in the older cohorts, whose encounters with negative labels associated with Indianness were direct and explicit, Reyna's experience attending a community-run school on the Navajo reservation had buffered her somewhat from similar experiences. Furthermore, growing up following the social movements of the 1960s and 1970s, Reyna had access to, and social support for articulating more positive constructions of pan-Indian identity.

Women in this youngest cohort tended to use different strategies from the older women to resist the marginalizing and stigmatizing potential of Indian identification.

For example, Sandra noted that "I know that it's really hard to be an Indian off the reservation. There's a lot of times when the stereotypes can really bring you down." She went on to describe her efforts to confront her college classmates' stereotypic portrayal of "all Indians (as) drunks, . . . or not smart, . . . or not responsible . . .":

> I always want to show them that those things are not true. . . . (I tell them) that my mom was a personnel manager of a school, and it's not like that at all. There are some people that are that way, and there are some people that aren't.
>
> (Sandra, age 27) . . .

The education and other life experiences of this youngest cohort were shaped by the shift to self-determination that emerged from the activism of the 1960s and 1970s (Senese 1991; Szasz 1973). As noted earlier, most women in this age range had attended elementary and often junior and senior high schools located on the Navajo Nation, and several had attended community-run schools with an explicit emphasis on bilingual education and Navajo history and culture. Many of these young women had grown up in homes where their parents, products of the boarding schools of the 1940s and 1950s, spoke English fluently and where they learned English as their first language. As a result, while their parents had struggled to learn the English language in school, many of these young women struggled to learn to speak, read, and write the Navajo language. Their sense of themselves as Navajo was developed in part through their interactions at school and the intentional efforts of teachers and parents to teach them what it meant to be Navajo. This exposure to Navajo language and identity differed from that of the older cohorts, who developed a Navajo identity through interactions with extended family members and encountered explicit efforts to eradicate it in the schools.

These results illuminate changes in the ways that women living on the Navajo Nation negotiated the meanings of Indianness, and the influence of social, economic, and political contexts on those negotiations. Women born prior to the social movements of the 1960s were more likely than younger women to distance themselves from an Indian identity, emphasizing or making salient their tribal affiliations and using them to disrupt dominant group images of Indians. Those who came of age after 1960 were much more likely to actively embrace an Indian identity. Women in both groups, however, sought to claim particular meanings associated with Indianness (e.g., citizens, human beings) and to disrupt others (e.g., drunken, illiterate, warriors), positioning themselves as different from Anglos but resisting the association of difference with deficiency. Thus, while the particular strategies and forms through which Indianness was contested changed, the contestation itself continued. . . .

CONCLUSION

The processes that shape ethnic group identities are complex and multifaceted, constructed against a backdrop of history, inter- and intra-group relationships. The analysis presented in the preceding pages emphasizes the interplay among multiple identities, and the historically as well as socially situated nature of those identities (Taylor 1989; Woodward 1997). As women describe their efforts to negotiate Indian and Navajo identities, the complexities of constructing dynamic and multidimensional personal identities within particular political and social contexts become visible.

The salience of different layers of identity varies situationally and contextually. Across the three cohorts, Indian identity was salient in relation to Anglos and reflected politicized relationships between the groups. As women spoke about what it meant to them to be Indian, their examples referenced interactions with Anglo foster families, teachers, school mates, tourists, and researchers, as well as their encounters with negative and stereotypic representations of Indians in film and literature. Throughout the interviews women attempted to interrupt power relations between Indians and Anglos by disrupting and reconstructing Indian identities and their associated meanings (Calhoun 1994; Cohen 1985; Cornell 1988; Nagel 1996).

The consistency of this resistance across all the cohorts in this study speaks to the importance of women's day-to-day resistance to the assimilationist efforts of the dominant group (Ward 1993; Yuval-Davis 1997). Women struggle to construct identities that affirm their cultural distinctiveness, but that do not accept the pervasive negative constructions of that difference; they assert and seek to maintain distinct, valued identities. The strategies they use toward this end vary according to the resources, both material and conceptual, available at different historical moments (Cornell 1988; Ortner 1989; Swidler 1986). These resources include language, history, and alternative identities, and access to them is influenced by historical patterns and events that shape the social and political contexts within which women's personal histories unfold.

Women who came of age prior to the civil rights movements of the 1960s, in general, encountered a social and political context in which American Indians were denigrated and efforts to promote economic and cultural assimilation were explicit. Those who had grown up in extended families used their Navajo identities as alternative frames to distance themselves from constructions of Indians as inferior, savage, or uncivilized. Arguing that they were labeled *by others* as Indian, women rejected both the label and the meanings associated with this externally imposed identity. Their claim to particular tribal identities subverted the homogenizing and marginalizing potential of Indianness.

For these women, the politics of identity involved not simply naming or labeling, but defining the meanings associated with the name and, perhaps most importantly, establishing who has claim to define those meanings (Barth 1969; Cohen 1985; Nagel 1996). The subjective experience of themselves as Navajo, Navajo language, and Navajo teachings or belief systems were resources upon which they drew to define themselves actively *against* the definitions of outsiders. They used Navajo identity to locate themselves within a history that they claimed, positioning themselves in relation to the Navajo, rather than in relation to Anglos. The themes of strength, resilience, and survival that were central to Navajo identity offered an alternative to the portrayal of Indians in Western history. As others have noted, access to such oppositional frames supports the ability to claim a separate and positive group identity that resists and disrupts pervasive negative group constructions (Groch 1994; Hall 1990; hooks 1990; Mulling 1992). As such, the ability to name and claim one's identity becomes an explicitly political act.

However, disruptions in patterns of practice that were central to Navajo identity occurred as post World War II era policy initiatives were implemented; it interrupted access to these frames for many women. Women like Nora were exposed to dominant group representations of Indians without access to the cultural resources with which to develop and articulate alternate constructions of identity. They described the impact of those representations on their sense of themselves. This does not mean that Navajo women who grew up away from the reservation did not resist representations of Indianness. They were, however, less likely to draw upon Navajo identities as they did so—in some cases perhaps because of the contested nature of those tribal identities. This suggests that social and historical patterns that distance Navajo women from language, patterns of practice, spiritual practices, and other dimensions of Navajo identity, as in relocation to urban communities, may interrupt the use of Navajo identity as a means to disrupt outsider notions of Indianness. Collective actions to maintain or re-create cultural practices are one strategy for countering these processes; for example, bilingual schools emerged on the Navajo Nation in the 1970s with an explicit agenda to teach Navajo language, culture, and history.

In contrast to women who grew up in the post-war era, those who came of age following the self-determination and civil rights movements of the 1960s actively embraced Indian identities. Like the women who grew up in the 1940s, these younger women also sought to redefine the content and meanings associated with Indian identity, claiming particular meanings associated with Indianness and disrupting others. Unlike the older women who grew up in extended families, and more like the older women who grew up among Anglos, the younger cohort did not draw upon their particular tribal identities to disrupt constructions of Indians. Rather, they sought to reconstruct Indianness by claiming the more pluralist constructions made available through the political struggles of the 1960s, and by directly challenging negative representations. These women drew upon frameworks that claimed Indian identity as one that distinguished them from Anglos, and actively worked to challenge prevailing constructions of that difference as deficiency.

. . . While women's specific strategies varied with the historical moment and also with individual life trajectories, women across all cohorts developed strategies to resist cultural annihilation. They illustrate both the constancy of the pressure toward incorporation and women's creativity as they draw upon available resources to maintain distinct—but not static—identities as Indian and as Navajo. The construction of identities that disrupt dominant group representations of Indians emerge as everyday acts of political resistance that challenge the continued pressures toward incorporation and the loss of a distinct and valued group identity.

NOTES

1. The gendered nature of ethnic or tribal identities is examined in detail elsewhere (Schulz 1994).
2. The specific wording of questions varied according to the particulars of the interview.

3. Data on Navajo women's education is from Table NN10: Social Characteristics of the Navajo Nation 1990. 1990 Census: Population and Housing Characteristics of the Navajo Nation (The Printing Company: Scottsdale, Arizona, 1993). See Schulz (1994) for a more complete description of the characteristics of study respondents as compared to all women living on the Navajo Nation in 1990.

4. Data on Navajo women's labor force participation is from Table NN11: Labor Force and Commuting Characteristics of the Navajo Nation: 1990. 1990 Census: Population and Housing Characteristics of the Navajo Nation. (The Printing Company: Scottsdale, Arizona, 1993).

5. Navajo spellings are from Garth A. Wilson's (1989) *Conversational Navajo Dictionary*.

REFERENCES

Anderson, Benedict. 1983. *Imagined Communities: Reflections on the Origins and Spread of Nationalism.* London:Verso Press.

Barth, Frederik. 1969. *Ethnic Groups and Boundaries: The Social Organization of Culture Difference.* Boston, MA: Little, Brown and Company.

Berkhofer, Robert F., Jr. 1979. *The White Man's Indian: Images of the American Indian from Columbus to the Present.* New York: Vintage Books.

Burt, Larry C. 1986. *Tribalism in Crisis: Federal Indian Policy 1953–1961.* Albuquerque, New Mexico: University of New Mexico Press.

Calhoun, Craig. 1994. "Social Theory and the Politics of Identity." In *Social Theory and the Politics of Identity,* ed. C. Calhoun, 9–36. Cambridge, MA: Blackwell.

Cohen, Anthony P. 1985. *The Symbolic Construction of Community.* London: Routledge.

Cooley, Charles H. 1902. *The Looking Glass Self: Human Nature and the Social Order.* New York: Charles Scribner Publishers.

Cornell, Stephen. 1988. *The Return of the Native: American Indian Political Resurgence.* New York: Oxford University Press.

Erikson, Kai. 1976. *Everything in Its Path: Destruction of Community in the Buffalo Creek Flood.* New York: Simon and Schuster.

Groch, Sharon A. 1994. "Oppositional consciousness: Its manifestation and development: The case of people with disabilities." *Sociological Inquiry* 64:369–395.

Hall, Stuart. 1990. "Cultural identity and the diaspora." In *Identity: Community, Culture, Difference,* ed. J. Rutherford, 222–237. London: Lawrence and Wishart.

hooks, bell. 1990. *Yearning: Race, Gender and Cultural Politics.* Boston, MA: South End Press.

Karp, David. 1996. *Speaking of Sadness.* New York: Oxford University Press.

Lemert, Charles. 1994. "Dark thoughts about the self." In *Social Theory and the Politics of Identity,* ed. C. Calhoun,100–130. Cambridge, MA: Blackwell.

Mbilinyi, Marjorie. 1989. "I'd have been a man: Politics and the labor process in producing personal narratives." In *Interpreting Women's Lives: Feminist Theory and Personal Narratives,* eds.The Personal Narratives Group, 204–227. Bloomington, IN: Indiana University Press.

Mulling, Leith. *Race, Class and Gender: Representations and Reality.* Center for Research on Women, Memphis, Tennessee: Memphis State University.

Nagel, Joane. 1996. *American Indian Ethnic Renewal: Red Power and the Resurgence of Identity and Power.* New York: Oxford University Press.

Navajo Government Publication. 1993. *Population and Housing Characteristics of the Navajo Nation, 1990.* Scottsdale, AZ: The Printing Company.

Oakley, Anne. 1981. "Interviewing women: A contradiction in terms." In *Doing Feminist Research,* ed. H. Roberts, 30–61. London: Routledge Kegan Paul.

Ortner, Sherry. 1989. *High Religion: A Cultural and Political History of Sherna Buddhism.* Princeton, NJ: Princeton University Press.

Schulz, Amy. 1994. "I raised my children to speak Navajo. . . . My Grandkids are all English speaking people": Identity, resistance and transformation among Navajo women. Doctoral Dissertation. (University of Michigan, Ann Arbor)

Senese, Guy B. 1991. "Self determination and the social education of Native Americans." In *Journal of Thought.* New York: Praeger Press.

Strauss, Anselm, and Juliet Corbin. 1990. *Basics of Qualitative Research: Grounded Theory Procedures and Techniques.* Newbury Park, CA: Sage.

Swidler, Anne. 1986. "Culture in action: Symbols and strategies." *American Sociological Review* 51:273–286.

Szasz, Margaret. 1973. *Education and the American Indian: On the Road to Self Determination, 1928–1973.* Albuquerque, NM: University of New Mexico Press.

Taylor, Charles. 1989. *Sources of the Self: The Making of Modern Identity.* Cambridge, MA: Harvard University Press.

Ward, Kathryn B. 1993. "Reconceptualizing world systems theory to include women." In *Theory of Gender/Feminism on Theory,* ed. Paula England, 43–68. New York: Aldine de Gruyter.

Waters, Mary C. 1990. *Ethnic Options: Choosing Identities in America.* Berkeley, CA: University of California Press.

Wiley, Norbert. 1994. "The politics of identity in American history." In *Social Theory and the Politics of Identity,* ed. C. Calhoun, 131–149. Cambridge, MA: Blackwell.

Wilson, Garth A. 1989. *Conversational Navajo Dictionary.* Blanding, UT: Conversational Navajo Publications.

Woodward, Kay. 1997. *Identity and Difference.* London: Sage.

Yuval-Davis, Nira. 1997. *Gender and Nation.* Thousand Oaks, CA: Sage.

DISCUSSION QUESTIONS

1. How does the history of these Navajo women specifically influence the development of their identity? How does gender influence the identities of these Navajo women?

2. How could you replicate this study among a different racial-ethnic group? What historical information is important to understanding the construction of identity among another racial-ethnic group?

11

Gender and Aging

KATHLEEN SLEVIN AND TONI CALASANTI

Slevin and Calasanti discuss how cultural views of aging and older people are predomi-
nantly negative. Aging is a difficult life change for all people, most especially women, people
of color, homosexuals, and the economically disadvantaged. This reading uncovers how gen-
der, race, class, and sexuality converge to create difficult life course change for older Americans.

AGEISM

. . . Ageism allows people to see the old as different, as "other," as not "like them,"
thereby making it easier to see the old as "not humans." Pink triangles (for gays and
lesbians) and Stars of David (for Jews) served that function in previous times by identifying those who were to be excluded but who might otherwise "pass" without visible means of differentiating. Constructions of old age and the visible markers of it
matter in this way. Old people mask the bodily changes with good reason. . . .

Similar to racism and sexism, ageism encourages viewing people not as individuals but as members of a social category. . . . Ageism is more than the attitudes and beliefs held by individuals in a particular society; it is also embedded in
patterns of behavior and it serves as a social organizing principle. . . . Further,
ageism takes new forms over time. Images of the old as vulnerable have recently
given way to images of the affluent "greedy geezer," whose selfish drain on society has left younger generations in poverty. . . .

We can see the influence of capitalism and culture on ageism when we look
at our "moral economy," the rationale underlying particular notions of distributive justice. . . . Within capitalism, U.S. cultural values dictate that an individual's worth rests on his or her productivity and ability to maintain independence.
This economic imperative devalues the old who cannot live up to its standards.

The institutionalization of retirement is a key illustration. The passage of the
Social Security Act in 1935 did more than provide a minimum of financial support
for (former) workers over the age of 65; it also created a new form of dependence
based on age. The demarcation of age 65 as the time when one would be eligible to
collect full Social Security benefits defined old age in terms of years since birth while
also equating it with dependence. Prior to the institutionalization of retirement, there

From: Toni M. Calasanti and Kathleen F. Slevin. 2001. *Gender, Social Inequalities, and*
Aging. Walnut Creek, CA: Alta Mira Press.

was no dependent group of adults based on age alone. Ironically, these new social policies also set the stage for the more recent claims that the old are an "economic burden." . . . This process of first creating and then reviling an old "dependent" population has been conditioned by the needs of a capitalist political economy. During the Depression of the 1930s, business owners wanted to control workers, to push out more highly paid employees, to lower unemployment artificially and stimulate consumption. At the same time, labor unions wanted a way to insure job stability and seniority, and establish a forty-hour work week; they also sought a national system of retirement that would keep management from using pensions as a form of labor control. . . . The result was the Social Security Act. . . .

While Social Security's institutionalization of retirement has created dependence among the old, ageism does not equally influence all those over age sixty-five, nor does it equally influence persons of the same age. Many can avoid ageism by "passing" for younger ages through the use of cosmetic surgery, hair dyes, exercise, and the like; others use their wealth or position to escape stereotypes. Still others seem to avoid much of the stigma by virtue of group membership; for example, their families and communities may see them as offering wisdom . . . in contrast with the larger social image. In other words, while the designation "old" often draws hostility, its application varies by other social locations.

Ageism and Inequality

. . . [W]e need to examine some of the sources of ageism. Ageism interacts with other types of discrimination—sexism, racism, classism, and heterosexism—in ways that make the experience of ageism both similar and different across various groups. Below, then, we explore some of the ways that ageism interacts with other forms of privilege and oppression in ways that vary the timing, sources, and experience of ageism.

Dependence

One of the most negative attributes of being marked as "old" is dependency. We know that class, racial and ethnic, gender, and sexual preference inequities throughout the life course translate into a variety of factors that lead to dependence sooner for some people than for others. . . . The most obvious is the way in which these relations exacerbate financial difficulties in old age. . . . At present, we simply note that the main sources of income in old age—Social Security, pensions, assets, and earnings—vary accordingly. For example, the original Social Security legislation did not cover all occupations, and so certain groups of the elderly—particularly those of working class and racial or ethnic minorities—had to depend upon different, means-tested sources of social support in old age, such as Supplemental Security Income. This latter form of economic support for the poor marks those who receive it as less "deserving" . . . than those who need not depend on welfare. Similarly, because later legislative revisions only allow "spouses" to collect benefits based on the earnings of a retired worker, those in same-sex partnerships cannot draw upon one another's benefits.

This is but one example of how class, race and ethnicity, sexual orientation, and gender can influence economic dependence in later life. Low lifetime earnings result in lower levels of Social Security; the inability to save or invest in a pension also reduces income in later life. Low lifetime earnings in turn are related to the intersections of these power relations. As a result, groups can see their disadvantages (or advantages) grow over their life courses. . . . For example, while a White, working-class man will often have lower lifetime earnings than a similarly situated middle-class man, a White working-class woman will not only have even lower benefits, due to lower earnings, but, if lesbian, will not be able to collect a pension or Social Security as spouse. Similarly, because of the types of jobs that Black men are more likely to occupy, they are likely to suffer both more spates of unemployment and higher rates of disability than White men. . . . Thus, they are more likely to rely on unemployment or disability payments than are White men. As a result, they will also have lower Social Security benefits, and their economic dependence upon the state will be increased. In turn, lack of economic resources will exacerbate their physical disabilities.

Therefore, the structure of Social Security, welfare, and private pensions casts some groups of old in the role of "dependents." Other groups are privileged by ownership and income, and will be able to avoid economic dependence indefinitely or even altogether. . . . Here, we can see that the intersections of gender, race and ethnicity, class, and sexual orientation all play a role in dependence in old age through such things as property, which translates into greater power and better life chances. . . .

Lifelong inequities for women become exacerbated in old age, leaving them with less financial security in retirement due to labor market discrimination or episodic employment patterns. In addition, race and ethnic relations play a part in making minority women poor earlier in life, reinforcing their vulnerability to poverty in old age and to different forms of ageism as they turn to the welfare state for such benefits as Supplemental Security Income. . . .

Ageism and Old Bodies

The discussion above focuses on the way in which power relations result in some groups being more likely to become "dependent" than others. However, this is not the only source of ageism, of designating someone as "old" and "other." Another critical source of ageism is physical appearance, and this too varies by the intersection of social locations.

Bodies serve as markers of age. Gray hair, wrinkles, brown spots—each of these denotes "old." Yet if we think about it, these traits are not universally judged to signify someone is old. Not all gray-haired people are seen to be old, nor are all who exhibit wrinkles. Most of us have heard of the "double standard" of aging, by which we usually mean that women are seen to be old at an earlier age than men. . . . Recent attitude polls confirm that the gray hair and wrinkles a woman experiences mark her as old sooner. . . . Why is this the case? How and why ageism based on physical appearance occurs is very much related to power relations. We begin with a focus on gender to make this clearer.

Why would people see an old woman wearing a miniskirt as deviant? Part of the reaction, and the rationale for regarding women as old earlier in their lives, arises from the fact that their value is based on their attractiveness to men and their reproductive abilities. Thus the old woman in the miniskirt is deviant for appearing sexual beyond her fertile years. By contrast, men's attractiveness stems from other sources not as quickly diminished. Indeed, sometimes age enhances men's attractiveness, especially if they are associated with public achievements, money, and power. . . . Women even "age" more quickly than men in the workplace, where they do deal with money, power, and public achievement. This is particularly true if they are engaged in jobs where "attractiveness" matters—such as jobs dealing with the public or working for (predominantly White) male supervisors. . . . For instance, when airline attendants in this country were almost exclusively women, their unions fought the airlines on a number of occasions where women were removed from their jobs because they were seen as "too old" (in other words, no longer attractive). However, as we have noted, such issues are shaped differently in different societies. Thus, for example, we find that in Finland youthfulness and attractiveness are not as important for women as they are in the United States. . . .

Having said this, however, we must note that the preceding scenario is too simplistic. What women are we talking about? Do physical signs of aging result in ageism for all women in similar ways? If we accept the fact that people see employed women as old sooner than men, and that this hinges at least in some part on their attractiveness to White men, we must question what this means for the aging of Black women, for example, in the labor force. Are they sexualized in the same way as White women, earlier or later in life? How about women who live openly as lesbians? Or working-class women?

Class plays an important role in another way as well. As was apparent in our discussion of dependence, class—through economic resources—can play a critical role in denying or providing resources that allow the old to choose the ways in which they will manage growing old. To the extent that outward signs of aging can be forestalled by such physical transformations as face-lifts, the well-to-do enjoy an obvious advantage. "Remaking" aging bodies is expensive and time consuming and, hence, beyond the reach of the working-class or poor. At the same time, which women do the remaking, and how, tells us about racial and ethnic relations. Not all women feel the "need" to hide gray hair or diminish wrinkles. . . .

Knowing the socially constructed and hence malleable nature of "old," we must still note that it remains negative across all groups. Whether the old appear in popular culture as vulnerable and potential victims or as affluent, the images are applied in a blanket fashion—that is, the old are seen to be homogenous— "all alike." This has led to additional problems for old women and members of racial and ethnic minority groups. For instance, the present ageist stereotype of the advantaged elder has obscured the class, racial, and gender stratification among the old. As a result, women, people of color, and working-class old remain disadvantaged, their problems ignored. . . .

DISCUSSION QUESTIONS

1. What stereotypes do you hold about older people? How did this reading inform your understanding of aging?

2. What are the gender differences in the aging experience? What are some class differences in aging?

http://infotrac.thomsonlearning.com

InfoTrac College Edition

BONUS READING

Giuliano, Traci A., Kathryn E. Popp, and Jennifer L. Knight. "Footballs versus Barbies: Childhood Play Activities as Predictors of Sport Participation by Women." *Sex Roles* (February 2000): 159–81.

The authors of this article researched how childhood play activities are agents of socialization. Stereotypically masculine play activities by girls led to greater likelihood of sports participation for these girls. Female sport socialization is influenced by the type of toys they played with as young girls. What toys and games did you play with as a child? How do these experiences relate to your current activities? Can you think of examples from your childhood of girls who were "tomboys" that grew up to play sports?

SEARCH TERMS

sports and socialization
Navajo and identity
ageism and society
college and identity

12

The Presentation of Self in Everyday Life

ERVING GOFFMAN

Erving Goffman likens social interaction to a "con game," in which we are consistently trying to put forward a certain impression or "self" in order to get something from others. Although many will not see human behavior so cynically, Goffman's analysis sheds light on how people try to manage the impression that others have of them.

When an individual plays a part he implicitly requests his observers to take seriously the impression that is fostered before them. They are asked to believe that the character they see actually possesses the attributes he appears to possess, that the task he performs will have the consequences that are implicitly claimed for it, and that, in general, matters are what they appear to be. In line with this, there is the popular view that the individual offers his performance and puts on his show "for the benefit of other people." It will be convenient to begin a consideration of performances by turning the question around and looking at the individual's own belief in the impression of reality that he attempts to engender in those among whom he finds himself.

At one extreme, one finds that the performer can be fully taken in by his own act; he can be sincerely convinced that the impression of reality which he stages is the real reality. When his audience is also convinced in this way about the show he puts on—and this seems to be the typical case—then for the moment at least, only the sociologist or the socially disgruntled will have any doubts about the "realness" of what is presented.

At the other extreme, we find that the performer may not be taken in at all by his own routine. This possibility is understandable, since no one is in quite as good an observational position to see through the act as the person who puts it on. Coupled with this, the performer may be moved to guide the conviction of his audience only as a means to other ends, having no ultimate concern in the conception that they have of him or of the situation. When the individual has no belief in his own act and no ultimate concern with the beliefs of his audience, we may call him cynical, reserving the term "sincere" for individuals who believe in the impression fostered by their own performance. It should be understood that the cynic, with all his professional disinvolvement, may obtain unprofessional

From: Erving Goffman. 1959. *The Presentation of Self in Everyday Life*. Garden City, NY: Anchor Doubleday, pp. 17–27.

pleasures from his masquerade, experiencing a kind of gleeful spiritual aggression from the fact that he can toy at will with something his audience must take seriously.

It is not assumed, of course, that all cynical performers are interested in deluding their audiences for purposes of what is called "self-interest" or private gain. A cynical individual may delude his audience for what he considers to be their own good, or for the good of the community, etc. For illustrations of this we need not appeal to sadly enlightened showmen such as Marcus Aurelius or Hsun Tzŭ. We know that in service occupations practitioners who may otherwise be sincere are sometimes forced to delude their customers because their customers show such a heartfelt demand for it. Doctors who are led into giving placebos, filling station attendants who resignedly check and recheck tire pressures for anxious women motorists, shoe clerks who sell a shoe that fits but tell the customer it is the size she wants to hear—these are cynical performers whose audiences will not allow them to be sincere. Similarly, it seems that sympathetic patients in mental wards will sometimes feign bizarre symptoms so that student nurses will not be subjected to a disappointingly sane performance. So also, when inferiors extend their most lavish reception for visiting superiors, the selfish desire to win favor may not be the chief motive; the inferior may be tactfully attempting to put the superior at ease by simulating the kind of world the superior is thought to take for granted.

I have suggested two extremes: an individual may be taken in by his own act or be cynical about it. These extremes are something a little more than just the ends of a continuum. Each provides the individual with a position which has its own particular securities and defenses, so there will be a tendency for those who have traveled close to one of these poles to complete the voyage. Starting with lack of inward belief in one's role, the individual may follow the natural movement described by Park:

> It is probably no mere historical accident that the word person, in its first meaning, is a mask. It is rather a recognition of the fact that everyone is always and everywhere, more or less consciously, playing a role . . . It is in these roles that we know each other; it is in these roles that we know ourselves.[1]

In a sense, and in so far as this mask represents the conception we have formed of ourselves—the role we are striving to live up to—this mask is our truer self, the self we would like to be. In the end, our conception of our role becomes second nature and an integral part of our personality. We come into the world as individuals, achieve character, and become persons.[2] . . .

Front, then, is the expressive equipment of a standard kind intentionally or unwittingly employed by the individual during his performance. For preliminary purposes, it will be convenient to distinguish and label what seem to be the standard parts of front.

First, there is the "setting," involving furniture, décor, physical layout, and other background items which supply the scenery and stage props for the spate of human action played out before, within, or upon it. . . .

If we take the term "setting" to refer to the scenic parts of expressive equipment, one may take the term "personal front" to refer to the other items of expressive equipment, the items that we most intimately identify with the performer himself and that we naturally expect will follow the performer wherever he goes. As part of personal front we may include: insignia of office or rank; clothing; sex, age, and racial characteristics; size and looks; posture; speech patterns; facial expressions; bodily gestures; and the like. Some of these vehicles for conveying signs, such as racial characteristics, are relatively fixed and over a span of time do not vary for the individual from one situation to another. On the other hand, some of these sign vehicles are relatively mobile or transitory, such as facial expression, and can vary during a performance from one moment to the next. . . .

In addition to the fact that different routines may employ the same front, it is to be noted that a given social front tends to become institutionalized in terms of the abstract stereotyped expectations to which it gives rise, and tends to take on a meaning and stability apart from the specific tasks which happen at the time to be performed in its name. The front becomes a "collective representation" and a fact in its own right.

When an actor takes on an established social role, usually he finds that a particular front has already been established for it. Whether his acquisition of the role was primarily motivated by a desire to perform the given task or by a desire to maintain the corresponding front, the actor will find that he must do both.

Further, if the individual takes on a task that is not only new to him but also unestablished in the society, or if he attempts to change the light in which his task is viewed, he is likely to find that there are already several well-established fronts among which he must choose. Thus, when a task is given a new front we seldom find that the front it is given is itself new.

NOTES

1. Robert Ezra Park, *Race and Culture* (Glencoe, IL: The Free Press, 1950), p. 249.
2. Ibid., 250.

DISCUSSION QUESTIONS

1. How many "selves" do you think you could "play" or "do" in order to accomplish something with another person? Discuss two such selves and try to get them to be quite different from each other.

2. "All the world's a stage," wrote William Shakespeare. So might Goffman have said this. How does his analysis of the presentation of self in everyday life suggest that life is a drama where we all play our parts?

13

Code of the Street

ELIJAH ANDERSON

Elijah Anderson's study of interaction on the street shows the vast array of implicit "codes" of behavior or rules that guide street interaction. His analysis helps explain the complexity of street interaction and provides a sociological explanation of street violence.

In some of the most economically depressed and drug- and crime-ridden pockets of the city, the rules of civil law have been severely weakened, and in their stead a "code of the street" often holds sway. At the heart of this code is a set of prescriptions and proscriptions, or informal rules, of behavior organized around a desperate search for respect that governs public social relations, especially violence, among so many residents, particularly young men and women. Possession of respect—and the credible threat of vengeance—is highly valued for shielding the ordinary person from the interpersonal violence of the street. In this social context of persistent poverty and deprivation, alienation from broader society's institutions, notably that of criminal justice, is widespread. The code of the street emerges where the influence of the police ends and personal responsibility for one's safety is felt to begin, resulting in a kind of "people's law," based on "street justice." This code involves a quite primitive form of social exchange that holds would-be perpetrators accountable by promising an "eye for an eye," or a certain "payback" for transgressions. In service to this ethic, repeated displays of "nerve" and "heart" build or reinforce a credible reputation for vengeance that works to deter aggression and disrespect, which are sources of great anxiety on the inner-city street. . . .

In approaching the goal of painting an ethnographic picture of these phenomena, I engaged in participant-observation, including direct observation, and conducted in-depth interviews. Impressionistic materials were drawn from various social settings around the city, from some of the wealthiest to some of the most economically depressed, including carryouts, "stop and go" establishments, laundromats, taverns, playgrounds, public schools, the Center City indoor mall known as the Gallery, jails, and public street corners. In these settings I encountered a wide variety of people—adolescent boys and young women (some incarcerated, some not), older men, teenage mothers, grandmothers, and male and female schoolteachers, black and white, drug dealers, and common criminals. To protect the privacy and confidentiality of my subjects, names and certain details have been disguised. . . .

From: Elijah Anderson. 1999. *Code of the Street*. New York: W. W. Norton, pp. 9–11, 32–34, 312–317. Reprinted with permission.

Of all the problems besetting the poor inner-city black community, none is more pressing than that of interpersonal violence and aggression. This phenomenon wreaks havoc daily on the lives of community residents and increasingly spills over into downtown and residential middle-class areas. Muggings, burglaries, carjackings, and drug-related shootings, all of which may leave their victims or innocent bystanders dead, are now common enough to concern all urban and many suburban residents.

The inclination to violence springs from the circumstances of life among the ghetto poor—the lack of jobs that pay a living wage, limited basic public services (police response in emergencies, building maintenance, trash pickup, lighting, and other services that middle-class neighborhoods take for granted), the stigma of race, the fallout from rampant drug use and drug trafficking, and the resulting alienation and absence of hope for the future. Simply living in such an environment places young people at special risk of falling victim to aggressive behavior. Although there are often forces in the community that can counteract the negative influences—by far the most powerful is a strong, loving, "decent" (as inner-city residents put it) family that is committed to middle-class values—the despair is pervasive enough to have spawned an oppositional culture, that of "the street," whose norms are often consciously opposed to those of mainstream society. These two orientations—decent and street—organize the community socially, and the way they coexist and interact has important consequences for its residents, particularly for children growing up in the inner city. Above all, this environment means that even youngsters whose home lives reflect mainstream values—and most of the homes in the community do—must be able to handle themselves in a street-oriented environment.

This is because the street culture has evolved a "code of the street," which amounts to a set of informal rules governing interpersonal public behavior, particularly violence. The rules prescribe both proper comportment and the proper way to respond if challenged. They regulate the use of violence and so supply a rationale allowing those who are inclined to aggression to precipitate violent encounters in an approved way. The rules have been established and are enforced mainly by the street-oriented; but on the streets the distinction between street and decent is often irrelevant. Everybody knows that if the rules are violated, there are penalties. Knowledge of the code is thus largely defensive, and it is literally necessary for operating in public. Therefore, though families with a decency orientation are usually opposed to the values of the code, they often reluctantly encourage their children's familiarity with it in order to enable them to negotiate the inner-city environment.

At the heart of the code is the issue of respect—loosely defined as being treated "right" or being granted one's "props" (or proper due) or the deference one deserves. However, in the troublesome public environment of the inner city, as people increasingly feel buffeted by forces beyond their control, what one deserves in the way of respect becomes ever more problematic and uncertain. This situation in turn further opens up the issue of respect to sometimes intense interpersonal negotiation, at times resulting in altercations. In the street culture, especially among young people, respect is viewed as almost an external entity, one that is hard-won but easily lost—and so must constantly be guarded. The rules of the code in fact provide a framework for negotiating respect. With the right amount of respect, individuals can avoid being bothered in public. This security is important, for if they *are* bothered, not only may they face physical danger, but they will have been disgraced or "dissed" (disrespected). Many of the forms dissing can take may seem petty to middle-class people (maintaining eye contact for too long, for example), but to those invested in the street code, these actions, a virtual slap in the face, become serious indications of the other person's intentions. Consequently, such people become very sensitive to advances and slights, which could well serve as a warning of imminent physical attack or confrontation.

The hard reality of the world of the street can be traced to the profound sense of alienation from mainstream society and its institutions felt by many poor inner-city black people, particularly the young. The code of the street is actually a cultural adaptation to a profound lack of faith in the police and the judicial system—and in others who would champion one's personal security. The police, for instance, are most often viewed as representing the dominant white society and as not caring to protect inner-city residents. When called, they may not respond, which is one reason many residents feel they must be prepared to take extraordinary measures to defend themselves and their loved ones against those who are inclined to aggression. Lack of police accountability has in fact been incorporated into the local status system: the person who is believed capable of "taking care of himself" is accorded a certain deference and regard, which translates into a sense of physical and psychological control. The code of the street thus emerges where the influence of the police ends and where personal responsibility for one's safety is felt to begin. Exacerbated by the proliferation of drugs and easy access to guns, this volatile situation results in the ability of the street-oriented minority (or those who effectively "go for bad") to dominate the public spaces. . . .

The attitudes and actions of the wider society are deeply implicated in the code of the street. Most people residing in inner-city communities are not totally invested in the code; it is the significant minority of hard-core street youth who maintain the code in order to establish reputations that are integral to the extant social order. Because of the grinding poverty of the communities these people inhabit, many have—or feel they have—few other options for expressing themselves. For them the standards and rules of the street code are the only game in town.

And as was indicated above, the decent people may find themselves caught up in problematic situations simply by being at the wrong place at the wrong time, which is why a primary survival strategy of residents here is to "see but don't see." The extent to which some children—particularly those who through upbringing have become most alienated and those who lack strong and conventional social support—experience, feel, and internalize racist rejection and contempt from mainstream society may strongly encourage them to express contempt for the society in turn. In dealing with this contempt and rejection, some youngsters consciously invest themselves and their considerable mental resources in what amounts to an oppositional culture, a part of which is the code of the street. They do so to preserve themselves and their own self-respect. Once they do, any respect they might be able to garner in the wider system pales in comparison with the respect available in the local system; thus they often lose interest in even attempting to negotiate the mainstream system.

At the same time, many less alienated young people have assumed a street-oriented demeanor as way of expressing their blackness while really embracing a much more moderate way of life; they, too, want a nonviolent setting in which to live and one day possibly raise a family. These decent people are trying hard to be part of the mainstream culture, but the racism, real and perceived, that they encounter helps legitimize the oppositional culture and, by extension, the code of the street. On occasion they adopt street behavior; in fact, depending on the demands of the situation, many people attempt to codeswitch, moving back and forth between decent and street behavior. . . .

In addition, the community is composed of working-class and very poor people since those with the means to move away have done so, and there has also been a proliferation of single-parent households in which increasing numbers of kids are being raised on welfare. The result of all this is that the inner-city community has become a kind of urban village, apart from the wider society and limited in terms of resources and human capital. Young people growing up here often receive only the truncated version of mainstream society that comes from television and the perceptions of their peers. . . .

According to the code, the white man is a mysterious entity, a part of an enormous monolithic mass of arbitrary power, in whose view black people are insignificant. In this system and in the local social context, the black man has very little clout; to salvage something of value, he must outwit, deceive, oppose, and ultimately "end-run" the system.

Moreover, he cannot rely on this system to protect him; the responsibility is his, and he is on his own. If someone rolls on him, he has to put his body, and often his life, on the line. The physicality of manhood thus becomes extremely important. And urban brinksmanship is observed and learned as a matter of course. . . .

Urban areas have experienced profound structural economic changes, as deindustrialization—the movement from manufacturing to service and high-tech—and the growth of the global economy have created new economic conditions. Job opportunities increasingly go abroad to Singapore, Taiwan, India, and Mexico, and to nonmetropolitan America, to satellite cities like King of Prussia, Pennsylvania. Over the last fifteen years, for example, Philadelphia has lost 102,500 jobs, and its manufacturing employment has declined by 53 percent. Large numbers of inner-city people, in particular, are not adjusting effectively to the new economic reality. Whereas low-wage jobs—especially unskilled and low-skill factory jobs—used to exist simultaneously with poverty and there was hope for the future, now jobs simply do not exist, the present economic boom notwithstanding. These dislocations have left many inner-city people unable to earn a decent living. More must be done by both government and business to connect inner-city people with jobs.

The condition of these communities was produced not by moral turpitude but by economic forces that have undermined black, urban, working-class life and by a neglect of their consequences on the part of the public. Although it is true that persistent welfare dependency, teenage pregnancy, drug abuse, drug dealing, violence, and crime reinforce economic marginality, many of these behavioral problems originated in frustrations and the inability to thrive under conditions of economic dislocation. This in turn leads to a weakening of social and family structure, so children are increasingly not being socialized into mainstream values and behavior. In this context, people develop profound alienation and may not know what to do about an opportunity even when it presents itself. In other words, the social ills that the companies moving out of these neighborhoods today sometimes use to justify their exodus are the same ones that their corporate predecessors, by leaving, helped to create.

Any effort to place the blame solely on individuals in urban ghettos is seriously misguided. The focus should be on the socioeconomic structure, because it was structural change that caused jobs to decline and joblessness to increase in many of these communities. But the focus also belongs on the public policy that has radically threatened the well-being of many citizens. Moreover, residents of these communities lack good education, job training, and job networks, or connections with those who could help them get jobs. They need enlightened employers able to understand their predicament and willing to give them a chance. Government, which should be assisting people to adjust to the changed economy, is instead cutting what little help it does provide. . . .

The emergence of an underclass isolated in urban ghettos with high rates of joblessness can be traced to the interaction of race prejudice, discrimination, and the effects of the global economy. These factors have contributed to the profound social isolation and impoverishment of broad segments of the inner-city black population. Even though the wider society and economy have been experiencing accelerated prosperity for almost a decade, the fruits of it often miss the truly disadvantaged isolated in urban poverty pockets.

In their social isolation an oppositional culture, a subset of which is the code of the street, has been allowed to emerge, grow, and develop. This culture is essentially one of accommodation with the wider society, but different from past efforts to accommodate the system. A larger segment of people are now not simply isolated but ever more profoundly alienated from the wider society and its institutions. For instance, in conducting the fieldwork for this book, I visited numerous inner-city schools, including elementary, middle, and high schools, located in areas of concentrated poverty. In every one, the so-called oppositional culture was well entrenched. In one elementary school, I learned from interviewing kindergarten, first-grade, second-grade, and fourth-grade teachers that through the first grade, about a fifth of the students were invested in the code of the street; the rest are interested in the subject matter and eager to take instruction from the teachers—in effect, well disciplined. By the fourth grade, though, about three-quarters of the students have bought into the code of the street or the oppositional culture.

As I have indicated throughout this work, the code emerges from the school's impoverished neighborhood, including overwhelming numbers of single-parent homes, where the fathers, uncles, and older brothers are frequently incarcerated—so frequently, in fact, that the word "incarcerated" is a prominent part of the young child's spoken vocabulary. In such communities there is not only a high rate of crime but also a generalized diminution of respect for law. As the residents go about meeting the exigencies of public life, a kind of people's law results, . . . Typically, the local streets are, as we saw, tough and dangerous places where people often feel very much on their own, where they themselves must be personally responsible for their own security, and where in order to be safe and to travel the public spaces unmolested, they must be able to show others that they are familiar with the code—that physical transgressions will be met in kind.

In these circumstances the dominant legal codes are not the first thing on one's mind; rather, personal security for self, family, and loved ones is. Adults, dividing themselves into categories of street and decent, often encourage their children in this adaptation to their situation, but at what price to the children and at what price to wider values of civility and decency? As the fortunes of the inner city continue to decline, the situation becomes ever more dismal and intractable. . . .

DISCUSSION QUESTIONS

1. List several ways that subtle or nonverbal behavior becomes important "on the street."

2. What specific ways does Anderson see street behavior as stemming from social structural conditions for African Americans?

14

Life Beyond the Screen:
Embodiment and Identity
Through the Internet

MICHAEL HARDEY

The use of the Internet for online communication has become increasingly popular and use-ful. In this reading, Hardey examines Internet dating sites and how people interact with others to establish the foundations for intimate relationships offline. Social interaction that takes place without physical presence has specific consequences and benefits.

. . . The purpose of this paper is to gain an understanding of how people con-struct and negotiate virtual identities and relationships within a digital space that offers opportunities to meet people on-line and move into relationships off-line. The analysis of internet dating sites provides an appropriate environment in which to examine how users negotiate the tensions between the development of virtual relationships, and the norms and conventions associated with the "inter-action order" of physical copresence (Goffman, 1983). It also serves to illustrate how virtual spaces may be shaped by and grounded in the social, bodily and cul-tural experiences of users. Giddens's (1992) conception of pure relationships will be used to situate relationships initiated through the internet within the broader context of sociability in late modernity. The paper finally explores the possible consequences of this study for those commentaries on cyberspace, which have variously celebrated the potentialities of the internet and lamented the effects it has on human life. First, however, it is necessary to briefly describe and contex-tualise internet dating within established mechanism for meeting strangers.

INFORMATION TECHNOLOGY
AND THE MEETING OF OTHERS

. . . Internet dating is characterised by a seamless movement between reading descriptions, writing responses and exchanging messages. Compared to the effort, awkwardness, risks and physical embarrassments often associated with "real world" dating, this points to some of the advantages of the internet. It should be remembered that what is described as "dating" covers a wide range of social activities. The concern here is to focus on the internet as a social space that may be used to meet others rather than the nature of any encounters that might follow. . . . Internet dating sites are similar to newspaper dating services in so far as they provide a medium through which individuals advertise themselves, yet their respective limitations vary greatly. Newspaper dating services usually require those advertising to restrict their text to twenty or thirty words, and often use a voice message system for those wishing to provide supplementary information. Internet sites, in contrast, are less prescriptive in content, and enable users to move seamlessly, and at no additional cost, between initial advertisement and contacting others using the service. Users are informed (as in print media) that the content of their text and pictures will be vetted and may be edited in order to achieve consistency within a site. Sites typically include an email system that allows frequent and lengthy correspondence between individuals and a "blocking" system whereby users can choose to receive no more correspondence from specified individuals. . . .

The form and content of sites vary, yet it is possible to describe an ideal type that provides a context for the interactions described in this paper. On entering a site users are presented with introductory information and a listing of new advertisers hyperlinked to longer self-description. Users can then create an advertisement, browse advertisements, or undertake searches. Advertisers may be encouraged to complete questions about their bodily appearance which is then included alongside their self-description. Above all, advertisers are recommended to be "realistic" and "truthful" about themselves and about what they hope for in a partner. In searching for potential partners, users can access "women advertising for men," or "men advertising for women," or whatever combination is made available by the site owners. Users are then presented with brief descriptions structured via categories that may include age, location, employment, interests and other specific attributes which are hyperlinked to longer descriptions of potential partners.

RESEARCHING THE VIRTUAL

The terms "dating," "dates," "romance" and "lonely heart" were entered into six common search engines and two multiple search engines. It was decided to focus on single heterosexuals who appear to constitute the largest sector of the dating market, both in print media (Jagger, 2001), and on the internet and to avoid sites which were otherwise exclusive with regard to such variables as religion. The owners of four internet dating sites originated in the UK were approached and agreed to collaborate in the research. The objectives of these sites that were displayed to users had a common theme of facilitating "partnership" and the identification of a "soulmate," in this the sites represent those that cater to people who are concerned to find a long-term relationship. However, does not preclude the use of the sites studied to find or initiate other forms of relationship. An advertisement for the research was displayed on the dating site from which users could follow a link to the study Web site (Jones, 1999). Here the project was described so that users were able to provide informed consent to participation and be assured that their identities would remain confidential (Mann and Stewart, 2000). While research on newsgroups commonly analyses and provides examples of posted exchanges, email correspondence within dating sites is not public and it would be difficult to establish the consent of users to access to such material. This paper utilises the responses from an email based questionnaire that included yes/no box questions and open-ended responses. It fell into four sections: the first asked general questions about relationships, dating sites and the internet, the following two parts explored experiences in contacting and meeting others and the final section collected profiling information.

A total of 437 completed questionnaires from users of these sites were received (men = 294, women = 143). All had visited several dating sites and many had used both free-to-users and fee based sites. They were resident in places across the UK and, when employed, about two thirds worked in occupations classified as falling into social classes I, II and III (non-manual). . . .

MEETINGS THROUGH THE INTERNET: PURE RELATIONSHIPS OR IMPOVERISHED MEETINGS?

The use of internet dating sites as a means to meet partners operates in stark contrast to traditional ideologies of romantic love (in which individuals physically meet and "fall in love" with each other). Giddens (1992) argues that the development of our "late modern" era is associated with the erosion of traditional forms of close personal relationships and the increased significance of "pure relationships." Entered into for their "own sake," for the intrinsic satisfactions they offer, pure relationships eschew tradition and contract, and are maintained only while they "deliver enough satisfactions" to induce individuals to "stay within" them (Giddens, 1992: 58). Characteristic of pure relationships is "confluent love," a contingent love based on "opening oneself out to the other" (ibid. 61). This form of intimacy involves the maintenance of "clear personal boundaries" rather than an absorption into the other (Bataille, 1962). It is "above all a matter of emotional communication, with others and with the self, in a context of interpersonal equality" (Giddens, 1992: 130). The consequent vision of a highly discursive, disembodied late modern intimacy based on talk rather than passion, negotiation rather than commitment, and the advancement of the self rather than the development of the couple suggest that the internet is uniquely placed to facilitate such relationships. . . .

ENTERING INTO VIRTUAL ENCOUNTERS

As we have noted internet sites provide users with a more or less open environment which they can tailor [to] meet their needs. For those that create descriptions of themselves in the form of advertisements this extends to deciding content, length and on some sites whether to include a small photograph. Reliance on only textual descriptions provides individuals with the potential to present themselves unhindered by visual images, and mostly unencumbered by the need to negotiate those shared "vocabularies of bodily idiom" that Goffman (1963, 1969) suggested is central to the "presentation of self" in public. This is because text renders invisible outward signs of dress, bearing, posture, movement, facial decorations and emotional expressions that are usually so important in determining how individuals respond to us, and how we come to perceive ourselves (Goffman, 1963: 33). As one man commented in an illustrative response about the perceived benefits of the digital medium:

> I value the flexibility and control it gives compared to newspaper advertisements . . . I would not put up a photograph because it cannot say much about me. Describing your life and yourself is not simple but gives a much better picture of the true person than any amount of pictures. It is also much easier to get to know someone on the Web and there is a record of everything that is written so I can look back on them. The main difference from paper dating is that I dip in and out of it whenever I want, have any number of conversations going at the same time and get to really know people. Reading what someone has written and getting into what amounts to sending letters is the best way to get close to someone quickly because you both have time to reflect and think. Very different to chat rooms! (questionnaire, 82)

For some users, and women in particular, the internet offers a space for emotional expression that is perceived to be unavailable elsewhere.

> One of the main things I found when I started using Interdate was that I could have conversations with men that would not have happened if I met them in person. I feel like you communicate on a different level on the Web. It allows you to get into emotional things that men often don't feel comfortable with unless they have known you for a long time. (questionnaire, 359)

Intimidation, harassment and flaming (abusing someone in public environments such as newsgroups) is common in some virtual spaces (Spender, 1995) but appears to [be] largely absent from exchanges within dating sites. Beth, who worked as a network manager, captures the way dating sites provide a space that is not conducive to harassment.

> I've been flamed in a group by men who are simply out to abuse women who use the WWW. Sexism is a common experience for women in the IT industry but it is not something that I have experienced when I've used Net Dating. Some guys post what they think are macho descriptions but you can choose to ignore those. All the men I've emailed have been really interesting and when I have decided to end a correspondence I have not encountered any abuse. (questionnaire, 97)

NEGOTIATING RELATIONSHIPS

Once a user has made contact with another member of a dating site through the internal email system a decision has to be made about whether to enter into an exchange of messages, or to simply ignore the invitation. This simple opportunity provides a sense of control experienced by users that some described as "liberating" them from what they see as the limitations or possible embarrassments of encounters in off-line. As one woman put it:

> It may sound odd but it is wonderful to be liberated from the usual guff that is involved in trying to meet someone new. I feel none of the inhibitions I am used to because I can choose when and if I reply or contact a guy. (questionnaire, 247)

While internet dating sites may free individuals from many of the encumbrances associated with copresent meetings, they also remove initial contacts from those multiple visual clues which are "given off" by copresent others (Goffman, 1967). In this context, it is revealing that users' experiences of establishing and maintaining interaction with others approximated much more closely to Goffman's view that interaction proceeds via rituals and norms that protect the self rather than to a vision of the internet as a revolutionary social space. Paul's comments are revealing in this respect:

> There's what you might call a set of unwritten rules or manners that you follow on these sites. It is not done on this site to come on with a lot of sexual content because that is not appropriate to what people expect or want from it. People who want that would be using a different bit of the WWW. You also work out with people how quickly they expect you to reply to them and the sort of depth you feel appropriate to get on with someone who starts out as a stranger. I'd say it was rather like dancing where you get two people who learn how to move together. (questionnaire, 101)

Goffman (1983) argued that the "interaction order," the domain of face-to-face relations of copresence, constituted the arena in which people developed and maintained a morally acceptable social self, and was informed by a battery of rules designed to protect individuals. These rules were necessary, and were widely shared, because copresent interaction makes people vulnerable to physical assault and to assaults on their sense of self (Goffman, 1983: 4). Interactional "rules" facilitate the building of "trust" between participants and the supporting and saving of "face" (Goffman, 1967). When they are broken, the identity and trustworthiness of the culpable party becomes "tainted"; their behaviour is seen as evidence of "weakness . . . moral guilt and other unenviable attributes" (Goffman, 1956: 266). Authenticity, reciprocal revelation of personal details, the building of trust, turn taking, and the dialogical establishment of intimacy may be characteristics of a "new" form of "pure relationship," then, but they have long been considered key to interaction rituals. . . .

Interactions are maintained only for as long as both parties agree so should they feel uncomfortable with a particular correspondence they can simply withdraw from it. As Mary explained:

> I get into quite a lot of emails from men because I don't have to worry about cutting them off or come to that if they cut me out. Not like reality where the ego might get a bit of a battering! (questionnaire, 347)

In communications which proceed smoothly, shared "likes" and "dislikes" are often exchanged as a means of establishing more regular and serious contact. A common theme concerns the "pace" of communication whereby a rhythm is established so that it is anticipated that an email will arrive, for example, every day. There is also a widely held expectation that there should be "turn taking" in email exchange whereby emails should be exchanged on a one-for-one basis. Should an email not be responded to after a "reasonable time" users see it as a signal that the relationship had come to end. This was accompanied by a balance between what is described as "coming on too strongly" and "appearing uninterested" that also underpins interaction rituals (Goffman, 1967, 1983). In email correspondences that continues, the discernible skepticism often characterising initial exchanges generally falls away the longer and apparently more self revealing the interchange becomes. The other becomes known and trusted through a mutual process of revealing the self. Penny provides a graphic illustration of how she perceives this process:

> If I had to sum up what it is like getting to know people here I would say it is like a striptease! What I mean is that I start with a full clothed version of me that I put up as an advertisement—makeup and posh frock to get men interested! Then as I write emails and get to know someone I reveal more of the real me and if we both seem to like what we see of each other we might arrange a meeting. If I decide I'm no longer interested I can easily say so and that is the end of it. (questionnaire, 87)

From the posting of a self description on a dating site to the exchange of email with others, users are concerned about the possible discrepancy between "cyberselves" and "real selves." Indeed the latter is often referred to as an important aspect of describing the self in email exchanges. . . .

In off-line meetings with strangers there is a prioritisation of bodily attraction that is replaced by an emphasis on textual adroitness in on-line meetings. As John's account suggests, the body no longer "sizes our attention (and that of others)" (Williams, 1998: 61) but becomes something that is defined and managed through textual interaction. How the body is written and read creates a space for negotiation and disjunctures between the lived body and how it is seen by others. The domain of internet dating is, therefore, a space in which individuals seek to close the gap between the embodied and disembodied self, the public and the private individual, and anonymity and intimacy. While many commentaries on the internet highlight the possibilities it facilitates for forging new selves and new relationships unencumbered by the constraints of time, place and body (e.g., Featherstone and Burrows, 1995), the users of these sites are concerned to translate virtual relationships into meetings between flesh and blood individuals.

FROM VIRTUAL INTERACTION TO COPRESENCE

In a study of internet communities Rheingold (1993) has commented that when community members meet in the real world, the relationship is, in a sense, "backwards" because they are already familiar with much of each others lives. . . . This sense of already "knowing the other" is an important aspect of how users negotiate the transition to a meeting. The process may also be accompanied by one or more phone conversations as Sally explains:

> In my experience there is mutual common ground in likes and dislikes and so on. It does not take many notes to see whether you get on with someone and then it naturally moves onto a meeting. John, who was a nice guy but there was no spark between us, wanted to go to the same exhibition as me so we swapped mobile numbers and fixed to meet for coffee before we went in . . . I've not felt any of the awkwardness I have felt when I've had completely blind dates. (questionnaire, 254)

The process of "getting-to-know" the other through email contributes to a sense of mutual trust. This reduces the risks and potential embarrassment of what would otherwise be a first-acquaintance meeting. Nevertheless, there is an inevitable gap between the virtual world of the internet and the reality of copresent meetings. There is a certain "shock" of presence involved in meeting physically an individual who was previously known only as a "sensitive," "funny," "open" writer of emails. No matter how open and honest individuals have been, meeting each other in the flesh was the crucial test for previously virtual relationships. The tensions of such meetings are reflected in the comments of Helen who had met two men:

> What strikes me about meeting in the flesh is the way no amount of description can prepare you for the real appearance. The two men I met did not look how I imagined. Not that they misled me, it is just that normally when you meet people you know what they look like at the same time. (questionnaire, 82) . . .

CONCLUSION

The internet has been interpreted as leading to the emergence of a distinction between the embodied self, and disembodied, multiple cyberselves. At their most utopian, these analyses suggest that on-line identities are "disengaged from gender, ethnicity and other problematic constructions" and "float free of biological and sociocultural determinant" (Dery, 1993: 560–1). This focus is less revealing than it might first appear, however, because the embodied lives, identities and material circumstances of users are themselves significant in affecting patterns of access to and use of the internet. As suggested in the introduction, the environment of internet dating provides us with an example of how the real world acts as a casing for this virtual media. Changing household patterns and class biased trends in internet use, for example, have fed a demand for this mode of dating and influenced who is able to make use of it.

It would be a mistake, however, to underestimate the changes facilitated by the internet. While Giddens's conception of "pure relationships" may ultimately constitute an unrealistic ideal, rather than a valid ideal type (Shilling, 1997), the internet does provide a medium in which individuals engage in a communicative process of building up trust, of self-disclosure, and of exploring the other in relation to one's own reflexively constructed needs and desires. While this paper is based on the predominate form of dating site it would be imprudent to ignore the diversity of dating and other sites through which people can arrange off-line meetings. There are, for example, a growing number of sites that promise to arrange relationships between people located in different countries that exploit economic deprivation in, for example, Eastern Europe and Asia. Within the gay community the internet is being increasingly used to meet people and arrange social events. However, there is the common use of the internet as a different starting point for off-line relationships. The general preference for text based description over photographs in the sites examined here suggests that, at least in the first stages of a relationship, communicative appeal is less subordinated to physical attraction than in other social contexts. However, these conditions do not mean the end of interaction rituals. Widely shared norms appear to have emerged that include turn taking in the sending of emails, reciprocity in disclosing details about the self, and respecting other people's presentations of self, that mirror those characteristic of daily life (Goffman, 1983). . . .

The environment discussed in this paper represents one of many areas within the internet where the self may be articulated and explored but is unlikely to be transcended or reconstructed into any number of virtual selves. The casing of the off-line world in such spaces remains important because it shapes whether, how and why people turn to the internet. Internet dating sites are but one example of a growing number of virtual places that are devised to have a potential impact on users' off-line lifestyles. Indeed many of the new resources that developed for the internet have been designed to address or fulfill off-line needs. The growing number of internet sites that provide health information, for example, do so by grounding information in the symptoms and experiences of real users in the real world (Hardey, 2002). The way participants in welfare related newsgroups provide advice and support has also been shown to be anchored in their own, as opposed to fantasy experiences, to the degree that users may flame anyone they see as inauthentic (Burrows et al., 2000; Burrows and Nettleton, 2000). . . . Rather than visions of another "life-world" . . . occupied by users with multiple identities (Haraway, 1985) the internet for many is just a different space where they may meet others and make use of a vast number of services and resources.

The nature of the internet and its social consequences discussed in this paper is, then, different from its common image as a realm dominated by the unreal, fantastic and imagined multiple selves (cf. Gibson, 1984; Turkle, 1995; Stone, 1996). Within the domains of MUDs and chat rooms people can remain anonymous and derive satisfactions from the disembodied interactions that take place, yet as the use of the internet has grown it has become increasingly used in ways that are grounded in pre-existing social and economic processes. The anonymity of individuals that characterises dating sites rarely seems to facilitate the construction of fantasy selves, but acts as a foundation for the building of trust and establishing real world relationships. Rather than forming a distinct cyberspace culture, the internet is opening up new opportunities to shape the extant contours and contents of social life. . . .

REFERENCES

Bataille, G. (1962), *Eroticism*, London: John Calder.

Burrows, R. and Nettleton, S. (2000), Reflexive modernization and the emergence of wired self-help, in Renninger, K. A. and Shumar, Q. (eds), *Building Virtual Communities: Learning and Change in Cyberspace,* New York: Cambridge University Press.

Burrows, R., Nettleton, S., Pleace, N., Loader, B., and Muncer, S. (2000), Virtual Community Care? Social Policy and the Emergence of Computer Mediated Social Support, *Information Communication and Society,* 3 (1): 23–31.

Dery, M. (1993), *Flame Wars: The Discourse of Cyberculture,* Durham, NC: Duke University Press.

Featherstone, M. and Burrows, R. (eds), (1995), *Cyberspace, Cyberbodies, Cyberpunk: Cultures of Technological Embodiment,* London: Routledge.

Gibson, W. (1984), *Neuromancer,* New York: Bantam Books.

Giddens, A. (1992), *The Transformation of Intimacy: Sexuality, Love, and Eroticism in Modern Societies,* Cambridge: Polity Press.

Goffman, E. (1956), 'Embarrassment and social organisation,' *American Journal of Sociology* LXII (3): 264–71.

Goffman, E. (1963), *Behaviour in Public Places,* New York: Free Press.

Goffman, E. (1967), *Interaction Ritual,* Harmondsworth: Penguin Books.

Goffman, E. (1969), *The Presentation of Self in Everyday Life,* Harmondsworth: Penguin Books.

Goffman, E. (1983), 'The interaction order,' *American Sociological Review,* 48: 1–17.

Haraway, D. (1985/1994), 'A manifesto for cyborgs: Science, technology and socialist feminism in the 1980s,' in Seidman, S. (ed.) *The Postmodern Turn,* Cambridge: Cambridge University Press.

Hardey, M. (2002), 'The story of my illness: personal accounts of illness on the Internet,' *Health: An Interdisciplinary Journal* 6 (1): 31–46.

Jagger, E. (2001), 'Marketing Molly and Melville: Dating in a postmodern, consumer society,' *Sociology,* 35 (1): 39–57.

Jones, S. (1999), Studying the net: Intricacies and issues, in Jones, S. (ed.) *Doing Internet Research: Critical Issues and Methods for Examining the Net,* London: Sage.

Mann, C. and Stewart, F. (2000), *Internet Communication and Qualitative Research: A Handbook for Researching Online,* London: Sage.

Rheingold, H. (1993), *The Virtual Community: Homesteading on the Electronic Frontier,* Reading, MA: Addison-Wesley.

Shilling, C. (1997), 'Emotions, embodiment and the sensation of society,' *The Sociological Review* 45 (2): 195–219.

Spender, D. (1995), *Nattering on the Net: Women Power and Cyberspace,* Melbourne: Spinifex Press.

Stone, A. R. (1996), *The War of Desire and Technology at the End of the Mechanical Age,* Cambridge, MA: MIT Press.

Turkle, S. (1995), *Life on the Screen: Identity in the Age of the Internet,* New York: Simon and Schuster.

Williams, S. J. (1998), Bodily dys-order: Desire, excess and the transgression of corporeal boundaries, *Body and Society,* 4 (2): 59–82.

DISCUSSION QUESTIONS

1. How does your communication change when using e-mail, instant messaging, or online chat rooms? Is the nature of that interaction different than in-person interaction? How?

2. What are the benefits to online interaction? What are the disadvantages to it?

http://infotrac.thomsonlearning.com

InfoTrac College Edition

BONUS READING

Gerschick, Thomas J. "Toward a Theory of Disability and Gender." *Signs* 25 (Summer 2000): 1263–68.

   This reading attempts to create a theory to explain how gender intersects with disability to create a unique experience for social interaction. Past gender scholarship established that gender is an interactional process, something we "do." This process involves the body. For men and women with disabilities, how they "do gender" is greatly influenced by their physical limitations. Consider how we use our bodies to establish gender in our interpersonal relationships. How would a disability alter this process?

SEARCH TERMS

impression management
urban violence
cyber relationships
cyber-interaction
disability and stigma

15

Clique Dynamics

PATRICIA ADLER AND PETER ADLER

Patricia Adler and Peter Adler take a look at clique formation and friendship groupings in schools. In their study of children's friendship groups, they analyze how cliques can generate tremendous power and influence over clique members.

A dominant feature of children's lives is the clique structure that organizes their social world. The fabric of their relationships with others, their levels and types of activity, their participation in friendships, and their feelings about themselves are tied to their involvement in, around, or outside the cliques organizing their social landscape. Cliques are, at their base, friendship circles, whose members tend to identify each other as mutually connected. Yet they are more than that; cliques have a hierarchical structure, being dominated by leaders, and are exclusive in nature, so that not all individuals who desire membership are accepted. They function as bodies of power within grades, incorporating the most popular individuals, offering the most exciting social lives, and commanding the most interest and attention from classmates. . . . As such they represent a vibrant component of the preadolescent experience, mobilizing powerful forces that produce important effects on individuals.

The research on cliques is cast within the broader literature on elementary school children's friendship groups. A first group of such works examines independent variables that can have an influence on the character of children's friendship groups. A second group looks at the features of children's inter- and intra-group relations. A third group concentrates on the behavioral dynamics specifically associated with cliques. Although these studies are diverse in their focus, they identify several features as central to clique functioning without thoroughly investigating their role and interrelation: boundary maintenance and definitions of membership (exclusivity); a hierarchy of popularity (status stratification and differential power), and relations between in-groups and out-groups (cohesion and integration).

In this [essay] we look at these dynamics and their association, at the way clique leaders generate and maintain their power and authority (leadership, power/dominance), and at what it is that influences followers to comply so readily with clique leaders' demands (submission). These interactional dynamics

From: Patricia Adler and Peter Adler. 1998. *Peer Power: Preadolescent Culture and Identity.* New Brunswick, NJ: Rutgers University Press, pp. 56–69. Reprinted with permission.

are not intended to apply to all children's friendship groups, only those (populated by one-quarter to one-half of the children) that embody the exclusive and stratified character of cliques.

TECHNIQUES OF INCLUSION

The critical way that cliques maintained exclusivity was through careful membership screening. Not static entities, cliques irregularly shifted and evolved their membership, as individuals moved away or were ejected from the group and others took their place. In addition, cliques were characterized by frequent group activities designed to foster some individuals' inclusion (while excluding others). Cliques had embedded, although often unarticulated, modes for considering and accepting (or rejecting) potential new members. These modes were linked to the critical power of leaders in making vital group decisions. Leaders derived power through their popularity and then used it to influence membership and social stratification within the group. This stratification manifested itself in tiers and subgroups within cliques composed of people who were hierarchically ranked into levels of leaders, followers, and wannabes. Cliques embodied systems of dominance, whereby individuals with more status and power exerted control over others' lives.

Recruitment

Initial entry into cliques often occurred at the invitation or solicitation of clique members. . . . Those at the center of clique leadership were the most influential over this process, casting their votes for which individuals would be acceptable or unacceptable as members and then having other members of the group go along with them. If clique leaders decided they liked someone, the mere act of their friendship with that person would accord them group status and membership. . . .

Potential members could also be brought to the group by established members who had met and liked them. The leaders then decided whether these individuals would be granted a probationary period of acceptance during which they could be informally evaluated. If the members liked them, the newcomers would be allowed to remain in the friendship circle, but if they rejected them, they would be forced to leave.

Tiffany, a popular, dominant girl, reflected on the boundary maintenance she and her best friend Diane, two clique leaders, had exercised in fifth grade:

Q: *Who defines the boundaries of who's in or who's out?*

TIFFANY: Probably the leader. If one person might like them, they might introduce them, but if one or two people didn't like them, then they'd start to get everyone up. Like in fifth grade, there was Dawn Bolton and she was new. And the girls in her class that were in our clique liked her, but Diane and I didn't like her, so we kicked her out. So then she went to the other clique, the Emily clique. . . .

Application

A second way for individuals to gain initial membership into a clique occurred through their actively seeking entry. . . . Several factors influenced the likelihood that a person would be accepted as a candidate for inclusion, as Darla, a popular fourth-grade girl described: "Coming in, it's really hard coming in, it's like really hard, even if you are the coolest person, they're still like, 'What is *she* doing [exasperated]?' You can't be too pushy, and like I don't know, it's really hard to get in, even if you can. You just got to be there at the right time, when they're nice, in a nice mood."

According to Rick, a fifth-grade boy who was in the popular clique but not a central member, application for clique entry was more easily accomplished by individuals than groups. He described the way individuals found routes into cliques: "It can happen any way. Just you get respected by someone, you do something nice, they start to like you, you start doing stuff with them. It's like you just kind of follow another person who is in the clique back to the clique, and he says, 'Could this person play?' So you kind of go out with the clique for a while and you start doing stuff with them, and then they almost like invite you in. And then soon after, like a week or so, you're actually in. It all depends. . . . But you can't bring your whole group with you, if you have one. You have to leave them behind and just go in on your own."

Successful membership applicants often experienced a flurry of immediate popularity. Because their entry required clique leaders' approval, they gained associational status.

Friendship Realignment

Status and power in a clique were related to stratification, and people who remained more closely tied to the leaders were more popular. Individuals who wanted to be included in the clique's inner echelons often had to work regularly to maintain or improve their position.

Like initial entry, this was sometimes accomplished by people striving on their own for upward mobility. In fourth grade, Danny was brought into the clique by Mark, a longtime member, who went out of his way to befriend him. After joining the clique, however, Danny soon abandoned Mark when Brad, the clique leader, took an interest in him. Mark discussed the feelings of hurt and abandonment this experience left him with: "I felt really bad, because I made friends with him when nobody knew him and nobody liked him, and I put all my friends to the side for him, and I brought him into the group, and then he dumped me. He was my friend first, but then Brad wanted him. . . . He moved up and left me behind, like I wasn't good enough anymore."

The hierarchical structure of cliques, and the shifts in position and relationships within them, caused friendship loyalties within these groups to be less reliable than they might have been in other groups. People looked toward those above them and were more susceptible to being wooed into friendship with individuals more popular than they. When courted by a higher-up, they could easily drop their less popular friends. . . .

Ingratiation

Currying favor with people in the group, like previous inclusionary endeavors, can be directed either upward (supplication) or downward (manipulation). . . . Note that children often begin their attempts at entry into groups with low-risk tactics; they first try to become accepted by more peripheral members, and only later do they direct their gaze and inclusion attempts toward those with higher status. The children we observed did this as well, making friendly overtures toward clique followers and hoping to be drawn by them into the center.

The more predominant behavior among group members, however, involved currying favor with the leader to enhance their popularity and attain greater respect from other group members. One way they did this was by imitating the style and interests of the group leader. Marcus and Adam, two fifth-grade boys, described the way borderline people would fawn on their clique and its leader to try to gain inclusion:

> MARCUS: Some people would just follow us around and say, "Oh yeah, whatever he says, yeah, whatever his favorite kind of music is, is my favorite kind of music."

> ADAM: They're probably in a position then they want to be more in because if they like what we like, then they think more people will probably respect them. Because if some people in the clique think this person likes their favorite groups, say it's REM, or whatever, so it's say Bud's [the clique leader's], this person must know what we like in music and what's good and what's not, so let's tell him that he can come up and join us after school and do something.

Fawning on more popular people not only was done by outsiders and peripherals but was common practice among regular clique members, even those with high standing. Darla, a second-tier fourth-grade girl, . . . described how, in fear, she used to follow the clique leader and parrot her opinions: "I was never mean to the people in my grade because I thought Denise might like them and then I'd be screwed. Because there were some people that I hated that she liked and I acted like I loved them, and so I would just be mean to the younger kids, and if she would even say, 'Oh she's nice,' I'd say, 'Oh yeah, she's really nice!'" Clique members, then, had to stay abreast of the leader's shifting tastes and whims if they were to maintain status and position in the group. Part of their membership work involved a regular awareness of the leader's fads and fashions, so that they would accurately align their actions and opinions with the current trends in timely manner. . . .

TECHNIQUES OF EXCLUSION

Although inclusionary techniques reinforced individuals' popularity and prestige while maintaining the group's exclusivity and stratification, they failed to contribute to other, essential, clique features such as cohesion and integration, the management of in-group and out-group relationships, and submission to clique leadership. These features are rooted, along with further sources of domination and power, in cliques' exclusionary dynamics.

Out-Group Subjugation

When they were not being nice to try to keep outsiders from straying too far from their realm of influence, clique members predominantly subjected outsiders to exclusion and rejection. They found sport in picking on these lower-status individuals. As one clique follower remarked, "One of the main things is to keep picking on unpopular kids because it's just fun to do." [Sociologist] Eder . . . notes that this kind of ridicule, where the targets are excluded and not enjoined to participate in the laughter, contrasts with teasing, where friends make fun of each other in a more lighthearted manner but permit the targets to remain included in the group by also jokingly making fun of themselves. Diane, a clique leader in fourth grade, described the way she acted toward outsiders: "Me and my friends would be mean to the people outside of our clique. Like, Eleanor Dawson, she would always try to be friends with us, and we would be like, 'Get away, ugly.' "

Interactionally sophisticated clique members not only treated outsiders badly but managed to turn others in the clique against them. Parker and Gottman . . . observe that one of the ways people do this is through gossip. Diane recalled the way she turned all the members of her class, boys as well as girls, against an outsider: "I was always mean to people outside my group like Crystal, and Sally Jones; they both moved schools. . . . I had this gummy bear necklace, with pearls around it and gummy bears. She [Crystal] came up to me one day and pulled my necklace off. I'm like, 'It was my favorite necklace,' and I got all of my friends, and all the guys even in the class, to revolt against her. No one liked her. That's why she moved schools, because she tore my gummy bear necklace off and everyone hated her. They were like, 'That was mean. She didn't deserve that. We hate you.' " . . .

In-Group Subjugation

Picking on people within the clique's confines was another way to exert dominance. More central clique members commonly harassed and were mean to those with weaker standing. Many of the same factors prompting the ill treatment of outsiders motivated high-level insiders to pick on less powerful insiders. Rick, a fifth-grade clique follower, articulated the systematic organization of downward harassment: "Basically the people who are the most popular, their life outside in the playground is picking on other people who aren't as popular, but are in the group. But the people just want to be more popular so they stay in the group, they just kind of stick with it, get made fun of, take it. . . . They come back everyday, you do more ridicule, more ridicule, more ridicule, and they just keep taking it because they want to be more popular, and they actually like you but you don't like them. That goes on a lot, that's the main thing in the group. You make fun of someone, you get more popular, because insults is what they like, they like insults."

The finger of ridicule could be pointed at any individual but the leader. It might be a person who did something worthy of insult, it might be someone who the clique leader felt had become an interpersonal threat, or it might be someone singled out for no apparent reason. . . . Darla, the second tier fourth grader discussed earlier, described the ridicule she encountered and her feelings of mortification when the clique leader derided her hair: "Like I remember, she embarrassed me so bad one day. Oh my God, I wanted to kill her! We were in music class and we were standing there and she goes, 'Ew! what's all that shit in your hair?' in front of the whole class. I was so embarrassed, 'cause, I guess I had dandruff or something."

Often, derision against insiders followed a pattern, where leaders started a trend and everyone followed it. This intensified the sting of the mockery by compounding it with multiple force. Rick analogized the way people in cliques behaved to the links on a chain: "Like it's a chain reaction, you get in a fight with the main person, then the person right under him will not like you, and the person under him won't like you, and et cetera, and the whole group will take turns against you. A few people will still like you because they will do their own thing, but most people will do what the person in front of them says to do, so it would be like a chain reaction. It's like a chain; one chain turns, and the other chain has to turn with them or else it will tangle."

Compliance

Going along with the derisive behavior of leaders or other high-status clique members could entail either active or passive participation. Active participation occurred when instigators enticed other clique members to pick on their friends. For example, leaders would often come up with the idea of placing phony phone calls to others and would persuade their followers to do the dirty work. They might start the phone call and then place followers on the line to finish it, or they might pressure others to make the entire call, thus keeping one step distant from becoming implicated, should the victim's parents complain.

Passive participation involved going along when leaders were mean and manipulative, as when Trevor submissively acquiesced in Brad's scheme to convince Larry that Rick had stolen his money. Trevor knew that Brad was hiding the money the whole time, but he watched while Brad whipped Larry into a frenzy, pressing him to deride Rick, destroy Rick's room and possessions, and threaten to expose Rick's alleged theft to others. It was only when Rick's mother came home, interrupting the bedlam, that she uncovered the money and stopped Larry's onslaught. The following day at school, Brad and Trevor could scarcely contain their glee. As noted earlier, Rick was demolished by the incident and cast out by the clique; Trevor was elevated to the status of Brad's best friend by his coconspiracy in the scheme. . . .

Stigmatization

Beyond individual incidents of derision, clique insiders were often made the focus of stigmatization for longer periods of time. Unlike outsiders who commanded less enduring interest, clique members were much more involved in picking on their friends, whose discomfort more readily held their attention.

Rick noted that the duration of this negative attention was highly variable: "Usually at certain times, it's just a certain person you will pick on all the time, if they do something wrong. I've been picked on for a month at a time, or a week, or a day, or just a couple of minutes, and then they will just come to respect you again." When people became the focus of stigmatization, as happened to Rick, they were rejected by all their friends. The entire clique rejoiced in celebrating their disempowerment. They would be made to feel alone whenever possible. Their former friends might join hands and walk past them through the play yard at recess, physically demonstrating their union and the discarded individual's aloneness.

Worse than being ignored was being taunted. Taunts ranged from verbal insults to put-downs to singsong chants. Anyone who could create a taunt was favored with attention and imitated by everyone. Even outsiders, who would not normally be privileged to pick on a clique member, were able to elevate themselves by joining in on such taunting. . . .

The ultimate degradation was physical. Although girls generally held themselves to verbal humiliation of their members, the culture of masculinity gave credence to boys' injuring each other. . . . Fights would occasionally break out in which boys were punched in the ribs or stomach, kicked, or given black eyes. When this happened at school, adults were quick to intervene. But after hours or on the school bus boys could be hurt. Physical abuse was also heaped on people's homes or possessions. People spit on each other or others' books or toys, threw eggs at their family's cars, and smashed pumpkins in front of their house.

Expulsion

While most people returned to a state of acceptance following a period of severe derision . . . this was not always the case. Some people became permanently excommunicated from the clique. Others could be cast out directly, without undergoing a transitional phase of relative exclusion. Clique members from any stratum of the group could suffer such a fate, although it was more common among people with lower status.

When Davey, mentioned earlier, was in sixth grade, he described how expulsion could occur as a natural result of the hierarchical ranking, where a person at the bottom rung of the system of popularity was pushed off. He described the ordinary dynamics of clique behavior:

Q: *How do clique members decide who they are going to insult that day?*

DAVEY: It's just basically everyone making fun of everyone. The small people making fun of smaller people, the big people making fun of the small people. Nobody is really making fun of people bigger than them because they can get rejected, because then they can say, "Oh yes, he did this and that, this and that, and we shouldn't like him anymore." And everybody else says, "Yeah, yeah, yeah," 'cause all the lower people like him, but all the higher people don't. So the lowercase people just follow the highercase people. If one person is doing something wrong, then they will say, "Oh yeah, get out, good-bye." . . .

DISCUSSION QUESTIONS

1. Take a look at your own friendship group in school. Which of the processes of both inclusion and exclusion do you observe?

2. What forms of negative sanction, or punishment, do the more powerful high status clique members deliver to others? List some, noting how they differ in severity.

16

Fraternities and Collegiate Rape Culture

Why Are Some Fraternities More Dangerous Places for Women?

A. AYRES BOSWELL
AND JOAN Z. SPADE

Boswell and Spade studied fraternity parties and bar settings as places where college women are at risk for rape. The authors point to the significance of the fraternity as a group that encourages the rape culture. The fraternity house is a setting that is ruled by the group norms of fraternity men.

Date rape and acquaintance rape on college campuses are topics of concern to both researchers and college administrators. Some estimate that 60 to 80 percent of rapes are date or acquaintance rape (Koss, Dinero, Seibel, and Cox 1988). Further, 1 out of 4 college women say they were raped or experienced an attempted rape, and 1 out of 12 college men say they forced a woman to have sexual intercourse against her will (Koss, Gidycz, and Wisniewski 1985).

Although considerable attention focuses on the incidence of rape, we know relatively little about the context or the *rape culture* surrounding date

From: A. Ayres Boswell and Joan Z. Spade. "Fraternities and Collegiate Rape Culture: Why Are Some Fraternities More Dangerous Places for Women? *Gender & Society*, Vol. 10, No. 2, April 1996: 133–47. © 1996 Sociologists for Women in Society. Reprinted with permission.

acquaintance rape. Rape culture is a set of values and beliefs that provide an environment conducive to rape (Buchwald, Fletcher, & Roth 1993; Herman 1984). The term applies to a generic culture surrounding and promoting rape, not the specific settings in which rape is likely to occur. We believe that the specific settings also are important in defining relationships between men and women.

Some have argued that fraternities are places where rape is likely to occur on college campuses (Martin and Hummer 1989; O'Sullivan 1993; Sanday 1990) and that the students most likely to accept rape myths and be more sexually aggressive are more likely to live in fraternities and sororities, consume higher doses of alcohol and drugs, and place a higher value on social life at college (Gwartney-Gibbs and Stockard 1989; Kalof and Cargill 1991). Others suggest that sexual aggression is learned in settings such as fraternities and is not part of predispositions or preexisting attitudes (Boeringer, Shehan, and Akers 1991). To prevent further incidences of rape on college campuses, we need to understand what it is about fraternities in particular and college life in general that may contribute to the maintenance of a rape culture on college campuses.

Our approach is to identify the social contexts that link fraternities to campus rape and promote a rape culture. Instead of assuming that all fraternities provide an environment conducive to rape, we compare the interactions of men and women at fraternities identified on campus as being especially *dangerous* places for women, where the likelihood of rape is high, to those seen as *safer* places, where the perceived probability of rape occurring is lower. Prior to collecting data for our study, we found that most women students identified some fraternities as having more sexually aggressive members and a higher probability of rape. These women also considered other fraternities as relatively safe houses, where a woman could go and get drunk if she wanted to and feel secure that the fraternity men would not take advantage of her. We compared parties at houses identified as high-risk and low-risk houses as well as at two local bars frequented by college students. Our analysis provides an opportunity to examine situations and contexts that hinder or facilitate positive social relations between undergraduate men and women.

The abusive attitudes toward women that some fraternities perpetuate exist within a general culture where rape is intertwined in traditional gender scripts. Men are viewed as initiators of sex and women as either passive partners or active resisters, preventing men from touching their bodies (LaPlante, McCormick, and Brannigan 1980). Rape culture is based on the assumptions that men are aggressive and dominant whereas women are passive and acquiescent (Buchwald et al. 1993; Herman 1984). What occurs on college campuses is an extension of the portrayal of domination and aggression of men over women that exemplifies the double standard of sexual behavior in U.S. society (Barthel 1988; Kimmel 1993). . . .

Whereas some researchers explain these attitudes toward sexuality and rape using an individual or a psychological interpretation, we argue that rape has a social basis, one in which both men and women create and re-create masculine and feminine identities and relations. Based on the assumption that rape is part of the social construction of gender, we examine how men and women "do gender" on a college campus (West and Zimmerman 1987). We focus on fraternities because they have been identified as settings that encourage rape (Sanday 1990). By comparing fraternities that are viewed by women as places where there is a high risk of rape to those where women believe there is a low risk of rape as well as two local commercial bars, we seek to identify characteristics that make some social settings more likely places for the occurrence of rape.

METHOD

We observed social interactions between men and women at a private coeducational school in which a high percentage (49.4 percent) of students affiliate with Greek organizations. The university has an undergraduate population of approximately 4,500 students, just more than one third of whom are women; the students are primarily from upper-middle-class families. The school, which admitted only men until 1971, is highly competitive academically.

We used a variety of data collection approaches: observations of interactions between men and women at fraternity parties and bars, formal interviews, and informal conversations. The first author, a former undergraduate at this school and a graduate student at the time of the study, collected the data. She knew about the social life at the school and had established rapport and trust between herself and undergraduate students as a teaching assistant in a human sexuality course.

. . . In our study, 40 women students identified fraternities that they considered to be high risk, or to have more sexually aggressive members and higher incidence of rape, as well as fraternities that they considered to be safe houses. The women represented all four years of undergraduate college and different living groups (sororities, residence halls, and off-campus housing). Observations focused on the four fraternities named most often by these women as high-risk houses and the four identified as low-risk houses.

Throughout the spring semester, the first author observed at two fraternity parties each weekend at two different houses (fraternities could have parties only on weekends at this campus). She also observed students' interactions in two popular university bars on weeknights to provide a comparison of students' behavior in non-Greek settings. The first local bar at which she observed was popular with seniors and older students; the second bar was popular with first-, second-, and third-year undergraduates because the management did not strictly enforce drinking age laws in this bar.

The observer focused on the social context as well as interaction among participants at each setting. In terms of social context, she observed the following: ratio of men to women, physical setting such as the party decor and theme, use and control of alcohol and level of intoxication, and explicit and implicit norms. She noted interactions between men and women (i.e., physical contact, conversational style, use of jokes) and the relations among men (i.e., their treatment of pledges and other men at fraternity parties). Other than the observer, no one knew the identity of the high- or low-risk fraternities. Although this may have introduced bias into the data collection, students on this campus who read this article before it was submitted for publication commented on how accurately the social scene is described. . . .

RESULTS

The Settings

Fraternity Parties We observed several differences in the quality of the interaction of men and women at parties at high-risk fraternities compared to those at low-risk houses. A typical party at a low-risk house included an equal number of women and men. The social atmosphere was friendly, with considerable interaction between women and men. Men and women danced in groups and in couples, with many of the couples kissing and displaying affection toward each other. Brothers explained that, because many of the men in these houses had girlfriends, it was normal to see couples kissing on the dance floor. Coed groups engaged in conversations at many of these houses, with women and men engaging in friendly exchanges, giving the impression that they knew each other well. Almost no cursing and yelling was observed at parties in low-risk houses; when pushing occurred, the participants apologized. Respect for women extended to the women's bathrooms, which were clean and well supplied.

At high-risk houses, parties typically had skewed gender ratios, sometimes involving more men and other times involving more women. Gender segregation also was evident at these parties, with the men on one side of a room or in the bar drinking while women gathered in another area. Men treated women differently in the high-risk houses. The women's bathrooms in the high-risk houses were filthy, including clogged toilets and vomit in the sinks. . . .

Men attending parties at high-risk houses treated women less respectfully, engaging in jokes, conversations, and behaviors that degraded women. Men made a display of assessing women's bodies and rated them with thumbs up or thumbs down for the other men in the sight of the women. One man attending a party at a high-risk fraternity said to another, "Did you know that this week is Women's Awareness Week? I guess that means we get to abuse them more this week." Men behaved more crudely at parties at high-risk houses. At one party, a brother dropped his pants, including his underwear, while dancing in front of several women. Another brother slid across the dance floor completely naked.

The atmosphere at parties in high-risk fraternities was less friendly overall. With the exception of greetings, men and women rarely smiled or laughed and spoke to each other less often than was the case at parties in low-risk houses. The few one-on-one conversations between women and men appeared to be strictly flirtatious (lots of eye contact, touching, and very close talking). It was rare to see a group of men and women together talking. Men were openly hostile, which made the high-risk parties seem almost threatening at times. For example, there was a lot of touching, pushing, profanity, and name calling, some done by women.

Students at parties at the high-risk houses seemed self-conscious and aware of the presence of members of the opposite sex, an awareness that was sexually charged. Dancing early in the evening was usually between women. Close to midnight, the sex ratio began to balance out with the arrival of more men or more women. Couples began to dance together but in a sexual way (close dancing with lots of pelvic thrusts). Men tried to pick up women using lines such as "Want to see my fish tank?" and "Let's go upstairs so that we can talk; I can't hear what you're saying in here." . . .

As others have found, fraternity brothers at high-risk houses on this campus told about routinely discussing their sexual exploits at breakfast the morning after parties and sometimes at house meetings (cf. Martin and Hummer 1989; O'Sullivan 1993; Sanday 1990). During these sessions, the brothers we interviewed said that men bragged about what they did the night before with stories of sexual conquests often told by the same men, usually sophomores. The women involved in these exploits were women they did not know or knew but did not respect, or *faceless victims.* Men usually treated girlfriends with respect and did not talk about them in these storytelling sessions. Men from low-risk houses, however, did not describe similar sessions in their houses.

The Bar Scene The bar atmosphere and social context differed from those of fraternity parties. The music was not as loud, and both bars had places to sit and have conversations. At all fraternity parties, it was difficult to maintain conversations with loud music playing and no place to sit. The volume of music at parties at high-risk fraternities was even louder than it was at low-risk houses, making it virtually impossible to have conversations. In general, students in the local bars behaved in the same way that students did at parties in low-risk houses with conversations typical, most occurring between men and women. . . .

On a couple of occasions, however, the atmosphere at the second bar became similar to that of a party at a high-risk fraternity. As the number of people in the bar increased, they removed chairs and tables, leaving no place to sit and talk. The music also was turned up louder, drowning out conversation. With no place to dance or sit, most people stood around but could not maintain conversations because of the noise and crowds. Interactions between women and men consisted mostly of flirting. Alcohol consumption also was greater than it was on the less crowded nights, and the number of visibly drunk people increased. The more people drank, the more conversation and socializing broke down. The only differences between this setting and that of a party at a high-risk house were that brothers no longer controlled the territory and bedrooms were not available upstairs.

Gender Relations

Relations between women and men are shaped by the contexts in which they meet and interact. As is the case on other college campuses, *hooking up* has replaced dating on this campus, and fraternities are places where many students hook up. Hooking up is a loosely applied term on college campuses that had different meanings for men and women on this campus. . . .

In contrast to hooking up, students also described monogamous relationships with steady partners. Some type of commitment was expected, but most people did not anticipate marriage. The term *seeing each other* was applied when people were sexually involved but free to date other people. This type of relationship involved less commitment than did one of boyfriend/girlfriend but was not considered to be a hook-up.

The general consensus of women and men interviewed on this campus was that the Greek system, called "the hill," set the scene for gender relations. The predominance of Greek membership and subsequent living arrangements segregated men and women. During the week, little interaction occurred between women and men after their first year in college because students in fraternities or sororities live and dine in separate quarters. In addition, many non-Greek upperclass students move off campus into apartments. Therefore, students see each other in classes or in the library, but there is no place where students can just hang out together.

Both men and women said that fraternities dominate campus social life, a situation that everyone felt limited opportunities for meaningful interactions. One senior Greek man said,

> This environment is horrible and so unhealthy for good male and female relationships and interactions to occur. It is so segregated and male dominated. . . .
> It is our party, with our rules and our beer. We are allowing these women and other men to come to our party. Men can feel superior in their domain. . . .

Some students claim that fraternities even control the dating relationships of their members. One senior woman said, "Guys dictate how dating occurs on this campus, whether it's cool, who it's with, how much time can be spent with the girlfriend and with the brothers." Couples either left campus for an evening or hung out separately with their own same-gender friends at fraternity parties, finally getting together with each other at about 2 A.M. Couples rarely went together to fraternity parties. Some men felt that a girlfriend was just a replacement for a hook-up. According to one junior man, "Basically a girlfriend is someone you go to at 2 A.M. after you've hung out with the guys. She is the sexual outlet that the guys can't provide you with."

Some fraternity brothers pressure each other to limit their time with and commitment to their girlfriends. One senior man said, "The hill [fraternities] and girlfriends don't mix." A brother described a constant battle between girlfriends and brothers over who the guy is going out with for the night, with the brothers usually winning. Brothers teased men with girlfriends with remarks such as "whipped" or "where's the ball and chain?" A brother from a high-risk house said that few brothers at his house had girlfriends; some did, but it was uncommon. One man said that from the minute he was a pledge he knew he would probably never have a girlfriend on this campus because "it was just not the norm in my house. No one has girlfriends; the guys have too much fun with [each other]."

The pressure on men to limit their commitment to girlfriends, however, was not true of all fraternities or of all men on campus. Couples attended low-risk fraternity parties together, and men in the low-risk houses went out on dates more often. A man in one low-risk house said that about 70 percent of the members of his house were involved in relationships with women, including the pledges (who were sophomores).

Treatment of Women

. . . Characteristic of rape culture, a double standard of sexual behavior for men versus women was prevalent on this campus. As one Greek senior man stated, "Women who sleep around are sluts and get bad reputations; men who do are champions and get a pat on the back from their brothers." Women also supported a double standard for sexual behavior by criticizing sexually active women. A first-year woman spoke out against women who are sexually active: "I think some girls here make it difficult for the men to respect women as a whole."

One concrete example of demeaning sexually active women on this campus is the "walk of shame." Fraternity brothers come out on the porches of their houses the night after parties and heckle women walking by. It is assumed that these women spent the night at fraternity houses and that the men they were with did not care enough about them to drive them home. Although sororities now reside in former fraternity houses, this practice continues and sometimes the victims of hecklings are sorority women on their way to study in the library. . . .

Fraternity men most often mistreated women they did not know personally. Men and women alike reported incidents in which brothers observed other brothers having sex with unknown women or women they knew only casually. . . .

DISCUSSION AND CONCLUSION

These findings describe the physical and normative aspects of one college campus as they relate to attitudes about and relations between men and women. Our findings suggest that an explanation emphasizing rape culture also must focus on those characteristics of the social setting that play a role in defining heterosexual relationships on college campuses (Kalof and Cargill 1991). The degradation of women as portrayed in rape culture was not found in all fraternities on this campus. Both group norms and individual behavior changed as students went from one place to another. Although individual men are the ones who rape, we found that some settings are more likely places for rape than are others. Our findings suggest that rape cannot be seen only as an isolated act and blamed on individual behavior and proclivities, whether it be alcohol consumption or attitudes. We also must consider characteristics of the settings that promote the behaviors that reinforce a rape culture.

Relations between women and men at parties in low-risk fraternities varied considerably from those in high-risk houses. Peer pressure and situational norms influenced women as well as men. Although many men in high- and low-risk houses shared similar views and attitudes about the Greek system, women on this campus, and date rape, their behaviors at fraternity parties were quite different. . . .

The social scene on this campus, and on most others, offers women and men few other options to socialize. Although there may be no such thing as a completely safe fraternity party for women, parties at low-risk houses and commercial bars encouraged men and women to get know each other better and decreased the probability that women would become faceless victims. Although both men and women found the social scene on this campus demeaning, neither demanded different settings for socializing, and attendance at fraternity parties is a common form of entertainment.

These findings suggest that a more conducive environment for conversation can promote more positive interactions between men and women. Simple changes would provide the opportunity for men and women to interact in meaningful ways such as adding places to sit and lowering the volume of music at fraternity parties or having parties in neutral locations, where men are not in control. The typical party room in fraternity houses includes a place to dance but not to sit and talk. The music often is loud, making it difficult, if not impossible, to carry on conversations; however, there were more conversations at the low-risk parties, where there also was more respect shown toward women. Although the number of brothers who had steady girlfriends in the low-risk houses as compared to those in the high-risk houses may explain the differences, we found that commercial bars also provided a context for interaction between men and women. At the bars, students sat and talked and conversations between men and women flowed freely, resulting in deep discussions and fewer hook-ups. . . .

In many ways, the fraternities on this campus determined the settings in which men and women interacted. As others before us have found, pressures for conformity to the norms and values exists at both high-risk and low-risk houses (Kalof and Cargill 1991; Martin and Hummer 1989; Sanday 1990). The desire to be accepted is not unique to this campus or the Greek system (Holland and Eisenhart 1990; Horowitz 1988; Moffat 1989). The degree of conformity required by Greeks may be greater than that required in most social groups, with considerable pressure to adopt and maintain the image of their houses. The fraternity system intensifies the "groupthink syndrome" (Janis 1972) by solidifying the identity of the in-group and creating an us/them atmosphere. Within the fraternity culture, brothers are highly regarded and women are viewed as outsiders. For men in high-risk fraternities, women threatened their brotherhood; therefore, brothers discouraged relationships and harassed those who treated women as equals or with respect. The pressure to be one of the guys and hang out with the guys strengthens a rape culture on college campus by demeaning women and encouraging the segregation of men and women.

Students on this campus were aware of the contexts in which they operated and the choices available to them. They recognized that, in their interactions, they created differences between men and women that are not natural, essential, or biological (West and Zimmerman 1987). Not all men and women accepted the demeaning treatment of women, but they continued to participate in behaviors that supported aspects of a rape culture. Many women participated in the hook-up scene even after they had been humiliated and hurt because they had few other means of initiating contact with men on campus. Men and women alike played out this scene, recognizing its injustices in many cases but being unable to change the course of their behaviors. . . .

Our findings indicate that a rape culture exists in some fraternities, especially those we identified as high-risk houses. College administrators are responding to this situation by providing counseling and educational programs that increase awareness of date rape including campaigns such as "No means no." These strategies are important in changing attitudes, values, and behaviors; however, changing individuals is not enough. The structure of campus life and the impact of that structure on gender relations on campus are highly determinative. To eliminate campus rape culture, student leaders and administrators must examine the situations in which women and men meet and restructure these settings to provide opportunities for respectful interaction. Change may not require abolishing fraternities; rather, it may require promoting settings that facilitate positive gender relations.

REFERENCES

Barthel, D. 1988. *Putting on appearances: Gender and advertising.* Philadelphia: Temple University Press.

Boeringer, S. B., C. L. Shehan, and R. L. Akers. 1991. Social contexts and social learning in sexual coercion and aggression. Assessing the contribution of fraternity membership. *Family Relations* 40:58–64.

Buchwald, E., R. Fletcher, and M. Roth. eds. 1993 *Transforming a rape culture.* Minneapolis, MN: Milkweed Editions.

Gwartney-Gibbs, P., and J. Stockard. 1989. Courtship aggression and mixed-sex peer groups. In *Violence in dating relationships: Emerging social issues,* edited by M. A. Pirog-Good and J. E. Stets. New York: Praeger.

Herman, D. 1984. The rape culture. In *Women: A feminist perspective,* edited by J. Freeman. Mountain View, CA: Mayfield.

Holland, D. C., and M. A. Eisenhart. 1990. *Educated in romance: Women, achievement, and college culture.* Chicago: University of Chicago Press.

Horowitz, H. L. 1988. *Campus life: Undergraduate cultures from the end of the 18th century to the present.* Chicago: University of Chicago Press.

Janis, I. L. 1972. *Victims of groupthink.* Boston: Houghton Mifflin.

Kalof, L., and T. Cargill. 1991. Fraternity and sorority membership and gender dominance attitudes. *Sex Roles* 25:417–23.

Kimmel, M. S. 1993. Clarence, William, Iron Mike, Tailhook, Senator Packwood, Spur Posse, Magic . . . and us. In *Transforming a rape culture,* edited by E. Buchwald, P. R. Fletcher, and M. Roth. Minneapolis, MN: Milkweed Editions.

Koss, M. P., T. E. Dinero, C. A. Seibel, and S. L. Cox. 1988. Stranger and acquaintance rape: Are there differences in the victim's experience? *Psychology of Women Quarterly* 12:1–24.

Koss, M. P., C. A. Gidyez, and N. Wisniewski. 1985. The scope of rape: Incidence and prevalence of sexual aggression and victimization in a national sample of higher education students. *Journal of Consulting and Clinical Psychology* 55:162–70.

LaPlante, M. N., N. McCormick, and G. G. Brannigan. 1980. Living the sexual script: College students' views of influence in sexual encounters. *Journal of Sex Research* 16:338–55.

Martin, P. Y., and R. Hummer. 1989. Fraternities and rape on campus. *Gender & Society* 3:457–73.

Moffat, M. 1989. *Coming of age in New Jersey: College life in American culture.* New Brunswick, NJ: Rutgers University Press.

O'Sullivan, C. 1993. Fraternities and the rape culture. In *Transforming a rape culture,* edited by E. Buchwald, P. R. Fletcher, and M. Roth. Minneapolis, MN: Milkweed Editions.

Sanday, P. R. 1990. *Fraternity gang rape: Sex, brotherhood, and privilege on campus.* New York: New York University Press.

West, C., and D. Zimmerman. 1987. Doing gender. *Gender & Society* 1:125–51.

DISCUSSION QUESTIONS

1. What characteristics described in the reading are similar to college parties you have attended? How do these characteristics represent the group or organization hosting the party?

2. What can be done on college campuses to change the rape culture in these group party settings?

17

Social Networks: The Value of Variety

BONNIE ERICKSON

Erickson's reading examines the importance of social networks in everything from gaining employment to maintaining good health. Social acquaintances are most helpful to individuals when they are many and varied. Diversity in the people you know provides greater networking opportunities.

From: Bonnie Erickson. "Social Networks: The Value of Variety." *Contexts* 2, no. 1 (Winter 2003): 25–31. Reprinted with permission.

S ociologists have measured acquaintance networks by focusing on occupations. People in different occupations differ from each other in many
important ways. The work we do reflects much of our pasts, such as schooling and family background, and shapes the ways we live, such as tastes and
lifestyles. Generally, someone who knows people in diverse kinds of jobs will
thereby know people who are diverse in many respects. The standard strategy is
to present a respondent with a list of occupations that range from very high to
very low in prestige, and ask whether the respondent knows anyone in each. The
greater the number of occupations within which a respondent has a contact, the
more the variety in the respondent's social network. . . .

Such acquaintances are a more diverse set than are the few people to whom
we feel really close—both because weak ties greatly outnumber strong ones,
and because our close ties are usually limited to people very much like ourselves. For example, when I studied the private security industry in Toronto I
asked whether people knew close friends, relatives, or anyone at all in each of
19 occupations. My respondents knew relatives, on average, in about two of
these occupations, close friends in about half a dozen and anyone at all in about
a dozen. . . .

NETWORKS AND JOBS

Diverse networks can help people to get good jobs. Having a variety of acquaintances improves a jobseeker's chances of having one really useful contact, and
variety itself is a qualification for some upper-end jobs.

People in North America find their jobs with the help of a contact roughly
half the time. We might assume that such helpers must be close friends and relatives willing to work hard for the jobhunter. But this is not the usual story in
Western nations. Close friends and kin want to help, but often cannot do very
much because they are too much alike: they move in the same social circles and
share information and influence, so they can do little for the candidate beyond
what he or she can do alone. But acquaintances are more varied, less like each
other, more likely to have new information and more likely to include people
highly-placed enough to influence hiring. Thus family and close friends provide
fewer jobs (and often worse jobs) than do people outside the intimate
circle. . . .

The strength of strong ties applies best to the few people at the top,
because they have highly-placed kin and friends who collect a lot of information and can exert a lot of influence. In general, more highly-placed people
can connect a jobseeker to more highly-placed jobs, and one big advantage of
having a diverse network is the improved chance of knowing such a useful
contact. . . .

Having a diverse set of acquaintances matters where there is a fairly free market in jobs and a fairly rich supply of jobs. If jobs are scarce, those in the know
will hoard access to good ones for people they care about the most, so strong ties

are more valuable in these circumstances. In non-market systems run by the state, the private use of personal contacts to get jobs may be risky: networking subverts state power and policy, and influential people may not want to be responsible for the occupational or political errors of acquaintances whom they help. Well-placed people still provide personal help, but mainly to jobseekers or intermediaries whom they know well and can trust. . . .

Diversified acquaintances are valuable as an ensemble when employers want to recruit both a person and the person's contacts, to make his or her network work for the organization. This is especially true for higher-level jobs because it is only higher-level jobs that include consequential responsibility for the "foreign affairs" of the organization.

When employers think of good contacts, what do they mean? In a word, variety. Employers named desirable contacts of many kinds (in their own industry, government, the police, senior management, etc.) and sometimes explicitly wanted variety as such ("all available"). The more varied a person's network, the more that network can do for the organization.

Employees with more network variety got jobs with higher rank and higher income. This was true whether or not people got those jobs through someone they knew. Again, a network of acquaintances is more useful than one of intimates, because acquaintances have the diversity employers seek.

Does all this add up to "it's not what you know, but who you know?" Not really. Sometimes what you know is critical. Even in the security industry, which has no formal certifications, employers often want to hire people with contacts and skills, not contacts instead of skills. Because employers look for both, using personal connections helps most to get a job at the top or bottom of the ladder, not in the middle. At the bottom, skill requirements are modest. Employers just want a reliable employee and jobseekers just want an adequate job. Using contacts is one cheap way to make this match. . . .

NETWORKS AND HEALTH

Knowing people is important in getting a job, but it also matters for other areas of our lives that are less obvious, such as good health. Research has long shown that having close friends and family is good for a person's health. People who say they have someone they can count on feel less depressed, get less physically ill and live longer than those who do not. The newer news is that having a variety of acquaintances also improves health. . . .

Acquaintances make more subtle contributions in small, invisible increments over the long run. One such contribution is a sense of control over one's life, a well-documented source of good health. People who feel more in control are less depressed just because of that, since feeling pushed around is a miserable and unwelcome experience. Moreover, having a sense of control encourages people to tackle problems they encounter, so they cope better with stress. This valuable sense of control grows with the diversity of acquaintances.

People with diverse contacts consciously adapt to different situations and manage conflicting obligations. They have to decide whom to see, how to act appropriately with others differing in their expectations, how to balance sometimes conflicting demands. As they navigate their intricate options, they develop a well-grounded sense of control over their lives. . . .

Acquaintance diversity also contributes to being better informed about health. People with wider networks are better informed about most things, but they may not realize how many of their good health practices go back to a thousand tiny nudges from casual conversations. They may know that they are committed to pushing down the broccoli and getting some exercise, while forgetting how many acquaintances mentioned the importance of such healthy habits. . . .

Feeling in control and being well-informed both flow from the diversity of the whole ensemble of acquaintances. But health, like work, sometimes benefits from a varied network because varied connections are more likely to include particular useful ones. For example, people who knew many kinds of people in the social movement group were much more likely to get some help with health (from organic vegetables to massage) from associates in the group. They knew what to look for and whom to trust to provide it.

Diverse networks also improve people's health indirectly, by helping them get ahead economically, and wealthier people tend to be healthier people. But the connection between wealth and health might suggest that all these benefits of having a variety of acquaintances might really just reflect the advantages of high social position. People with more network variety, better jobs, more feelings of control and better health may be that way because they come from more privileged circumstances. . . .

WHAT NEXT?

Other possible benefits of network variety are yet to be studied. Students of politics have speculated that interacting with a range of people expands one's sources of political information and activity, and increases tolerance for others different from oneself. . . .

Another critical avenue for future work is the way in which we think about and measure network diversity. At present, almost all studies focus on the variety of occupations within which a respondent knows someone. This works very well, because occupation goes with so many important differences of resources, views, lifestyles and so on. But occupation is not the only way in which the social world is carved up into different kinds of peoples—gender and ethnicity also shape networks.

For instance, men occupy more powerful positions in organizations, so knowing a variety of men may help one's job search more than knowing a wide variety of women. But women take more responsibility for health, including the health of others, so knowing a good range of women may be better for one's health than knowing many kinds of men. In countries like the United States or

Canada, ethnic groups have distinctive cultures and, sometimes, even labor markets. Knowing a variety of people in an ethnic group may lead to better jobs within the ethnic economy, to richer knowledge of the ethnic culture, to better access to alternative medicines and to feeling better about the group. At the same time, having acquaintances exclusively in an ethnic group may cut one off from broader social benefits.

Indeed, there are many kinds of network variety: variety of occupation, gender, ethnicity and much more. Each probably goes with a somewhat different menu of benefits. Future research should elaborate on the finding that, not only is knowing people good for you, but knowing many different kinds of people is especially good for you.

DISCUSSION QUESTIONS

1. Think of your own friendship groups and acquaintances. How diverse are the people you interact with? Do your social networks provide a variety of opportunities?

2. According to Erickson, how does diversity in the people you interact with provide advantages? What is the disadvantage in having only a few close friends?

http://infotrac.thomsonlearning.com

InfoTrac College Edition

BONUS READING

Editorial. "When Agencies Sleep." *The Christian Science Monitor,* July 11, 2003, 10.

This brief editorial points out the danger in organizational culture that does not encourage alertness and vigilance from the individuals who are a part of it. Corporate, bureaucratic cultures are difficult to change even when complacency is problematic for the organization. In what ways can a large organization, like NASA, improve the organizational culture to allow for individual vigilance? What other organizations can you think of that face this same issue?

SEARCH TERMS

stigma and groups
rape culture
social networks and the labor market
organizational culture and work

18

The Functions of Crime

EMILE DURKHEIM

This classic essay, written in 1895 and translated many times since, points to crime as an inevitable part of society. Durkheim's main functionalist thesis that criminal behavior exists in all social settings is still the theoretical basis for many sociological inquiries into crime and deviance.

If there is a fact whose pathological nature appears indisputable, it is crime. All criminologists agree on this score. Although they explain this pathology differently, they none the less unanimously acknowledge it. However, the problem needs to be treated less summarily.

. . . Crime is not only observed in most societies of a particular species, but in all societies of all types. There is not one in which criminality does not exist, although it changes in form and the actions which are termed criminal are not everywhere the same. Yet everywhere and always there have been men who have conducted themselves in such a way as to bring down punishment upon their heads. If at least, as societies pass from lower to higher types, the crime rate (the relationship between the annual crime figures and population figures) tended to fall, we might believe that, although still remaining a normal phenomenon, crime tended to lose that character of normality. Yet there is no single ground for believing such a regression to be real. Many facts would rather seem to point to the existence of a movement in the opposite direction. From the beginning of the century statistics provide us with a means of following the progression of criminality. It has everywhere increased, and in France the increase is of the order of 300 percent. Thus there is no phenomenon which represents more incontrovertibly all the symptoms of normality, since it appears to be closely bound up with the conditions of all collective life. To make crime a social illness would be to concede that sickness is not something accidental, but on the contrary derives in certain cases from the fundamental constitution of the living creature. This would be to erase any distinction between the physiological and the pathological. It can certainly happen that crime itself has normal forms; this is what happens, for instance, when it reaches an excessively high level. There is no doubt that this excessiveness is pathological in nature. What is normal is simply that criminality exists, provided that for each social type it does not reach or

From: Emile Durkheim. 1982. *The Rules of Sociological Method*. Steven Lukes (Ed.) Translated by W. D. Halls, New York: The Free Press. A division of Macmillan, pp. 64–75.

go beyond a certain level which it is perhaps not impossible to fix in conformity with the previous rules.

We are faced with a conclusion which is apparently somewhat paradoxical. Let us make no mistake: to classify crime among the phenomena of normal sociology is not merely to declare that it is an inevitable though regrettable phenomenon arising from the incorrigible wickedness of men; it is to assert that it is a factor in public health, an integrative element in any healthy society. At first sight this result is so surprising that it disconcerted even ourselves for a long time. However, once that first impression of surprise has been overcome it is not difficult to discover reasons to explain this normality and at the same time to confirm it.

In the first place, crime is normal because it is completely impossible for any society entirely free of it to exist.

Crime consists of an action which offends certain collective feelings which are especially strong and clear-cut. In any society, for actions regarded as criminal to cease, the feelings that they offend would need to be found in each individual consciousness without exception and in the degree of strength requisite to counteract the opposing feelings. Even supposing that this condition could effectively be fulfilled, crime would not thereby disappear; it would merely change in form, for the very cause which made the well-springs of criminality to dry up would immediately open up new ones.

Indeed, for the collective feelings, which the penal law of a people at a particular moment in its history protects, to penetrate individual consciousnesses that had hitherto remained closed to them, or to assume greater authority—whereas previously they had not possessed enough—they would have to acquire an intensity greater than they had had up to then. The community as a whole must feel them more keenly, for they cannot draw from any other source the additional force which enables them to bear down upon individuals who formerly were the most refractory. . . .

In order to exhaust all the logically possible hypotheses, it will perhaps be asked why this unanimity should not cover all collective sentiments without exception, and why even the weakest sentiments should not evoke sufficient power to forestall any dissentient voice. The moral conscience of society would be found in its entirety in every individual, endowed with sufficient force to prevent the commission of any act offending against it, whether purely conventional failings or crimes. But such universal and absolute uniformity is utterly impossible, for the immediate physical environment in which each one of us is placed, our hereditary antecedents, the social influences upon which we depend, vary from one individual to another and consequently cause a diversity of consciences. It is impossible for everyone to be alike in this matter, by virtue of the fact that we each have our own organic constitution and occupy different areas in space. This is why, even among lower peoples where individual originality is very little developed, such originality does however exist. Thus, since there cannot be a society in which individuals do not diverge to some extent from the collective type, it is also inevitable that among these deviations some assume a criminal character. What confers upon them this character is not the intrinsic importance of the acts but the importance which the common consciousness ascribes to them. Thus if the latter is stronger and possesses sufficient authority to make these divergences very weak in absolute terms, it will also be more sensitive and exacting. By reacting against the slightest deviations with an energy which it elsewhere employs against those that are more weighty, it endues them with the same gravity and will brand them as criminal.

Thus crime is necessary. It is linked to the basic conditions of social life, but on this very account is useful, for the conditions to which it is bound are themselves indispensable to the normal evolution of morality and law.

Indeed today we can no longer dispute the fact that not only do law and morality vary from one social type to another, but they even change within the same type if the conditions of collective existence are modified. Yet for these transformations to be made possible, the collective sentiments at the basis of morality should not prove unyielding to change, and consequently should be only moderately intense. If they were too strong, they would no longer be malleable. Any arrangement is indeed an obstacle to a new arrangement; this is even more the case the more deep-seated the original arrangement. The more strongly a structure is articulated, the more it resists modification; this is as true for functional as for anatomical patterns. If there were no crimes, this condition would not be fulfilled, for such a hypothesis presumes that collective sentiments would have attained a degree of intensity unparalleled in history. Nothing is good indefinitely and without limits. The authority which the moral consciousness enjoys must not be excessive, for otherwise no one would dare to attack it and it would petrify too easily into an immutable form. For it to evolve, individual originality must be allowed to manifest itself. But so that the originality of the idealist who dreams of transcending his era may display itself, that of the criminal, which falls short of the age, must also be possible. One does not go without the other.

Nor is this all. Beyond this indirect utility, crime itself may play a useful part in this evolution. Not only does it imply that the way to necessary changes remains open, but in certain cases it also directly prepares for these changes. Where crime exists, collective sentiments are not only in the state of plasticity necessary to assume a new form, but sometimes it even contributes to determining beforehand the shape they will take on. Indeed, how often is it only an anticipation of the morality to come, a progression towards what will be! . . . The freedom of thought that we at present enjoy could never have been asserted if the rules that forbade it had not been violated before they were solemnly abrogated. However, at the time the violation was a crime, since it was an offence against sentiments still keenly felt in the average consciousness. Yet this crime was useful since it was the prelude to changes which were daily becoming more necessary. . . .

From this viewpoint the fundamental facts of criminology appear to us in an entirely new light. Contrary to current ideas, the criminal no longer appears as an utterly unsociable creature, a sort of parasitic element, a foreign, unassimilable body introduced into the bosom of society. He plays a normal role in social life. For its part, crime must no longer be conceived of as an evil which cannot be circumscribed closely enough. Far from there being cause for congratulation when it drops too noticeably below the normal level, this apparent progress assuredly coincides with and is linked to some social disturbance. Thus the number of crimes of assault never falls so low as it does in times of scarcity. Consequently, at the same time, and as a reaction, the theory of punishment is revised, or rather should be revised. If in fact crime is a sickness, punishment is the cure for it and cannot be conceived of otherwise; thus all the discussion aroused revolves round knowing what punishment should be to fulfill its role as a remedy. But if crime is in no way pathological, the object of punishment cannot be to cure it and its true function must be sought elsewhere. . . .

DISCUSSION QUESTIONS

1. According to Durkheim's theory, criminal behavior exists in all societies. Consider the possibility of a society without the ability to punish criminal behavior (no prisons, no courts). How would individuals respond to crime? What informal social control mechanisms would help to maintain order?

2. How could you use Durkheim's theory as the basis for a research project on deviant behavior? What hypotheses could you test that would challenge or support the functionalist view of crime?

19

The Medicalization of Deviance

PETER CONRAD AND JOSEPH W. SCHNEIDER

This essay outlines the social construction of social deviance. The authors specifically refer to the medical profession as redefining certain deviant behaviors as "illness," rather than as "badness." They argue that the "medicalization of deviance changes the social response to such behavior to one of treatment rather than punishment."

From: Peter Conrad and Joseph W. Schneider. 1992. *Deviance and Medicalization: From Badness to Sickness.* Philadelphia: Temple University Press, pp. 28–37.

Consider the following situations. A woman rides a horse naked through the streets of Denver claiming to be Lady Godiva and after being apprehended by authorities, is taken to a psychiatric hospital and declared to be suffering from a mental illness. A well-known surgeon in a Southwestern city performs a psychosurgical operation on a young man who is prone to violent outbursts. An Atlanta attorney, inclined to drinking sprees, is treated at a hospital clinic for his disease, alcoholism. A child in California brought to a pediatric clinic because of his disruptive behavior in school is labeled hyperactive and is prescribed methylphenidate (Ritalin) for his disorder. A chronically overweight Chicago housewife receives a surgical intestinal bypass operation for her problem of obesity. Scientists at a New England medical center work on a million-dollar federal research grant to discover a heroin-blocking agent as a "cure" for heroin addiction. What do these situations have in common? In all instances medical solutions are being sought for a variety of deviant behaviors or conditions. We call this "the medicalization of deviance" and suggest that these examples illustrate how medical definitions of deviant behavior are becoming more prevalent in modern industrial societies like our own. The historical sources of this medicalization, and the development of medical conceptions and controls for deviant behavior, are the central concerns of our analysis.

Medical practitioners and medical treatment in our society are usually viewed as dedicated to healing the sick and giving comfort to the afflicted. No doubt these are important aspects of medicine. In recent years the jurisdiction of the medical profession has expanded and encompasses many problems that formerly were not defined as medical entities. . . . There is much evidence for this general viewpoint—for example, the medicalization of pregnancy and childbirth, contraception, diet, exercise, child development norms—but our concern here is more limited and specific. Our interests focus on the medicalization of deviant behavior: the defining and labeling of deviant behavior as a medical problem, usually an illness and mandating the medical profession to provide some type of treatment for it. Concomitant with such medicalization is the growing use of medicine as an agent of social control, typically as medical intervention. Medical intervention as social control seeks to limit, modify, regulate, isolate, or eliminate deviant behavior with medical means and in the name of health. . . .

Conceptions of deviant behavior change, and agencies mandated to control deviance change also. Historically there have been great transformations in the definition of deviance—from religious to state-legal to medical-scientific. Emile Durkheim (1893/1933) noted in *The Division of Labor in Society* that as societies develop from simple to complex, sanctions for deviance change from repressive to restitutive or, put another way, from punishment to treatment or rehabilitation. Along with the change in sanctions and social control agent there is a corresponding change in definition or conceptualization of deviant behavior. For example, certain "extreme" forms of deviant drinking (what is now called alcoholism) have been defined as sin, moral weakness, crime, and most recently illness. . . . In modern industrial society there has been a substantial growth in the prestige, dominance, and jurisdiction of the medical profession (Freidson, 1970).

It is only within the last century that physicians have become highly organized, consistently trained, highly paid, and sophisticated in their therapeutic techniques and abilities. . . . The medical profession dominates the organization of health care and has a virtual monopoly on anything that is defined as medical treatment, especially in terms of what constitutes "illness" and what is appropriate medical intervention. . . . Although Durkheim did not predict this medicalization, perhaps in part because medicine of his time was not the scientific, prestigious, and dominant profession of today, it is clear that medicine is the central restitutive agent in our society.

EXPANSION OF MEDICAL JURISDICTION
OVER DEVIANCE

When treatment rather than punishment becomes the preferred sanction for deviance, an increasing amount of behavior is conceptualized in a medical framework as illness. As noted earlier, this is not unexpected, since medicine has always functioned as an agent of social control, especially in attempting to "normalize" illness and return people to their functioning capacity in society. Public health and psychiatry have long been concerned with social behavior and have functioned traditionally as agents of social control (Foucault, 1965; Rosen, 1972). What is significant, however, is the expansion of this sphere where medicine functions in a social control capacity. In the wake of a general humanitarian trend, the success and prestige of modern biomedicine, the technological growth of the 20th century, and the diminution of religion as a viable agent of control, more and more deviant behavior has come into the province of medicine. In short, the particular, dominant designation of deviance has changed; much of what was badness (i.e., sinful or criminal) is now sickness. Although some forms of deviant behavior are more completely medicalized than others (e.g., mental illness), recent research has pointed to a considerable variety of deviance that has been treated within medical jurisdiction: alcoholism, drug addiction, hyperactive children, suicide, obesity, mental retardation, crime, violence, child abuse, and learning problems, as well as several other categories of social deviance. Concomitant with medicalization there has been a change in imputed responsibility for deviance: with badness the deviants were considered responsible for their behavior, with sickness they are not, or at least responsibility is diminished (see Stoll, 1968). The social response to deviance is "therapeutic" rather than punitive. Many have viewed this as "humanitarian and scientific" progress; indeed, it often leads to "humanitarian and scientific" treatment rather than punishment as a response to deviant behavior. . . .

A number of broad social factors underlie the medicalization of deviance. As psychiatric critic Thomas Szasz (1974) observes, there has been a major historical shift in the manner in which we view human conduct:

With the transformation of the religious perspective of man into the scientific, and in particular the psychiatric, which became fully articulated during the nineteenth century, there occurred a radical shift in emphasis away from viewing man as a *responsible agent acting in and on the world* and toward viewing him *as a responsive organism being acted upon* by biological and social "forces." (p. 149)

This is exemplified by the diffusion of Freudian thought, which since the 1920s has had a significant impact on the treatment of deviance, the distribution of stigma, and the incidence of penal sanctions.

Nicholas Kittrie (1971), focusing on decriminalization, contends that the foundation of the therapeutic state can be found in determinist criminology, that it stems from the *parens patriae* power of the state (the state's right to help those who are unable to help themselves), and that it dates its origin with the development of juvenile justice at the turn of the century. He further suggests that criminal law has failed to deal effectively (e.g., in deterrence) with criminals and deviants, encouraging a use of alternative methods of control. Others have pointed out that the strength of formal sanctions is declining because of the increase in geographical mobility and the decrease in strength of traditional status groups (e.g., the family) and that medicalization offers a substitute method for controlling deviance (Pitts, 1968). The success of medicine in areas like infectious disease has led to rising expectations of what medicine can accomplish. In modern technological societies, medicine has followed a technological imperative—that the physician is responsible for doing everything possible for the patient—while neglecting such significant issues as the patient's rights and wishes and the impact of biomedical advances on society (Mechanic, 1973). Increasingly sophisticated medical technology has extended the potential of medicine as social control, especially in terms of psychotechnology (Chorover, 1973). Psychotechnology includes a variety of medical and quasimedical treatments or procedures: psychosurgery, psychoactive medications, genetic engineering, disulfiram (Antabuse), and methadone. Medicine is frequently a pragmatic way of dealing with a problem (Gusfield, 1975). Undoubtedly the increasing acceptance and dominance of a scientific world view and the increase in status and power of the medical profession have contributed significantly to the adoption and public acceptance of medical approaches to handling deviant behavior.

THE MEDICAL MODEL AND "MORAL NEUTRALITY"

The first "victories" over disease by an emerging biomedicine were in the infectious diseases in which specific causal agents—germs—could be identified. An image was created of disease as caused by physiological difficulties located *within* the human body. This was the medical model. It emphasized the internal and biophysiological environment and deemphasized the external and social psychological environment.

There are numerous definitions of "the medical model." . . . We adopt a broad and pragmatic definition: the medical model of deviance locates the source of deviant behavior within the individual, postulating a physiological, constitutional, organic, or, occasionally, psychogenic agent or condition that is assumed to cause the behavioral deviance. The medical model of deviance usually, although not always, mandates intervention by medical personnel with medical means as treatment for the "illness." Alcoholics Anonymous, for example, adopts a rather idiosyncratic version of the medical model—that alcoholism is a chronic disease caused by an "allergy" to alcohol—but actively discourages professional medical intervention. But by and large, adoption of the medical model legitimates and even mandates medical intervention.

The medical model and the associated medical designations are assumed to have a scientific basis and thus are treated as if they were morally neutral (Zola, 1975). They are not considered moral judgments but rational, scientifically verifiable conditions. . . . Medical designations *are* social judgments, and the adoption of a medical model of behavior, a political decision. When such medical designations are applied to deviant behavior, they are related directly and intimately to the moral order of society. In 1851 Samuel Cartwright, a well-known Southern physician, published an article in a prestigious medical journal describing the disease "drapetomania," which only affected slaves and whose major symptom was running away from the plantations of their white masters (Cartwright, 1851). Medical texts during the Victorian era routinely described masturbation as a disease or addiction and prescribed mechanical and surgical treatments for its cure (Comfort, 1967; Englehardt, 1974). Recently many political dissidents in the Soviet Union have been designated mentally ill, with diagnoses such as "paranoia with counterrevolutionary delusions" and "manic reformism," and hospitalized for their opposition to the political order (Conrad, 1977). Although these illustrations may appear to be extreme examples, they highlight the fact that all medical designations of deviance are influenced significantly by the moral order of society and thus cannot be considered morally neutral. . . .

Even after a social definition of deviance becomes accepted or legitimated, it is not evident what particular type of problem it is. Frequently there are intellectual disputes over the causes of the deviant behavior and the appropriate methods of control. These battles about deviance designation (is it sin, crime, or sickness?) and control are battles over turf: Who is the appropriate definer and treater of the deviance? Decisions concerning what is the proper deviance designation and hence the appropriate agent of social control are settled by some type of political conflict.

How one designation rather than another becomes dominant is a central sociological question. In answering this question, sociologists must focus on claims-making activities of the various interest groups involved and examine how one or another attains ownership of a given type of deviance or social problem and thus generates legitimacy for a deviance designation. Seen from this perspective, public facts, even those which wear a "scientific" mantle are treated as products of the groups or organizations that produce or promote them rather than as accurate reflections of "reality." The adoption of one deviance designation or another has consequences beyond settling a dispute about social control turf.

. . . When a particular type of deviance designation is accepted and taken for granted, something akin to a paradigm exists. There have been three major deviance paradigms: deviance as sin, deviance as crime, and deviance as sickness. When one paradigm and its adherents become the ultimate arbiter of "reality" in society, we say a hegemony of definitions exists. In Western societies, and American society in particular, anything proposed in the name of science gains great authority. In modern industrial societies, deviance designations have become increasingly medicalized. We call the change in designations from badness to sickness the medicalization of deviance. . . .

REFERENCES

Cartwright, S. W. Report on the diseases and physical peculiarities of the negro race. *N. O. Med. Surg. J.,* 1851, 7, 691–715.

Chorover, S. "Big Brother and psychotechnology." *Psychol. Today,* 1973, 7, 43–54 (Oct.).

Comfort, A. *The anxiety makers.* London: Thomas Nelson & Sons, 1967.

Conrad, P. Soviet dissidents, ideological deviance, and mental hospitalization. Presented at Midwest Sociological Society Meetings, Minneapolis, 1977.

Durkheim, E. *The division of labor in society.* New York: The Free Press, 1933. (Originally published 1893.)

Englehardt, H. T. Jr. The disease of masturbation: Values and the concept of disease. *Bull. Hist. Med.,* 1974, 48, 234–48 (Summer).

Foucault, M. *Madness and civilization.* New York: Random House, Inc. 1965.

Freidson, E. *Profession of medicine.* New York: Harper & Row Publishers Inc. 1970.

Gusfield, J. R. Categories of ownership and responsibility in social issues: Alcohol abuse and automobile use. *J. Drug Issues,* 1975, 5, 285–303 (Fall).

Kittrie, N. *The right to be different: Deviance and enforced therapy.* Baltimore: Johns Hopkins University Press, 1971.

Mechanic, D. Health and illness in technological societies. *Hastings Center Stud.,* 1973, 1(3), 7–18.

Pitts, J. Social control: The concept. In D. Sills (Ed.) *International Encyclopedia of Social Sciences.* (Vol. 14). New York: Macmillan Publishing Co., Inc. 1968.

Rosen, G. The evolution of social medicine. In H. E. Freeman, S. Levine, and L. Reeder (Eds.) *Handbook of medical sociology* (2nd ed.). Englewood Cliffs, NJ: Prentice-Hall, Inc. 1972.

Stoll, C. S. Images of man and social control. *Soc. Forces,* 1968, 47, 119–127 (Dec.).

Szasz, T. *Ceremonial chemistry.* New York: Anchor Books, 1974.

Zola, I. K. In the name of health and illness: On some socio-political consequences of medical influence. *Soc. Sci. Med.,* 1975, 9, 83–87.

DISCUSSION QUESTIONS

1. Alcoholism is an example of a deviant behavior being medicalized. How has this altered the understanding of and treatment of alcoholism? How does the involvement of health professionals in the treatment of alcoholism influence societal reaction to excessive drinking?

2. Some argue that rapists should be castrated. How does this illustrate the transformation of understanding rape as a move "from badness to sickness"? What assumptions guide the suggestion that rapists should be castrated as a way of stopping rape?

20

The Rich Get Richer and the Poor Get Prison?

JEFFREY H. REIMAN

This essay challenges the reader to view the criminal justice system from a radically different angle. Specifically, Jeffrey Reiman argues that the corrections system and broader criminal justice policy in the United States simply provide the illusion of fighting crime. In reality, he argues that criminal justice policies reinforce public fears of crimes committed by the poor. These policies, in turn, help to maintain a "criminal class" of disadvantaged people.

A criminal justice system is a mirror in which a whole society can see the darker outlines of its face. Our ideas of justice and evil take on visible form in it, and thus we see ourselves in deep relief. Step through this looking glass to view the American criminal justice system—and ultimately the whole society it reflects—from a radically different angle of vision.

In particular, entertain the idea that the goal of our criminal justice system is not to eliminate crime or to achieve justice, *but to project to the American public a visible image of the threat of crime as a threat from the poor.* To do this, the justice system must maintain the existence of a sizable population of poor criminals. To do this, it must fail in the struggle to eliminate the crimes that poor people commit, or even to reduce their number dramatically. Crime may, of course, occasionally decline, as it has recently—*but not because of criminal justice policies.* . . .

In recent years, we have tripled our prison population and, in cities like New York, allowed the police new freedom to stop and search people they suspect. No one can deny that if you lock enough people up, and allow the police greater and greater power to interfere with the liberty and privacy of citizens,

From: Jeffrey H. Reiman. 1997. *The Rich Get Richer and the Poor Get Prison?: Ideology, Class, and Criminal Justice,* 5th ed., Boston, MA: Allyn & Bacon, pp. 1–9. Copyright ©1998 by Pearson Education. Reprinted with permission.

you will eventually prevent some crime that might otherwise have taken place. . . . When I say that criminal justice policy is failing to reduce crime, I mean that it is failing to reduce it in any substantial way, that it is failing to make more than a marginal difference. But it is failing nonetheless, because our rates of crime remain extremely high, our crime-reduction strategies do not touch on the social causes of crime, and our citizens remain extremely fearful about criminal victimization, even after the recent declines. . . .

Some years ago, I taught a seminar for graduate students titled "The Philosophy of Punishment and Rehabilitation." Many of the students were already working in the field of corrections as probation officers or prison guards or halfway-house counselors. First we examined the various philosophical justifications for legal punishment, and then we directed our attention to the actual functioning of our correctional system. For much of the semester we talked about the myriad inconsistencies and cruelties and overall irrationality of the system. We discussed the arbitrariness with which offenders are sentenced to prison and the arbitrariness with which they are treated once there. We discussed the lack of privacy and the deprivation of sources of personal identity and dignity, the everpresent physical violence, as well as the lack of meaningful counseling or job training within prison walls. We discussed the harassment of parolees, the inescapability of the "ex-con" stigma, the refusal of society to let a person finish paying his or her "debt to society," and the nearly total absence of meaningful noncriminal opportunities for the ex-prisoner. We confronted time and again the bald irrationality of a society that builds prisons to prevent crime knowing full well that they do not and that does not even seriously try to rid its prisons and post release practices of those features that guarantee a high rate of *recidivism:* the return to crime by prison alumni. How could we fail so miserably? We are neither an evil nor a stupid nor an impoverished people. How could we continue to bend our energies and spend our hard-earned tax dollars on cures we know are not working?

Toward the end of the semester I asked the students to imagine that, instead of designing a correctional system to reduce and prevent crime, we had to design one that would maintain a stable and visible "class" of criminals. What would it look like? The response was electrifying. In briefer and somewhat more orderly form, here is a sample of the proposals that emerged in our discussion:

First. It would be helpful to have laws on the books against drug use or prostitution or gambling—laws that prohibit acts that have no unwilling victim. This would make many people "criminals" for what they regard as normal behavior and would increase their need to engage in *secondary* crime (the drug addict's need to steal to pay for drugs, the prostitute's need for a pimp because police protection is unavailable, and so on).

Second. It would be good to give police, prosecutors, and/or judges broad discretion to decide who got arrested, who got charged, and who got sentenced to prison. This would mean that almost anyone who got as far as prison would know of others who committed the same crime but who either were not arrested or were not charged or were not sentenced to prison. This would assure us that a good portion of the prison population would experience their confinement as arbitrary and unjust and thus respond with rage, which would make them more "antisocial," rather than respond with remorse, which would make them feel more bound by social norms.

Third. The prison experience should be not only painful but also demeaning. The pain of loss of liberty might deter future crime. But demeaning and emasculating prisoners by placing them in an enforced childhood characterized by no privacy and no control over their time and actions, as well as by the constant threat of rape or assault, is sure to overcome any deterrent effect by weakening whatever capacities a prisoner had for self-control. Indeed, by humiliating and brutalizing prisoners we can be sure to increase their potential for aggressive violence.

Fourth. Prisoners should neither be trained in a marketable skill nor provided with a job after release. Their prison records should stand as a perpetual stigma to discourage employers from hiring them. Otherwise, they might be tempted *not* to return to crime after release.

Fifth. Ex-offenders' sense that they will always be different from "decent citizens," that they can never finally settle their debt to society, should be reinforced by the following means. They should be deprived for the rest of their lives of rights, such as the right to vote. They should be harassed by police as "likely suspects" and be subject to the whims of parole officers who can at any time threaten to send them back to prison for things no ordinary citizens could be arrested for, such as going out of town or drinking or fraternizing with the "wrong people."

And so on.

In short, *asked to design a system that would maintain and encourage the existence of a stable and visible "class of criminals," we "constructed" the American criminal justice system.*

. . . The practices of the criminal justice system keep before the public the *real* threat of crime and the *distorted* image that crime is primarily the work of the poor. The value of this *to those in positions of power* is that it deflects the discontent and potential hostility of Middle America away from the classes above them and toward the classes below them. If this explanation is hard to swallow, it should be noted in its favor that it not only explains our dismal failure to make a significant dent in crime but also explains why the criminal justice system functions in a way that is biased against the poor at every stage from arrest to conviction. Indeed, even at the earlier stage, when crimes are defined in law, the system primarily concentrates on the predatory acts of the poor and tends to exclude or deemphasize the equally or more dangerous predatory acts of those who are well off. In sum, I will argue that *the criminal justice system fails to reduce crime substantially while making it look as if crime is the work of the poor.* It does this in a way that conveys the image that the real danger to decent, law-abiding Americans comes from below them, rather than from above them, on the economic ladder. This image sanctifies the status quo with its disparities of wealth, privilege, and opportunity and thus serves the interests of the rich and powerful in America—the very ones who could change criminal justice policy if they were really unhappy with it.

Therefore, it seems appropriate to ask you to look at criminal justice "through the looking glass." On the one hand, this suggests a reversal of common expectations. Reverse your expectations about criminal justice and entertain the notion that the system's real goal is the very reverse of its announced goal. On the other hand, the figure of the looking glass suggests the prevalence of image over reality. My argument is that the system functions the way it does *because it maintains a particular image of crime: the image that it is a threat from the poor.* Of course, for this image to be believable there must be a reality to back it up. The system must actually fight crime—or at least some crime—but only enough to keep it from getting out of hand and to keep the struggle against crime vividly and dramatically in the public's view—never enough to substantially reduce or eliminate crime.

I call this outrageous way of looking at criminal justice policy the *Pyrrhic defeat* theory. A "Pyrrhic victory" is a military victory purchased at such a cost in troops and treasure that it amounts to a defeat. The Pyrrhic defeat theory argues that the failure of the criminal justice system yields such benefits to those in positions of power that it amounts to success. . . .

The Pyrrhic defeat theory has several components. Above all, it must provide an explanation of *how* the failure to reduce crime substantially could benefit anyone—anyone other than criminals, that is. I argue that the failure to reduce crime substantially broadcasts a potent *ideological* message to the American people, a message that benefits and protects the powerful and privileged in our society by legitimating the present social order with its disparities of wealth and privilege and by diverting public discontent and opposition away from the rich and powerful and onto the poor and powerless.

To provide this benefit, however, not just any failure will do. It is necessary that the failure of the criminal justice system take a particular shape. *It must fail in the fight against crime while making it look as if serious crime and thus the real danger to society is the work of the poor.* The system accomplishes this both by what it does and by what it refuses to do. I argue that the criminal justice system refuses to label and treat as a crime a large number of acts that produces as much or more damage to life and limb as the crimes of the poor. Even among the acts treated as crimes, the criminal justice system is biased from start to finish in a way that guarantees that *for the same crimes* members of the lower classes are much more likely than members of the middle and upper classes to be arrested, convicted, and imprisoned—thus providing living "proof" that crime is a threat from the poor.

Our criminal justice system is characterized by beliefs about what is criminal, and beliefs about how to deal with crime, that predate industrial society. Rather than being anyone's conscious plan, the system reflects attitudes so deeply embedded in tradition as to appear natural. To understand why it persists even though it fails to protect us, all that is necessary is to recognize that, on the one hand, those who are the most victimized by crime are not those in positions to make and implement policy. Crime falls more frequently and more harshly on the poor than on the better off. On the other hand, there are enough benefits to the wealthy from the identification of crime with the poor and system's failure to reduce crime that those with the power to make profound changes in the system feel no compulsion nor see any incentive to make them. In short, the criminal justice system came into existence in an earlier epoch and persists in the present because, even though it is failing, indeed because of the way it fails, it generates no effective demand for change. When I speak of the criminal justice system as "designed to fail," I mean no more than this. I call this explanation of the existence and persistence of our failing criminal justice system the *historical inertia* explanation. . . .

DISCUSSION QUESTIONS

1. What does Reiman mean in arguing that the current criminal justice system works to maintain a class of criminals? Do you agree or disagree that our corrections system fails to rehabilitate and fails to deter crime?

2. If you had the power to change the corrections system in the United States, what changes would you make to help reduce and prevent crime?

http://infotrac.thomsonlearning.com

InfoTrac College Edition

BONUS READING

Mauer, Marc. 2002. "The Social Cost of America's Race to Incarcerate."
Phi Kappa Phi Forum 82 (Winter): 28–32.

This reading summarizes the problems associated with criminal justice policies designed to deal with drug use and its related violence. Mauer argues that
our punitive approach to putting drug offenders in prison with harsh sentences
creates a multitude of problems for society. What are some of the societal consequences of having massive prison populations? How, according to Mauer, does
this compare with how we handled the AIDS epidemic? What might be a better
approach to dealing with crack cocaine and drug violence?

SEARCH TERMS

criminal justice policies
crime and function
medicalization
social constructionism
disability and identity

21

The Communist Manifesto

KARL MARX AND FREDERICH ENGELS

The analysis of the class system under capitalism, as developed by Marx and Engels, continues to influence sociological understanding of the development of capitalism and the structure of the class system. In this classic essay, first published in 1848, Marx and Engels define the class system in terms of the relationships between capitalism, the bourgeoisie, and the proletariat. Their analysis of the growth of capitalism and its influence on other institutions continues to provide a compelling portrait of an economic system based on the pursuit of profit.

BOURGEOIS AND PROLETARIANS

The history of all hitherto existing society is the history of class struggles. . . .

Modern industry has established the world market, for which the discovery of America paved the way. This market has given an immense development to commerce, to navigation, to communication by land. This development has, in its turn, reacted on the extension of industry; and in proportion as industry, commerce, navigation, railways extended, in the same proportion the bourgeoisie developed, increased its capital, and pushed into the background every class handed down from the Middle Ages.

We see, therefore, how the modern bourgeoisie is itself the product of a long course of development, of a series of revolutions in the modes of production and of exchange. . . .

The bourgeoisie has at last, since the establishment of modern industry and of the world market, conquered for itself, in the modern representative state, exclusive political sway. The executive of the modern state is but a committee for managing the common affairs of the whole bourgeoisie.

The bourgeoisie, historically, has played a most revolutionary part.

The bourgeoisie, wherever it has got the upper hand, has put an end to all feudal, patriarchal, idyllic relations. It has pitilessly torn asunder the motley feudal ties that bound man to his "natural superiors," and has left remaining no other nexus

From: Karl Marx and Frederick Engels. 1998. *Manifesto of the Communist Party*. With introduction by Eric Hobsbawm. New York: Verso, pp. 33–51.

between man and man than naked self-interest, than callous "cash payment." It has drowned the most heavenly ecstasies of religious fervour, of chivalrous enthusiasm, of philistine sentimentalism, in the icy water of egotistical calculation. It has resolved personal worth into exchange value, and in place of the numberless inde-feasible chartered freedoms, has set up that single, unconscionable freedom—free trade. In one word, for exploitation, veiled by religious and political illusions, it has substituted naked, shameless, direct, brutal exploitation.

The bourgeoisie has stripped of its halo every occupation hitherto honoured and looked up to with reverent awe. It has converted the physician, the lawyer, the priest, the poet, the man of science, into its paid wage labourers.

The bourgeoisie has torn away from the family its sentimental veil, and has reduced the family relation to a mere money relation. . . .

The need of a constantly expanding market for its products chases the bour-geoisie over the whole surface of the globe. It must nestle everywhere, settle everywhere, establish connections everywhere.

The bourgeoisie has through its exploitation of the world market given a cosmopolitan character to production and consumption in every country. To the great chagrin of reactionists, it has drawn from under the feet of industry the national ground on which it stood. All old-established national industries have been destroyed or are daily being destroyed. They are dislodged by new indus-tries, whose introduction becomes a life and death question for all civilized nations, by industries that no longer work up indigenous raw material, but raw material drawn from the remotest zones; industries whose products are con-sumed, not only at home, but in every quarter of the globe. In place of the old wants, satisfied by the productions of the country, we find new wants, requiring for their satisfaction the products of distant lands and climes. In place of the old local and national seclusion and self-sufficiency, we have intercourse in every direction, universal interdependence of nations. And as in material, so also in intellectual production. The intellectual creations of individual nations become common property. National one-sidedness and narrow-mindedness become more and more impossible, and from the numerous national and local literatures, there arises a world literature.

The bourgeoisie, by the rapid improvement of all instruments of production, by the immensely facilitated means of communication, draws all, even the most barbarian, nations into civilization. The cheap prices of its commodities are the heavy artillery with which it batters down all Chinese walls, with which it forces the barbarians' intensely obstinate hatred of foreigners to capitulate. It compels all nations, on pain of extinction, to adopt the bourgeois mode of production; it compels them to introduce what it calls civilization into their midst, i.e., to become bourgeois themselves. In one word, it creates a world after its own image.

The bourgeoisie has subjected the country to the rule of the towns. It has created enormous cities, has greatly increased the urban population as compared with the rural, and has thus rescued a considerable part of the population from the idiocy of rural life. Just as it has made the country dependent on the towns, so it has made barbarian and semi-barbarian countries dependent on the civi-lized ones, nations of peasants on nations of bourgeois, the East on the West.

The bourgeoisie keeps more and more doing away with the scattered state of the population, of the means of production, and of property. It has agglomerated population, centralized means of production, and has concentrated property in a few hands. The necessary consequence of this was political centralization. Independent, or but loosely connected provinces, with separate interests, laws, governments and systems of taxation, became lumped together into one nation, with one government, one code of laws, one national class interest, one frontier and one customs tariff. . . .

The weapons with which the bourgeoisie felled feudalism to the ground are now turned against the bourgeoisie itself.

But not only has the bourgeoisie forged the weapons that bring death to itself; it has also called into existence the men who are to wield those weapons—the modern working class—the proletarians.

In proportion as the bourgeoisie, i.e., capital, is developed, in the same proportion is the proletariat, the modern working class, developed—a class of labourers, who live only so long as they find work, and who find work only so long as their labour increases capital. These labourers, who must sell themselves piecemeal, are a commodity, like every other article of commerce, and are consequently exposed to all the vicissitudes of competition, to all the fluctuations of the market.

Owing to the extensive use of machinery and to division of labour, the work of the proletarians has lost all individual character, and, consequently, all charm for the workman. He becomes an appendage of the machine, and it is only the most simple, most monotonous, and most easily acquired knack, that is required of him. Hence, the cost of production of a workman is restricted, almost entirely, to the means of subsistence that he requires for his maintenance, and for the propagation of his race. But the price of a commodity, and therefore also of labour, is equal to its cost of production. In proportion, therefore, as the repulsiveness of the work increases, the wage decreases. Nay more, in proportion as the use of machinery and division of labour increases, in the same proportion the burden of toil also increases, whether by prolongation of the working hours, by increase of the work exacted in a given time or by increased speed of the machinery, etc.

Modern industry has converted the little workshop of the patriarchal master into the great factory of the industrial capitalist. Masses of labourers, crowded into the factory, are organized like soldiers. As privates of the industrial army they are placed under the command of a perfect hierarchy of officers and sergeants. Not only are they slaves of the bourgeois class, and of the bourgeois state; they are daily and hourly enslaved by the machine, by the overseer, and, above all, by the individual bourgeois manufacturer himself. The more openly this despotism proclaims gain to be its end and aim, the more petty, the more hateful and the more embittering it is.

The less the skill and exertion of strength implied in manual labour, in other words, the more modern industry becomes developed, the more is the labour of men superseded by that of women. Differences of age and sex have no longer any distinctive social validity for the working class. All are instruments of labour, more or less expensive to use, according to their age and sex. . . .

But with the development of industry the proletariat not only increases in number; it becomes concentrated in greater masses, its strength grows, and it feels that strength more. The various interests and conditions of life within the ranks of the proletariat are more and more equalized, in proportion as machinery obliterates all distinctions of labour, and nearly everywhere reduces wages to the same low level. The growing competition among the bourgeois, and the resulting commercial crises, make the wages of the workers ever more fluctuating. The unceasing improvement of machinery, ever more rapidly developing, makes their livelihood more and more precarious; the collisions between individual work-men and individual bourgeois take more and more the character of collisions between two classes. Thereupon the workers begin to form combinations (trade unions) against the bourgeois. . . .

This organization of the proletarians into a class, and consequently into a political party, is continually being upset again by the competition between the workers themselves. But it ever rises up again, stronger, firmer, mightier. It compels legislative recognition of particular interests of the workers, by taking advantage of the divisions among the bourgeoisie itself. . . .

Altogether, collisions between the classes of the old society further, in many ways, the course of development of the proletariat. The bourgeoisie finds itself involved in a constant battle: at first with the aristocracy; later on, with those portions of the bourgeoisie itself, whose interests have become antagonistic to the progress of industry; at all times, with the bourgeoisie of foreign countries. In all these battles it sees itself compelled to appeal to the proletariat, to ask for its help, and thus to drag it into the political arena. The bourgeoisie itself, therefore, supplies the proletariat with its own elements of political and general education, in other words, it furnishes the proletariat with weapons for fighting the bourgeoisie.

Further, as we have already seen, entire sections of the ruling classes are, by the advance of industry, precipitated into the proletariat, or are at least threatened in their conditions of existence. These also supply the proletariat with fresh elements of enlightenment and progress.

Finally, in times when the class struggle nears the decisive hour, the process of dissolution going on within the ruling class, in fact within the whole range of old society, assumes such a violent, glaring character, that a small section of the ruling class cuts itself adrift, and joins the revolutionary class, the class that holds the future in its hands. Just as, therefore, at an earlier period, a section of the nobility went over to the bourgeoisie, so now a portion of the bourgeoisie goes over to the proletariat, and in particular, a portion of the bourgeois ideologists, who have raised themselves to the level of comprehending theoretically the historical movement as a whole.

Of all the classes that stand face to face with the bourgeoisie today, the proletariat alone is a really revolutionary class. The other classes decay and finally disappear in the face of modern industry; the proletariat is its special and essential product. . . .

DISCUSSION QUESTIONS

1. What evidence do you see in contemporary society of Marx and Engels' claim that the need for a constantly expanding market means that capitalism "nestles everywhere"?

2. How do Marx and Engels depict the working class and what evidence do you see of their argument by looking at the contemporary labor market?

22

Great Divides

THOMAS M. SHAPIRO

Thomas Shapiro provides an overview of the dimensions of social stratification commonly studied by sociologists. In contrast to the "American Dream" suggesting that everyone has an equal chance to succeed, class, race, and gender shape the opportunities that people have in the U. S. system of stratification. Shapiro also identifies some of the social changes that are currently influencing the structure of opportunity in the United States.

We all know that "the American Dream" means economic opportunity, social mobility, and material success. It means that all of us do better than our parents did; that our standard of living improves over the course of our lifetimes; and that we can own our homes. It also means that regardless of background or origin, all of us have a chance to succeed. As Americans, we do not favor the idea that some people or groups are privileged by their background, while others are systematically blocked from success. Economic opportunity and social equality are thus twin pillars of the American belief system.

In the decades following the end of World War II—from 1945 to the early 1970s—the American Dream became a reality for millions of families and individuals in our society. People were able to buy homes, purchase new automobiles, take vacations, and see their standard of living steadily rise. At the same time, our society made strides toward improved social conditions for minorities and greater equality among groups of people. For these changes, many observers credit pressure from the civil rights movement, the women's movement, labor

From: Shapiro, Thomas M., ed. *Great Divides: Readings in Social Inequality in the United States,* 2nd ed. 1–6. Mountain View, CA: Mayfield, 2004. Reprinted with permission.

unions, and other social organizations and movements, along with changes in public policy and a growing, prosperous economy. A better standard of living and a narrowing gap between rich and poor—together, these changes reinforced vital elements of the American credo. Many Americans assumed that material life would continue to improve indefinitely and that the economic and social gaps among groups would continue to narrow.

In the 1970s, however, the economy slowed down and began to stagnate. The standard of living began to fall, poverty began to increase, and membership in the middle class became more tenuous. Just to stay in the same place, many families had to adapt, as more women sought work outside the home, living spaces became smaller, and time for leisure activities and for families getting together was reduced as people worked harder for longer hours. The 1980s saw a return to some of the inequalities of the past, as measured by comparisons of the relative material positions of African and European Americans and by comparisons of women's income with men's. Many observers began to feel that the United States was losing ground in its struggle against social inequality.

The changes we are seeing in our society during the 1990s and at the beginning of 2000—stagnating living standards, increasing poverty, a precarious middle class, and a growing gap between rich and poor—are probably the result of the specific way that economic restructuring is taking place in the United States. Economic restructuring is one way in which corporations, businesses, bureaucracies, individuals, and various levels of government respond to the challenge of keeping the United States a preeminent nation in the emerging global economy. Restructuring includes a variety of strategies, policies, and practices—including corporate and government downsizing, deindustrialization, movement of production from the central city to the suburbs, and changes from a permanent to a contingent workforce, from local to global production, from Fordist (mass assembly line) production to more flexible and decentralized production, and from a manufacturing-based to a service-based economy. Together with other factors, including the increasing diversity of the American population, economic restructuring is giving a new shape to social and economic inequality in our society. . . .

SOCIAL STRATIFICATION

Strata means layers, or hierarchy; *social stratification* is a process or system by which groups of people are arranged into a hierarchical social structure. Dimensions of power and powerlessness undergird this hierarchy and influence subsequent opportunities for rewards. Consequently, people have differential access to—and control over—prospects, rewards, and whatever is of value in society at any given time, based on their hierarchical positions, primarily because of social factors. Social stratification, an expression of social inequality, is so pervasive in American society that an entire field of sociology is devoted to its study.

Social stratification systems are based primarily on either ascribed status or achieved status. *Ascribed status* is a social position typically designated or given to each person at birth. In a society with ascribed status, differential opportunities, rewards, privileges, and power are provided to individuals according to criteria fixed at birth. *Achieved status* is a social position gained as a result of ability or effort. This type of stratification is evident in all industrial societies, including the United States.

Given that social stratification is an expression of social inequality, how does inequality result, in turn, from social stratification? One leading scholar has proposed that inequality is produced by two different kinds of matching processes: "The jobs, occupations, and social roles in society are first matched to 'reward packages' of unequal value"; then, individuals are sorted and matched to particular jobs, occupations, and social roles through training and other institutional processes (Grusky, 1994, p. 3). Both parts of this matching process have been the subject of much investigation in sociology, as many inquiries have probed these two questions (Fischer et al., 1996, p. 7): (1) What determines how much people get for performing various economic roles and tasks? (2) What social and institutional processes determine who gets ahead and who falls behind in the competition for positions of unequal value?

Other questions sociologists ask about stratification include the following: How is ascribed status constructed over time? What are the institutional processes and practices that shape ascriptive stratification? To what extent does an ascribed status circumscribe people's opportunities and rewards? To what extent is achieved status fixed or open? What determines how and whether individuals are able to move through the occupational and wage structure in a system characterized by achieved status? When people can move through such a structure, how do they move? . . .

Social Stratification in the United States

Ascriptive stratification based on gender is found in nearly every society, and racial and ethnic stratification is almost as widespread. Nevertheless, social stratification in the United States is based primarily on achieved status, at least in theory. When this nation was formed, the founders deliberately distinguished it from nations where life chances and social rank were determined by birth. A core element of the American credo is that life chances are determined largely by talent, skill, hard work, and achievement. We believe that everyone has a fair shot at whatever is valued or prized and that no individual or group is unfairly advantaged or disadvantaged.

This belief does not mean that we expect everyone to achieve equal results; rather, we expect that everyone is starting with the same opportunities for achieving these different outcomes. Indeed, we tend to see differences in material success as the legitimate result of playing by the agreed-on rules. Although our national history is ambiguous about our implementation of social inequality, we normally take great exception when systemic and systematic differences in achievement clearly and directly result from public policy, varying or hidden rules, discrimination, or differential rewards for similar accomplishments. These pernicious factors produce what we think of as inequality.

Despite our egalitarian values and beliefs, social inequality has been an enduring fact of life and politics in the United States. Some groups of people have sufficient power—through family, neighborhood, school, or community—to maintain higher economic class positions and higher social status in American society. People in these groups have the ability to get and stay ahead in the competition for success. Further, social inequality has always been integrally bound up with three dimensions of social stratification: socioeconomic class, race and ethnicity, and gender. Divisions based on these three dimensions are deeply embedded in the social structures and institutions that define our lives, so these three constructs must be at the center of any analysis of social inequality. The integration of these constructs is not simple, however, because we lack both a common understanding of them and an agreement as to their significance in the structure of social inequality.

An example illustrates this lack of common ground. Whenever I ask my students what they mean by *class,* they say they are sure that classes exist in the United States and that a lot of economic inequality, privilege, and disadvantage results from class structure. However, they become much less certain when I ask them what determines class status. A recent group of students suggested a number of ways to determine class status, including income, wealth, education, job, and neighborhood, as well as how many members of a given family were working. Even when we focused on one criterion for class that is often used—income—they could not agree on how much income put people into which class. This example suggests the lack of common understanding about class. As this article shows, the difficulties involved in analyzing the influence of ethnicity (and race) and gender on social inequality probably are even greater than those for analyzing class. This analysis is more difficult because there is little agreement regarding the existence or significance of ethnic (and racial) and gender inequality.

DIMENSIONS OF INEQUALITY IN THE UNITED STATES

Class, race and ethnicity, and gender shape the history, experiences, and opportunities of people in the United States. As a leading social theorist indicated, we should view class, race, and gender as different and interrelated, with interlocking levels of domination, not as discrete dimensions of stratification (Collins, 1990). Thus, even though the following discussion introduces class, race and ethnicity, and gender as separate concepts, many of the readings and discussions in this book examine these dimensions as simultaneous, interrelated, and interlocking means of configuring people's social relations and life opportunities.

Class

A *class* is a group of people who share the same economic or social status, life chances, and outlook on life. A *class system* is a system of social stratification in which social status is determined by the ownership and control of resources and by the kinds of work people do. The two major sociological explanations of class derive from its two most influential contributors, Karl Marx and Max Weber. In Marx's theory, social classes are defined by their distinct relationship to the means of production—that is, by whether people own the means of production (the capitalists) or sell their labor to earn a living (the workers). People's role in social life and their place in society are fixed by their place in the system of production. In Marx's theoretical perspective, the classes that dominate production also dominate other institutions in society, from schools and the mass media to the institutions that make and enforce rules.

German sociologist Max Weber also believed that divisions between capitalists and workers and their assigned classes were the driving force of social organization. For Weber, however, Marx's theory of social stratification was too strongly driven by the single motor of economics and by where an individual was positioned in the production process. In addition to a person's economic position, Weber included social status and *party* (i.e., coordinated political action) as different bases of power, independent of (but closely related to) economics. Weber's multidimensional perspective examines wealth, prestige, and power.

In the social sciences, the debate over class has not been whether classes exist; rather, essential theoretical perspectives flow from these two different ways (Marx's or Weber's) of understanding and constructing class. These two theoretical perspectives on class are fundamental to an understanding of social stratification and social inequality. . . .

Race and Ethnicity

Like class, race and ethnicity are important dimensions of social stratification and social inequality. Although the terms *race* and *ethnicity* are often used interchangeably, they do not refer to the same thing. Both concepts are complex, and both defy easy definition. In the past, *race* was usually defined as a category of people sharing genetically transmitted traits deemed significant by society. However, this simple view does not hold up when we take into consideration the complex biology of genetic inheritance, migration, intermarriage, and the resulting wide variation within so-called racial groups. Today, most social scientists view *race* as a far more subjective (and shifting) social category than the fixed definitions of the past, wherein people are labeled by themselves or by others as belonging to a group based on some physical characteristic, such as skin color or facial features. Racial-formation theory, which emphasizes the shifting meanings and power relationships inherent in notions of race, defines *race* "as a concept that signifies and symbolizes sociopolitical conflicts and interests in reference to different types of human bodies" (Winant, 1994, p. 115). The concept of race, then, has both biological and social components. Examples of race that have been used in the past are Caucasian, Asian, and African.

The concept of ethnicity is closely related to that of race. An *ethnic group* can be defined as a category of people distinguished by their ancestry, nationality, traditions, or culture. Examples of ethnic groups in the United States are Puerto Ricans, Japanese Americans, Cuban Americans, Irish Americans, and Lebanese Americans. Ethnicity is a cultural and social construct, and people's ethnic categories may be either self-chosen or assigned by outsiders to the group. Because characteristics such as culture, traditions, religion, and language are less visible and more changeable than skin color or facial features, ethnicity is even more arbitrary and subjective than is race.

The distinction between race and ethnicity is important. The basis for the social construction of race is primarily (though not entirely) biological; for ethnicity, it is primarily (though not entirely) cultural and social. Race is usually visible to an observer; ethnicity is usually a guess. The historical discourse about race in the United States is charged with notions of difference, of superiority and inferiority, of domination and subordination. Ideas and practices surrounding race—especially the deeply embedded divisions between African Americans and Caucasian (European) Americans—go to the core of the American experience of social inequality. . . .

Gender

Gender, perhaps the oldest and deepest division in social life, may be defined as the set of social and cultural characteristics associated with biological sex—being female or male—in a particular society. Like race and ethnicity, gender is socially constructed (whereas biological sex is not). It is rooted in society's belief that females and males are naturally distinct and opposed social beings. These beliefs are translated into reality when people are assigned to different and often unequal political, social, and economic positions based on their sex. It is common for societies to separate adult work, family, and civic roles by gender and to prepare girls and boys differently for those roles. The result is socially discrete gender roles.

Our society provides a great deal of occupational segregation by gender. The division of labor by gender is hierarchical, such that males occupy positions accorded higher prestige and value than females do. So-called women's work is matched with inferior reward packages and low status.

Many feminists consider the sexual division of labor to be the primary source of gender stratification, along with the socialization and institutional processes that prepare females and males for different lives. These processes include the institutions of *patriarchy,* the institutional arrangements that bestow power and privilege on male roles and occupations, thereby allowing men to perpetuate their political, social, and economic advantages. Put another way, patriarchy is the ability of men to control the laws and institutions of society and to command superior status and reward packages. . . .

EXPLORING SOCIAL INEQUALITY IN AN INSTITUTIONAL CONTEXT: EDUCATION

We have introduced some basic ideas about the three major dimensions of social inequality and social stratification in the United States—class, race and ethnicity, and gender. Discussing them one at a time is a simple and orderly way to begin, but it is critical to keep in mind that these dimensions do not function in isolation. Rather, their dynamic interaction is highly complex. Class, race and ethnicity, and gender are simultaneous, intersecting, and sometimes crosscutting systems of relationships and meanings.

For example, even though we might use the term *middle class* to refer to all people of a certain income level (or amount of wealth, or occupation, or educational level), African Americans of that income level—the black middle class—occupy a far more precarious position in our society than do members of the white middle class. In considering these two dimensions, we may ask, Do all middle-class people have the same interests? Is their social status a function more of class or of race? How do they identify themselves—as middle class or as African or European Americans?

Other examples of the intersection of the three dimensions abound. In a study conducted in several nations, employed married women identified their socioeconomic class positions more on the basis of their spouses' jobs than on their own familial or educational backgrounds or their own occupations (Erickson and Goldthorpe, 1992). Another study showed that European American women in the United States tend to mention gender alone when asked how they identify themselves, whereas African American women tend to emphasize both race and gender (Rubin, 1994).

Evidence points to growing conflict among different ethnic minorities of the same social class or community in the United States, perhaps as a result of contracting opportunities and limited resources. For example, in Los Angeles and in New York City, many Korean immigrants run grocery, liquor, produce, and fish retail businesses that are heavily concentrated in ethnic-minority neighborhoods. These Korean American merchants often act as an intermediate ethnic minority interposed between low-income Latino and African American minority consumers and high-income European American owners of large companies. The Korean Americans encounter conflicts with both their ethnic-minority customers and their ethnic-majority suppliers. As a result, intergroup conflicts, tension, misunderstandings, and violence erupt all too often in Los Angeles, New York, and other ethnically diverse cities (Min, 1996). . . .

. . . [E]ducation has traditionally been seen as a remedy for inequality in the United States. Americans believe that if society creates more educational opportunities for the disadvantaged, greater equality will eventually result. Thus, people who seek greater social equality often turn to education as an arena in which to institute change. Similarly, those who seek to maintain our society's system of social stratification may also try to manipulate the educational system to preserve the status quo. Hence, education becomes an arena for conflict between groups seeking change toward greater social equality and groups seeking to maintain the current social inequalities. . . .

REFERENCES

Collins, Patricia Hill. 1990. *Black Feminist Thought*. New York: Routledge.

Erickson, Robert, and John H. Goldthorpe. 1992. "Individual or Family? Results from Two Approaches to Class Assignment." *Acta Sociologica* 35: 95–105.

Fischer, Claude S., Michael Hout, Martin Sanchez Jankowski, Samuel R. Lucas, Ann
    Swidler, and Kim Voss. 1996. *Inequality by Design: Cracking the Bell Curve Myth.*
    Princeton, NJ: Princeton University Press.

Grusky, David B., ed. 1994. *Social Stratification: Class, Race, and Gender in Sociological
    Perspective.* Boulder, CO: Westview Press.

Min, Pyong Gap. 1996. *Caught in the Middle: Korean Merchants in America's Multiethnic Cities.*
    Berkeley and Los Angeles: University of California Press.

Rubin, Lillian B. 1994. *Families on the Fault Line.* New York: HarperCollins.

Winant, Howard. 1994. *Racial Conditions.* Minneapolis: University of Minnesota Press.

DISCUSSION QUESTIONS

1. What specific changes in the late twentieth century does Shapiro cite as
   evidence of economic restructuring in the United States? How do you see
   these changes as affecting the opportunities of diverse groups in this country?

2. What does Shapiro mean when he says that class, race, ethnicity, and gender
   are crosscutting systems of relationships and meanings? How is this
   exemplified by the middle class?

23

Wealth Matters

DALTON M. CONLEY

*By contrasting the experiences of two different families, Dalton Conley illustrates the great
difference in socioeconomic status that differences in wealth, even at a modest level, can produce. Wealth and the ability to pass it on also creates a cumulative difference in the status
of different groups in society. Conley shows how understanding differences in wealth
between racial groups is an important part of understanding the relationship between race
and class inequality.*

From: Conley, Dalton M. "Wealth Matters." In *Being Black, Living in the Red: Race,
Wealth, and Social Policy in America,* 1–7. Berkeley: University of California Press, 1999.
Reprinted with permission.

Property is theft.

Pierre Joseph Proudhon, 1809–65

. . . Contrast the situations of two hypothetical families. Let's say that both households consist of married parents, in their thirties, with two young children.[1] Both families are low-income—that is, the total household income of each family is approximately the amount that the federal government has "declared" to be the poverty line for a family of four (with two children). In 1996, this figure was $15,911.*

Brett and Samantha Jones (family 1) earned about $12,000 that year. Brett earned this income from his job at a local fast-food franchise (approximately two thousand hours at a rate of $6 per hour). He found himself employed at this low-wage job after being laid off from his relatively well-paid position as a sheet metal worker at a local manufacturing plant, which closed because of fierce competition from companies in Asia and Latin America. After six months of unemployment, the only work Brett could find was flipping burgers alongside teenagers from the local high school.

Fortunately for the Jones family, however, they owned their own home. Fifteen years earlier, when Brett graduated from high school, married Samantha, and landed his original job as a sheet metal worker, his parents had lent the newlyweds money out of their retirement nest egg that enabled Brett and Samantha to make a 10 percent down payment on a house. With Samantha's parents cosigning—backed by the value of their own home—the newlyweds took out a fifteen-year mortgage for the balance of the cost of their $30,000 home. Although money was tight in the beginning, they were nonetheless thrilled to have a place of their own. During those initial, difficult years, an average of $209 of their $290.14 monthly mortgage payment was tax deductible as a home mortgage interest deduction. In addition, their annual property taxes of $800 were completely deductible, lowering their taxable income by a total of $3,308 per year. This more than offset the payments they were making to Brett's parents for the $3,000 they had borrowed for the down payment.

After four years, Brett and Samantha had paid back the $3,000 loan from his parents. At that point, the total of their combined mortgage payment ($290.14), monthly insurance premium ($50), and monthly property tax payment ($67), minus the tax savings from the deductions for mortgage interest and local property taxes, was less than the $350 that the Smiths (family 2) were paying to rent a unit the same size as the Joneses' house on the other side of town.

That other neighborhood, on the "bad" side of town, where David and Janet Smith lived, had worse schools and a higher crime rate and had just been chosen as a site for a waste disposal center. Most of the residents rented their housing units from absentee landlords who had no personal stake in the community other than profit. A few blocks from the Smiths' apartment was a row of public

---

*Editors' note: In 2002, the poverty line for a family of four was $18,307.

housing projects. Although they earned the same salaries and paid more or less the same monthly costs for housing as the Joneses did, the Smiths and their children experienced living conditions that were far inferior on every dimension, ranging from the aesthetic to the functional (buses ran less frequently, large supermarkets were nowhere to be found, and class size at the local school was well over thirty).

Like Brett Jones, David Smith had been employed as a sheet metal worker at the now-closed manufacturing plant. Unfortunately, the Smiths had not been able to buy a home when David was first hired at the plant. With little in the way of a down payment, they had looked for an affordable unit at the time, but the real estate agents they saw routinely claimed that there was just nothing available at the moment, although they promised to "be sure to call as soon as something comes up. . . ." The Smiths never heard back from the agents and eventually settled into a rental apartment.

David spent the first three months after the layoffs searching for work, drawing down the family's savings to supplement unemployment insurance—savings that were not significantly greater than those of the Joneses, since both families had more or less the same monthly expenses. After several months of searching, David managed to land a job. Unfortunately, it was of the same variety as the job Brett Jones found:working as a security guard at the local mall, for about $12,000 a year. Meanwhile, Janet Smith went to work part time, as a nurse's aide for a home health care agency, grossing about $4,000 annually.

After the layoffs, the Joneses experienced a couple of rough months, when they were forced to dip into their small cash savings. But they were able to pay off the last two installments of their mortgage, thus eliminating their single biggest living expense. So, although they had some trouble adjusting to their lower standard of living, they managed to get by, always hoping that another manufacturing job would become available or that another company would buy out the plant and reopen it. If worst came to worst, they felt that they could always sell their home and relocate in a less expensive locale or an area with a more promising labor market.

The Smiths were a different case entirely. As renters, they had no latitude in reducing their expenses to meet their new economic reality, and they could not afford their rent on David's reduced salary. The financial strain eventually proved too much for the Smiths, who fought over how to structure the family budget. After a particularly bad row when the last of their savings had been spent, they decided to take a break; both thought life would be easier and better for the children if Janet moved back in with her mother for a while, just until things turned around economically—that is, until David found a better-paying job. With no house to anchor them, this seemed to be the best course of action.

Several years later, David and Janet Smith divorced, and the children began to see less and less of their father, who stayed with a friend on a "temporary" basis. Even though together they had earned more than the Jones family (with total incomes of $16,000 and $12,000, respectively), the Smiths had a rougher financial, emotional, and family situation, which,we may infer, resulted from a lack of property ownership.

What this comparison of the two families illustrates is the inadequacy of relying on income alone to describe the economic and social circumstances of families at the lower end of the economic scale. With a $16,000 annual income, the Smiths were just above the poverty threshold. In other words, they were not defined as "poor," in contrast to the Joneses, who were.[2] Yet the Smiths were worse off than the Joneses, despite the fact that the U. S. government and most researchers would have classified the Jones family as the one who met the threshold of neediness, based on that family's lower income.

These income-based poverty thresholds differ by family size and are adjusted annually for changes in the average cost of living in the United States. In 1998, more than two dozen government programs—including food stamps, Head Start, and Medicaid—based their eligibility standards on the official poverty threshold. Additionally, more than a dozen states currently link their needs standard in some way to this poverty threshold. The example of the Joneses and the Smiths should tell us that something is gravely wrong with the way we are measuring economic hardship—poverty—in the United States. By ignoring assets, we not only give a distorted picture of life at the bottom of the income distribution but may even create perverse incentives.

Of course, we must be cautious and remember that the Smiths and the Joneses are hypothetically embellished examples that may exaggerate differences. Perhaps the Smiths would have divorced regardless of their economic circumstances. The hard evidence linking modest financial differences to a propensity toward marital dissolution is thin; however, a substantial body of research shows that financial issues are a major source of marital discord and relationship strain.[3] It is also possible that the Smiths, with nothing to lose in the form of assets, might have easily slid into the world of welfare dependency. A wide range of other factors, not included in our examples, affect a family's well-being and its trajectory. For example, the members of one family might have been healthier than those of the other, which would have had important economic consequences and could have affected family stability. Perhaps one family might have been especially savvy about using available resources and would have been able to take in boarders, do under-the-table work, or employ another strategy to better its standard of living. Nor do our examples address educational differences between the two households.

But I have chosen not to address all these confounding factors for the purpose of illustrating the importance of asset ownership *per se.* Of course, home-ownership, savings behavior, and employment status all interact with a variety of other measurable and unmeasurable factors. This interaction, however, does not take away from the importance of property ownership itself.

. . . In order to understand a family's well-being and the life chances of its children—in short, to understand its class position—we not only must consider income, education, and occupation but also must take into account accumulated wealth (that is, property, assets, or net worth). . . . As you might have guessed, an important detail is missing from the preceding descriptions of the two families: the Smiths are black and have fewer assets than the Joneses, who are white.

At all income, occupational, and education levels, black families on average have drastically lower levels of wealth than similar white families. The situation of the Smiths may help us to understand the reason for this disparity of wealth between blacks and whites. For the Smiths, it was not discrimination in hiring or education that led to a family outcome vastly different from that of the Joneses; rather, it was a relative lack of assets from which they could draw. In contemporary America, race and property are intimately linked and form the nexus for the persistence of black-white inequality.

Let us look again at the Smith family, this time through the lens of race. Why did real estate agents tell the Smiths that nothing was available, thereby hindering their chances of finding a home to buy? This well-documented practice is called "steering," in which agents do not disclose properties on the market to qualified African American home seekers, in order to preserve the racial makeup of white communities—with an eye to maintaining the property values in those neighborhoods. Even if the Smiths had managed to locate a home in a predominantly African American neighborhood, they might well have encountered difficulty in obtaining a home mortgage because of "redlining," the procedure by which banks code such neighborhoods "red"—the lowest rating—on their loan evaluations, thereby making it next to impossible to get a mortgage for a home in these districts. Finally, and perhaps most important, the Smiths' parents were more likely to have been poor and without assets themselves (being black and having been born early in the century), meaning that it would have been harder for them to amass enough money to loan their children a down payment or to cosign a loan for them. The result is that while poor whites manage to have, on average, net worths of over $10,000, impoverished blacks have essentially no assets whatsoever.[4]

Since wealth accumulation depends heavily on intergenerational support issues such as gifts, informal loans, and inheritances, net worth has the ability to pick up both the current dynamics of race and the legacy of past inequalities that may be obscured in simple measures of income, occupation, or education. This thesis has been suggested by the work of sociologists Melvin Oliver and Thomas Shapiro in their recent book *Black Wealth/White Wealth*.[5] They claim that wealth is central to the nature of black-white inequality and that wealth—as opposed to income, occupation, or education—represents the "sedimentation" of both a legacy of racial inequality as well as contemporary, continuing inequities. . . .

NOTES

1. These family descriptions were extrapolated from profiles of specific families who were interviewed for this study. The age, racial, income, family size, wealth, housing tenure, and divorce descriptions of these families come directly from cases 4348 and 1586 of the PSID 1984 wave (inflation-adjusted to 1996 dollars). The names and other details are fictitious but are in line with previous research that would suggest such profiles.

2. Neither family received health insurance from an employer. Since the Smiths' income was under 185 percent of the poverty line, their children were eligible for Medicaid. (In most states, the Joneses' children would also have been eligible for Medicaid since that family's wealth was in the form of a home, which is excluded from the asset limits of many states.)

3. See, e.g., G. Levinger and O. Moles, eds. *Divorce and Separation: Contexts, Causes, and Consequences,* (New York: Basic Books, 1979); and R. Conger, G. H. Elder, et al., "Linking Economic Hardship to Marital Quality and Instability," *Journal of Marriage and the Family 52* (1990):643–56.

4. The terms "black" and "African American" are used interchangeably, as are the terms "Hispanic" and "Latino." Black people of Caribbean origin make up a negligible portion of the data sample.

5. M. Oliver and T. Shapiro, *Black Wealth/White Wealth: A New Perspective on Racial Inequality,* (New York: Routledge, 1995).

DISCUSSION QUESTIONS

1. What is the difference in *income* and *wealth* and why is this significant in understanding stratification by race?

2. Sociologists have long debated the relationship between race and class. What relationship between race and class is shown by the evidence in Dalton Conley's research?

24

Studying the Quagmire of Welfare Reform

SHARON HAYS

In this reading, Hays discusses the experiences of women resulting from welfare reform. Her research reveals that, despite the desire of poor women (most of whom are single mothers) to be off government dependence, the reality of economic inequality prevents any measurable improvement in financial independence.

It was a lot tougher than I expected it to be, studying welfare reform. I knew I had to examine the impact of the 1996 Personal Responsibility Act. I had listened closely to the politicians as they hammered out the language and logic of this law. I analyzed the newly established work rules. I read carefully the law's

From: Hays, Sharon. "Studying the Quagmire of Welfare Reform." *The Chronicle of Higher Education,* October 17, 2003, B7–B9.

condemnation of single parenting. I pondered the larger social significance of removing the 61-year-old guarantee of a safety net for the nation's most desperately poor women and children. And I tried to take seriously the Congressional proclamation that welfare reform would "end the dependence of needy parents on government benefits by promoting job preparation, work, and marriage."

I was skeptical. All that was needed, this law seemed to be saying, was to educate the poor in "mainstream" American values of work and family life. But I had to wonder, were welfare recipients really just suffering from bad values? Did they need new ones? Would they find them in the welfare office?

In any case, it was apparent to me that the majority of Congressional policy makers had never spent much time in the world of welfare; never followed the routines of a welfare office; never lived in ghetto poverty; never raised a child without the help of nannies, au pairs, housekeepers, or (at least) the best in child-care centers; never spent a month trying to make ends meet on a $350 welfare check. But then again, neither had I. To make sense of this law, to see how it would all unfold, I wanted to be inside the welfare office.

I thought I was well prepared for the task. As a scholar of gender, work, and family life, and a theorist of American culture, I felt confidently armed with the proper intellectual tools. And having spent my first years out of college as a street-level social worker, I figured that the structure of the welfare office would hold few surprises, and I was sure that I knew how to handle myself in poor neighborhoods. After all, in those younger days I'd spent time with drug addicts, murderers, and thieves; I'd worked with people who abused their kids, visited heroin-shooting galleries, and watched dogfights staged on the street for the sake of gambling and sport. Nothing in the world of welfare could rattle me.

Or so I thought.

To write my book, *Flat Broke With Children: Women in the Age of Welfare Reform,* I spent three years, from December 1997 to January 2001, visiting welfare offices and the homes of welfare families. Most of my time was spent in and around two welfare offices—one in a medium-size town in the Southeast I call Arbordale, another in a large metropolitan area in the West that I call Sunbelt City. I interviewed clients and caseworkers, hung out in waiting rooms, attended all the training sessions and caseworker meetings that welfare recipients are required to attend, went through all the forms they have to fill out. I visited the homes of welfare mothers and talked to them there. I spent a lot of time in housing projects. . . .

From the moment I set foot in the welfare office, I knew I had entered a quagmire. By early 1998, the White House was already applauding the results of reform, and the nightly news was already offering "human interest" coverage of smiling former welfare mothers gainfully employed in local supermarkets and small businesses. But the welfare caseworkers and clients I met were a good deal less sanguine. Caseworkers in Arbordale and Sunbelt City were still frantically trying to decipher all the rules and regulations of reform, still worried about whether they'd be able to keep up with the federal demand that they place an ever-increasing proportion of their clients in jobs, still spending their lunch breaks complaining about multiple glitches in the new computer systems, and still wondering whether they'd be able to convince their clients that the newly instituted lifetime limits on welfare benefits were both absolute and here to stay.

Welfare clients, for their part, noticed immediately that something big was happening. And, like caseworkers, they were struggling to determine just what the nation was asking of them, and wondering if they'd be able to manage. To understand the experience of welfare clients, and to make sense of their responses, it's important to know what I now know about this group of Americans. The vast majority of adult welfare recipients—over 90 percent—are mothers. Nearly all are raising their children alone. Over 80 percent have work experience, but more than half are without high-school diplomas, and nearly as many suffer from physical or mental-health disabilities that affect their ability to work. National studies suggest that about half have been the victims of domestic violence or sexual abuse. And welfare mothers have, on average, two children to worry about when they consider the costs and benefits of the jobs available to them.

The primary message these women heard when they arrived at the newly reformed Arbordale and Sunbelt City welfare offices was simple—you must get a job, get it fast, and accept whatever wages or hours you can get. The pressure was intense. It began with the mandatory "job search" requiring 40 verifiable job contacts in 30 days and the employability and skills assessments and the three- to five-day "job readiness" workshops that offered tips on how to dress for an interview, balance one's budget, manage child-care arrangements, cope with stress, and speak proper English rather than street slang.

Most of the welfare mothers who hadn't found a job within the first 30 days were placed in full-time employment-training programs. In Arbordale and Sunbelt City, nearly all these training programs were geared to low-wage jobs: nursing assistants, cook's helpers, introductory computer skills, child care, and janitorial work. Those welfare clients who were still unemployed at the completion of training, or those for whom training was deemed inappropriate, were assigned an unpaid "workfare" placement—sweeping city streets, serving food at school cafeterias, sorting papers for a county agency—working at least 30 hours a week in return for their welfare checks. Throughout, all welfare mothers were required to meet regularly with their welfare caseworkers to report on their overall progress toward "self-sufficiency."

While welfare mothers were spending 30 to 40 hours a week in all those seminars, training programs, and workfare placements, they also had to find some place to put their kids. If they were lucky, well organized, flexible, patient, and persistent, they could hope to be among the less than one-third of all welfare mothers who actually receive the federal child-care subsidies for which all poor families are technically eligible. If they were not so lucky, they'd have to somehow manage child care on their $350-a-month (average) welfare income.

Inside the welfare office, the demanding, lock-step work requirements of reform were enforced not just through the time limits on benefits, but also through the system of "sanctions." Welfare mothers who failed to attend job-readiness or employment-training sessions, failed to make a sufficient number of job contacts, failed to follow through on a workfare placement, or failed to report changes in their employment circumstances were sanctioned. These sanctioned welfare clients lost all or part of their family's welfare check for a specified number of months—they lost, in other words, their primary source of income. Nationwide, sanction rates have more than doubled since reform. . . .

At the same time that welfare reform instituted a whole new set of impossibly demanding and often punitive rules, it also brought with it a smaller, but significant, set of positive "supportive services." Those services included not just that (grossly insufficient) assistance in paying for child care, but also help in covering the cost of transportation, clothing, and supplies for work and, in special cases, aid in covering expenses like rent and utility payments, repairing one's car, purchasing eyeglasses, and paying for dental work. Welfare reform also allowed those mothers whose earnings from work were very low to continue to receive welfare benefits. This support was limited, and all of it was tied to the work requirements. Nonetheless, most of the welfare mothers I met were grateful for this assistance.

Further clouding my darker portrait of reform was the stunning fact that the majority of welfare mothers I met actually liked welfare reform. That is, even though many welfare clients disliked (or abhorred) the endless stream of rules and regulations that came with this law, most remained in favor of the Personal Responsibility Act. They wanted to work, they wanted their children to see them working, they wanted to be free of the welfare office, and they wanted financial independence.

Confusing matters even further, the more I talked to my colleagues, friends, and neighbors about my research, the more it became clear to me that everyone had his or her own agenda when it came to interpreting the significance of welfare reform. It was about poverty. It was about race. It was about women and motherhood. It was about work. It was about cutting costs. Some of my colleagues thought that international comparisons of state welfare policies were the only worthy approach to analyzing the significance of reform. Others thought the most important factor in assessing the Personal Responsibility Act was the racial discrimination that lies behind its more punitive and unforgiving features. A good number of my neighbors, on the other hand, remained convinced that welfare mothers were morally "undeserving"—lazy, promiscuous, and prone to illegal behavior and abuse of the system. These women, they said, needed to be taught a lesson.

I was dizzy.

Before I could regain my balance, I was faced with the "success" of welfare reform. By the time I was completing my research, the American public had been so frequently assured that the goals of reform had been met that they had turned their attention elsewhere. The welfare rolls had been cut by more than half (from 12 million recipients in 1996 to 5 million in 2002), and most former welfare mothers were employed. So it seemed that, even if there were still problems with the system, and even if there were still questions left unanswered about work, family, race, poverty, motherhood, and morality, wasn't it true that the Personal Responsibility Act had done its job? Wasn't it time to celebrate?

From where I sat, there were still plenty of things to worry about. Just as there was a vast difference between policy makers' goals and the more disturbing realities

I saw inside the welfare office, there was also a wide gap between the political cele-brations of the success of welfare reform and the hardship I encountered. And if you read the statistics carefully, you could see that. By 2002, nearly half (40 percent) of the women and children who had left the welfare rolls had no discernible source of income—no work, no welfare, nowhere to go. Of the 60 percent of for-mer welfare recipients who were employed, half were still living in poverty. Their average annual earnings were estimated at $8,000 to $10,000 a year, and most had jobs without medical insurance, sick leave, or retirement benefits.

Thus, in the context of the then-booming economy, more than two-thirds of the women and children who had left the welfare office as a result of reform were still living well below the poverty level. Although the welfare rolls had been cut by more than half, the number of families living in dire (welfare-level) poverty had declined by only 15 percent. What this meant, and what I knew to be true, was that millions of poor families were now more reluctant than ever to seek out help from the welfare office. State governments and local charities were already reporting rising rates of hunger and homelessness. And the economy had begun to sour.

As I read those numbers, I could not forget that they were referring to real people, to mothers and children. I also could not forget that there were millions of poor families still on the welfare rolls. New ones were coming in each day, others were coming back again. I knew all the stories about problems in manag-ing the graveyard shift, what it was like to find child care when one of the kids was sick, what it meant to have the phone turned off, how it felt to be unable to afford birthday gifts or winter coats, the major difficulties caused by flat tires or buses that didn't run on schedule, the distress of women forced to leave their children at home alone, the overdue power bills, the slum landlords, the problems with depression and stress.

In the end, it was clear to me that, although the outcomes of welfare reform would emerge slowly, and although there were a good number of real success sto-ries among the millions who had left the rolls, in the long run this law would not only result in rising rates of hunger and homelessness, but also rising rates of crime, rising numbers of women in mental-health facilities and domestic-violence shelters; rising numbers of children in foster care, in substandard child care, or left to fend for themselves; and rising numbers of working-poor families stretched to the breaking point. . . .

Looking at the results of reform, I also saw that worthy ideals and good values were not enough. To make those good intentions a reality, what was needed were decent jobs that paid a family wage, serious education and training programs that offered access to real career ladders, guarantees of quality child care, affordable housing, and reliable medical coverage. What was needed were programs that included poor men as well as women, and included the working poor as well as the nonworking poor. What was needed, in short, was a real commitment to the "mainstream" American values that the Personal Responsibility Act had claimed to champion.

In thinking about this, I also came to recognize that all the seemingly dis-parate agendas of my friends, neighbors, and colleagues were on target, at least to some extent. Welfare reform is simultaneously about work, family, poverty, race, capitalism, motherhood, and morality. It is about problems of gender and race inequality, the widening gap between rich and poor, widespread difficulties in juggling work and family commitments, and the significant question of how this nation will respond to its most disadvantaged members. . . .

All those concerns seem even more pressing to me now. And I am absolutely convinced that welfare reform, in all its complexity and with all the proclamations of its "success," remains one of the most urgent and important issues of our time.

DISCUSSION QUESTIONS

1. Hays gives some basic statistics about the people who make up welfare recipients. What surprised you about these numbers? Did you have a different image of a typical welfare recipient?

2. What would be a critical response to the "conclusion" that welfare reform is working? What evidence do we have that it is, in fact, not improving the economic situations of single mothers?

http://infotrac.thomsonlearning.com

InfoTrac College Edition

BONUS READING

Edin, Kathyrn. "Few Good Men: Why Poor Women Don't Remarry." *The American Prospect* 11, no. 4 (January 3, 2000): 26–31.

    This reading looks at the reasons low-income single mothers choose not to remarry. Contrary to common misperceptions, marrying a financially unstable man does more to hinder than help economic advancement for these women. What are the central reasons why poor women do not remarry? Given Edin's research, what can be our response to a government administration that encourages marriage as a step toward financial stability for welfare recipients?

SEARCH TERMS

welfare reform and race
proletarianization
Black middle class
living wage campaign
race and home ownership

25

The Garment Industry in the Restructuring Global Economy

EDNA BONACICH, LUCIE CHENG, NORMA CHINCHILLA,
NORA HAMILTON, AND PAUL ONG

These authors describe the consequences of the new processes in the global economy, particularly as they affect different classes of people in different international locales. The authors note the different interpretations that various scholars have given to the process of globalization—some seeing its positive effects, others being more critical of the impact of globalization on women workers, immigrants, and the working class.

Global integration, a long-standing feature of the world economy, is currently undergoing a restructuring. Generally, until after World War II, the advanced industrial countries of western Europe and the United States dominated the world economy and controlled most of its industrial production. The less-developed countries tended to concentrate in the production of raw materials. Since the late 1950s, and accelerating rapidly in the 1980s, however, industrial production has shifted out of the West, initially to Japan, then to the Asian NICs (newly industrializing countries—namely, Hong Kong, Taiwan, South Korea, and Singapore), and now to almost every country of the world. Less-developed countries are not manufacturing mainly for the domestic market or following a model of "import substitution"; rather, they are manufacturing for export, primarily to developed countries, and pursuing a development strategy of export-led industrialization. What we are witnessing has been termed by some a "new international division of labor" (Fröbel, Heinrichs, and Kreye 1980).

The developed countries are faced with the problem of "deindustrialization" in terms of traditional manufacturing, as their manufacturing base is shifted to other, less-developed countries (Bluestone and Harrison 1988). At the same time, they are faced with a massive rise in imports that compete with local industries' products, moving to displace them. This shift is accompanied by the rise of a new kind of transnational corporation (TNC). Of course, TNCs have existed since the beginning of the European expansion, but they concentrated mainly on the production of agricultural goods and raw materials and, in the postwar period, on manufacturing for

From: "The Garment Industry in the Restructuring Global Economy," by Edna Bonacich, Lucie Cheng, Norma Chinchilla, Nora Hamilton, and Paul Ong. As it appears in *Global Production: The Apparel Industry in the Pacific Rim.* Edited by Edna Bonacich, Lucie Cheng, Norma Chinchilla, Nora Hamilton, and Paul Ong. Reprinted by permission of Temple University Press © 1994 by Temple University. All rights reserved.

the host country market. The new TNCs are global firms that are able to use advanced communications and transportation technology to coordinate manufacturing in multiple locations simultaneously. They engage in "off-shore sourcing" to produce primarily for the home market (Grunwald and Flamm 1985; Sklair 1989).

TNCs sometimes engage in direct foreign investment, but globalized production does not depend on it. They can arrange for production in numerous locations through other, looser connections, such as subcontracting and licensing. In other words, TNCs can set up complex networks of global production without owning or directly controlling their various branches.

The nation-state has increasingly declined as an economic unit, with the result that states are often unable to control the actions of powerful TNCs. The TNCs are supragovernmental actors that make decisions on the basis of profit-making criteria without input from representative governments. Of course, strong states are still able to exercise considerable influence over trade policies and over the policies of the governments of developing countries.

Some scholars have used the concept of "commodity chains" to describe the new spatial arrangements of production (Gereffi and Korzeniewicz 1994). The concept shows how design, production, and distribution are broken down and geographically dispersed, with certain places serving as centers within the chain. Power is differentially allocated along the chain, and countries and firms vie to improve their position in the chain.

Focusing on the geographic aspects of global production also has led to the concept of "global cities" (Sassen 1991). These are coordination centers for the global economy, where planning takes place. They house the corporate headquarters of TNCs, as well as international financial services and a host of related business services. These cities have become the "capitals" of the new global economy.

Another way to view the restructuring is to see it as the proletarianization of most of the world. People who had been engaged primarily in peasant agriculture or in other forms of noncapitalist production are now being incorporated into the industrial labor force. Many of these people are first-generation wage-workers, and a disproportionate number of them are women. These "new" workers sometimes retain ties to noncapitalist sectors and migrate between them and capitalist employment, making their labor cheaper than that of fully proletarianized workers. But even if they are not attached to noncapitalist sectors, first-generation workers tend to be especially vulnerable to exploitative conditions. Thus, an important feature of the new globalization is that TNCs are searching the world for the cheapest available labor and are finding it in developing countries.

Countries pursuing export-led industrialization typically follow strategies that encourage the involvement of foreign capital. They offer incentives, including tax holidays and the setting up of export processing zones (EPZs), where the bureaucracy surrounding importing and exporting is curtailed; sometimes they also promise cheap and controllable labor. Countries using this development strategy do not plan to remain the providers of cheap labor for TNCs, however: they hope to move up the production ladder, gaining more economic power and control. They want to shift from labor-intensive manufactures to capital-intensive, high-technology goods. They hope to follow the path of Japan and the Asian NICs and become major economic players in the global economy.

Sometimes participation in global capitalist production is foisted on nations by advanced-industrial countries and/or suprastate organizations such as the World Bank and the International Monetary Fund (IMF), where advanced countries wield a great deal of influence. The United States, in particular, has backed regimes that support globalized production and has pushed for austerity programs that help to make labor cheap. At the same time, developed countries, including the United States, have been affected by the restructured global economy. Accompanying the rise in imports and deindustrialization has been a growth in unemployment and a polarization between the rich and the poor (Harrison and Bluestone 1988). This trend has coincided with increased racial polarization, as people of color have faced a disproportionate impact from these developments.

A rise in immigration from less-developed to more-developed countries has also accompanied globalization. The United States, for example, has experienced large-scale immigration from the Caribbean region and from Asia, two areas pursuing a manufacturing-for-export development strategy. At least part of this immigration is a product of globalization, as people are dislocated by the new economic order and are forced to emigrate for survival (Sassen 1988). Dislocations occur not only because global industries displace local ones (as in the case of agribusiness displacing peasants), but also because austerity programs exacerbate the wage gap between rich and poor countries (making the former ever more desirable). Political refugees, often from countries where the United States has supported repressive regimes, have added to the rise in immigration as well. Finally, some immigration results when people move to service global enterprise as managers, trade representatives, or technicians.

In the advanced countries, the immigration of workers has created a "Third World within." In this case, the newly created proletariat is shifting location. These immigrants play a part in the efforts of the advanced countries to hold on to their industries, by providing a local source of cheap labor to counter the low labor standards in competing countries.

In sum, we are seeing a shakeup of the old world economic order. Some countries have used manufacturing for export as a way to become major economic powers (Appelbaum and Henderson 1992; Gereffi and Wyman 1990). These countries now threaten U.S. dominance. Other countries are trying to pursue this same path, but it is not clear whether they will succeed. Meanwhile, despite the fact that the United States is suffering some negative consequences from the global restructuring, certain U.S.-based TNCs are deeply implicated in the process and benefit from it.

CONTRASTING VIEWS OF RESTRUCTURING

The new globalization receives different interpretations and different evaluations (Gondolf, Marcus, and Dougherty 1986). Some focus on the positive side; they see global production as increasing efficiency by allowing each country to specialize in its strengths. Less-developed countries are able to provide low-cost, unskilled labor while developed countries provide management, technical, and financial resources. Together they are able to maximize the efficient use of resources. The result is that more goods and services are produced more cheaply, to the benefit of all. Consumers, in particular, are seen as the great beneficiaries of globalized production, because of the abundance of low-cost, higher-quality goods from which to choose.

Globalization can be seen as part of the new system of flexible specialization (Piore and Sabel 1984). Consumer markets have become more differentiated, making the old, industrial system of mass production in huge factories obsolete. To be competitive today, a firm must be able to produce small batches of differentiated goods for diverse customers. Globalization contributes to this process by enabling firms to produce a vast range of products in multiple countries simultaneously.

Another aspect of the positive view is to see the entrance of less-developed countries into manufacturing for export as a step toward their industrialization and economic development. Although countries may enter the global economy at a tremendous disadvantage, by participating in exports they are able to accumulate capital and gradually increase their power and wealth. Japan and the Asian NICs have demonstrated the possibilities; now other countries can follow a similar path.

Although workers in the advanced countries may suffer some dislocation by the movement of industry abroad, in the long run they are seen to be beneficiaries of this process. While lower-skilled, more labor-intensive jobs will move to the developing countries, the advanced countries will gain higher-technology jobs, as well as jobs in coordinating and managing the global economy. Thus workers in the advanced countries will be "pushed up" to more middle-class positions, servicing and directing the workers in the rest of the world. Moreover, as other countries develop, their purchasing power will increase, leading to larger markets for the products of the developed countries. Growth in exports means growth in domestic production, and thus growth in domestic employment.

Those who favor globalization also note its inevitability. The economic logic that is propelling global production is immensely powerful. Technology allows globalization, and competition forges it; there is really no stopping the process, so the best one can do is adapt on the most favorable terms possible. Nations feel they must get into the game quickly so as not to be left behind.

A favorable standpoint on globalization is typically coupled with an optimistic view of the effects of immigration. Like new nations entering the global economy, immigrant workers are seen as having to suffer in the short run in order to make advances in the future. Instead of being viewed as exploited, the immigrants are seen as being granted an opportunity—one that they freely choose—to better their life circumstances. They may start off being paid low wages because they lack marketable skills, but with time, they or their children will acquire such skills and will experience upward mobility.

In general, a positive view of globalization is accompanied by a belief in the benefits of markets and free trade. The market, rather than political decision making, should, it is felt, be the arbiter of economic decision making. This favorable and inevitable view of globalization is by far the most predominant approach. It is promoted by the U.S. government, by the TNCs, by many governments in developing countries, and by various international agencies. This position receives considerable support from academics, especially economists, who provide governmental agencies with advice. It is the dominant world policy.

There is, however, a less sanguine interpretation of globalization voiced by U.S. trade unionists and many academics who study development, labor, women, inequality, and social class (Castells and Henderson 1987; Kamel 1990; Kolko 1988; Peet 1987; Ross and Trachte 1990; Sklair 1989). In general, their view is that globalization has a differential class impact: globalization is in the interests of capitalists, especially capitalists connected with TNCs, and of sectors of the capitalist class in developing nations. But the working class in both sets of countries is hurt, especially young women workers, who have become the chief employees of the TNCs (Fernandez-Kelly 1983; Fuentes and Ehrenreich 1983; Mies 1986; Nash and Fernandez-Kelly 1983).

Some argue that globalization is part of a response to a major crisis that has emerged in the advanced capitalist countries. In particular, after the post–World War II boom, the economies of these countries stagnated and profits declined; stagnation was blamed on the advances made by workers under the welfare state. Capital's movement abroad, which was preceded in the United States by regional relocation, is an effort to cut labor costs, weaken unions, and restore profitability. Put generally, globalization can be seen, in part, as an effort to discipline labor.

Globalization enables employers to pit workers from different countries against one another. Regions and nations must compete to attract investment and businesses. Competitors seek to undercut one another by offering the most favorable conditions to capital. Part of what they seek to offer is quality, efficiency, and timeliness, but they also compete in terms of providing the lowest possible labor standards: they promise a low-cost, disciplined, and unorganized work force. Governments pledge to ensure these conditions by engaging in the political repression of workers' movements (Deyo 1989).

The disciplining of the working class that accompanies globalization is not limited to conditions in the workplace. It also involves a cutback in state social programs. For example, in the United States, under the Reagan-Bush administrations, efforts were made to curtail multiple programs protecting workers' standard of living; these tax-based programs were seen as hindering capital accumulation. The argument was made that if these funds were invested by the private sector, everyone would benefit, including workers. This same logic has been imposed on developing countries; they have been granted aid and loans on the condition that they engage in austerity programs that cut back on social spending. The impact of such cutbacks is that workers are less protected from engaging in bargains of desperation when they enter the work force.

This view of globalization is accompanied by a pessimism about the policy of export-led development. Rather than believing that performing assembly for TNCs will lead to development, critics fear that it is another form of dependency, with the advanced capitalist countries and their corporations retaining economic (and political) control over the global economy (Bello and Rosenfeld 1990).

Critics also note a negative side to immigrants' experiences (Mitter 1986; Sassen 1988). They see the immigration of workers as, in part, a product of globalization and TNC activity, as workers in less-developed countries find their means of livelihood disrupted by capitalist penetration. Immigrants are thus not just people seeking a better life for themselves, but often those "forced" into moving because they have lost the means to survive. On arrival in the more-advanced economies, they are faced with forms of coercion, including immigration regulations, racism, and sexism, that keep them an especially disadvantaged work force. Especially coercive is the condition of being an undocumented immigrant. Critics point out that those who favor globalization promote the free movement of commodities and capital, but not the free movement of labor, in the form of open borders. Political restrictions on workers add to the weakening of the working class.

In sum, the critical perspective sees globalization as an effort to strengthen the hand of capital and weaken that of labor. The favorable view argues that the interests of capital and labor are not antagonistic and that everyone benefits from capital accumulation, investment, economic growth, and the creation of jobs. Critics, on the other hand, contend that certain classes benefit at the expense of others, and that, even if workers in poor countries do get jobs, these jobs benefit the capitalists much more than they do the workers, and also hurt the workers in the advanced capitalist countries through deindustrialization.

Where does the truth lie? . . .

To a certain extent, one's point of view depends on geographic location. Generally, Asian countries, especially the NICs, appear to be transforming themselves from dependencies into major actors and competitors in the global economy, leading to an optimism about the effects of globalization. This optimism, however, blots out the suffering and labor repression that is still occurring for some workers in these countries, despite the rise in standard of living for the majority.

On the other hand, the Caribbean region generally faces a harsher reality, in part because the closeness and dominance of the United States pose special problems for these countries. They are more likely to get caught in simple assembly for the TNCs, raising questions about whether manufacturing for export will be transformable into broader economic development. Of course, some in these countries are firm believers in this policy and are pursuing it avidly, but there are clear signs that many workers are severely exploited in the process. . . .

Other confusing issues remain. For example, do women benefit from their movement into the wage sector (proletarianization) as a result of globalization? A case can be made that working outside the home and earning money gives women new-found power in their relations with men. It can also be argued, however, that these women remain under patriarchal control, but that now, in addition to their fathers and husbands, they are under the control of male bosses. They have double and even triple workloads, as they engage in wage labor, domestic labor, and often industrial homework and other forms of informalized labor.

The two points of view lead to different politics. Those who hold the favorable outlook advocate working for the breakdown of all trade and investment barriers and to pushing rapidly ahead toward global integration. Critics are not trying to stem these forces completely, but rather, are attempting to set conditions on them. For example, globalization should be allowed only if labor and environmental standards are protected in the process. Similarly, the rights of workers to form unions should be safeguarded, so that business cannot wantonly pit groups of workers against one another. . . .

REFERENCES

Appelbaum, Richard P., and Jeffrey Henderson, eds. 1992. *States and Development in the Asian Pacific Rim*. Newbury Park, CA: Sage.

Bello, Walden, and Stephanie Rosenfeld. 1990. *Dragons in distress: Asia's miracle economies in crisis*. San Francisco: Institute for Food and Development Policy.

Castells, Manuel, and Jeffrey Henderson. 1987. "Technoeconomic restructuring, sociopolitical processes, and spatial transformation: A global perspective." In *Global restructuring and territorial development*, ed. Jeffrey Henderson and Manuel Castells, 1–17. London: Sage.

Deyo, Frederic C. 1989. *Beneath the Miracle: Labor subordination in the new Asian industrialism*. Berkeley: University of California Press.

Fernandez-Kelly, M. Patricia. 1983. *For we are sold, I and my people: Women and industry in Mexico's frontier*. Albany: State University of New York Press.

Fröbel, Folker, Jürgen Heinrichs, and Otto Kreye. 1980. *The new international division of labour: Structural unemployment in industrialised countries and industrialisation in developing countries*. Cambridge: Cambridge University Press.

Fuentes, Annette, and Barbara Ehrenreich. 1983. *Women in the global factory*. Boston: South End Press.

Gereffi, Gary, and Miguel Korzeniewicz, eds. 1994. *Commodity chains and global capitalism*. Westport, CT: Greenwood Press.

Gereffi, Gary, and Donald L. Wyman, eds. 1990. *Manufacturing miracles: paths of industrialization in Latin America and East Asia*. Princeton: Princeton University Press.

Gondolf, Edward W., Irwin M. Marcus, and James P. Daugherty. 1986. *The global economy: Divergent perspectives on economic change*. Boulder, CO: Westview Press.

Grunwald, Joseph, and Kenneth Flamm. 1985. *The global factory: Foreign assembly in international trade*. Washington, DC: Brookings Institution.

Harrison, Bennett, and Barry Bluestone, 1988. *The great U-turn: Corporate restructuring and the polarizing of America*. New York: Basic Books.

Kamel, Rachael. 1990. *The global factory: Analysis and action for a new economic era*. Philadelphia: American Friends Service Committee.

Kolko, Joyce. 1988. *Restructuring the world economy*. New York: Pantheon.

Mies, Maria. 1986. *Patriarchy and accumulation on a world scale: Women in the international division of labor*. London: Zed Books.

Mitter, Swasti. 1986. *Common fate, common bond: Women in the global economy*. London: Pluto Press.

Nash, June, and M. Patricia Fernandez-Kelly, eds. 1983. *Women, men, and the international division of labor*. Albany: State University of New York Press.

Peet, Richard, ed. 1987. *International capitalism and industrial restructuring*. Boston: Unwin Hyman.

Piore, Michael J., and Charles F. Sabel. 1984. *The second industrial divide: Possibilities for prosperity*. New York: Basic Books.

Ross, Robert J. S., and Kent C. Trachte. 1990. *Global capitalism: The new leviathan.* Albany: State University of New York Press.

Sassen, Saskia. 1988. *The mobility of labor and capital: A study in international investment and labor flow.* Cambridge: Cambridge University Press.

———. 1991. *The global city: New York, London, Tokyo.* Princeton: Princeton University Press.

Sklair, Leslie. 1989. *Assembling for development: The Maquila industry in Mexico and the United States.* London: Unwin Hyman.

DISCUSSION QUESTIONS

1. What is a *commodity chain*? What evidence do you see of commodity chains in the wardrobe that you wear?

2. Compare and contrast the two perspectives on the new globalization that the authors describe. How do proponents of the positive and critical views of globalization view immigration?

26

The Nanny Chain

ARLIE RUSSELL HOCHSCHILD

Arlie Hochschild identifies the "nanny chain" as a global system of work in which women workers from poor nations provide the "care work" for more privileged workers in other parts of the world. This pattern of labor is transforming social relations of care worldwide and, according to Hochschild, makes care and love a commodity that is transferred and exchanged in the world market.

Vicky Diaz, a 34-year-old mother of five, was a college-educated school-teacher and travel agent in the Philippines before migrating to the United States to work as a housekeeper for a wealthy Beverly Hills family and as a nanny for their two-year-old son. Her children, Vicky explained to Rhacel Parrenas,

> were saddened by my departure. Even until now my children are trying to convince me to go home. The children were not angry when I left because

From: Arlie Russell Hochschild. "The Nanny Chain.." 2000. *The American Prospect,* 3 (January 2000), pp. 33–36.

they were still very young when I left them. My husband could not get angry either because he knew that was the only way I could seriously help him raise our children, so that our children could be sent to school. I send them money every month.

In her book *Servants of Globalization,* Parrenas, an affiliate of the Center for Working Families at the University of California, Berkeley, tells an important and disquieting story of what she calls the "globalization of mothering." The Beverly Hills family pays "Vicky" (which is the pseudonym Parrenas gave her) $400 a week, and Vicky, in turn, pays her own family's live-in domestic worker back in the Philippines $40 a week. Living like this is not easy on Vicky and her family. "Even though it's paid well, you are sinking in the amount of your work. Even while you are ironing the clothes, they can still call you to the kitchen to wash the plates. It . . . [is] also very depressing. The only thing you can do is give all your love to [the two-year-old American child]. In my absence from my children, the most I could do with my situation is give all my love to that child."

Vicky is part of what we could call a global care chain: a series of personal links between people across the globe based on the paid or unpaid work of caring. A typical global care chain might work something like this: An older daughter from a poor family in a third world country cares for her siblings (the first link in the chain) while her mother works as a nanny caring for the children of a nanny migrating to a first world country (the second link) who, in turn, cares for the child of a family in a rich country (the final link). Each kind of chain expresses an invisible human ecology of care, one care worker depending on another and so on. A global care chain might start in a poor country and end in a rich one, or it might link rural and urban areas within the same poor country. More complex versions start in one poor country and extend to another slightly less poor country and then link to a rich country.

Global care chains may be proliferating. According to 1994 estimates by the International Organization for Migration, 120 million people migrated—legally or illegally—from one country to another. That's 2 percent of the world's population. How many migrants leave loved ones behind to care for other people's children or elderly parents, we don't know. But we do know that more than half of legal migrants to the United States are women, mostly between ages 25 and 34. And migration experts tell us that the proportion of women among migrants is likely to rise. All of this suggests that the trend toward global care chains will continue.

How are we to understand the impact of globalization on care? If, as globalization continues, more global care chains form, will they be "good" care chains or "bad" ones? Given the entrenched problem of third world poverty—which is one of the starting points for care chains—this is by no means a simple question. But we have yet to fully address it, I believe, because the world is globalizing faster than our minds or hearts are. We live global but still think and feel local.

FREUD IN A GLOBAL ECONOMY

Most writing on globalization focuses on money, markets, and labor flows, while giving scant attention to women, children, and the care of one for the other. Most research on women and development, meanwhile, draws a connection between, say, World Bank loan conditions and the scarcity of food for women and children in the third world, without saying much about resources expended on caregiving. Much of the research on women in the United States and Europe focuses on a chainless, two-person picture of "work-family balance" without considering the child care worker and the emotional ecology of which he or she is a part. Fortunately, in recent years, scholars such as Ernestine Avila, Evelyn Nakano Glenn, Pierette Hondagneu-Sotelo, Mary Romero, and Rhacel Parrenas have produced some fascinating research on domestic workers. Building on this work, we can begin to focus on the first world end of the care chain and begin spelling out some of the implications of the globalization of love.

One difficulty in understanding these implications is that the language of economics does not translate easily into the language of psychology. How are we to understand a "transfer" of feeling from one link in a chain to another? Feeling is not a "resource" that can be crassly taken from one person and given to another. And surely one person can love quite a few people; love is not a resource limited the same way oil or currency supply is. Or is it?

Consider Sigmund Freud's theory of displacement, the idea that emotion can be redirected from one person or object to another. Freud believed that if, for example, Jane loves Dick but Dick is emotionally or literally unavailable, Jane will find a new object (say, John, Dick and Jane's son) onto which to project her original feeling for Dick. While Freud applied the idea of displacement mainly to relations within the nuclear family, the concept can also be applied to relations extending far outside it. For example, immigrant nannies and au pairs often divert feelings originally directed toward their own children toward their young charges in this country. As Sau-ling C. Wong, a researcher at the University of California, Berkeley, has put it, "Time and energy available for mothers are diverted from those who, by kinship or communal ties, are their more rightful recipients."

If it is true that attention, solicitude, and love itself can be "displaced" from one child (let's say Vicky Diaz's son Alfredo, back in the Philippines) onto another child (let's say Tommy, the son of her employers in Beverly Hills), then the important observation to make here is that this displacement is often upward in wealth and power. This, in turn, raises the question of the equitable distribution of care. It makes us wonder, is there—in the realm of love—an analogue to what Marx calls "surplus value," something skimmed off from the poor for the benefit of the rich?

Seen as a thing in itself, Vicky's love for the Beverly Hills toddler is unique, individual, private. But might there not be elements in this love that are borrowed, so to speak, from somewhere and someone else? Is time spent with the first world child in some sense "taken" from a child further down the care chain? Is the Beverly Hills child getting "surplus" love, the way immigrant farm workers

give us surplus labor? Are first world countries such as the United States importing maternal love as they have imported copper, zinc, gold, and other ores from
third world countries in the past?

This is a startling idea and an unwelcome one, both for Vicky Diaz, who
needs the money from a first world job, and for her well-meaning employers,
who want someone to give loving care to their child. Each link in the chain feels
she is doing the right thing for good reasons—and who is to say she is not?

But there are clearly hidden costs here, costs that tend to get passed down
along the chain. One nanny reported such a cost when she described (to Rhacel
Parrenas) a return visit to the Philippines: "When I saw my children, I thought,
'Oh children do grow up even without their mother.' I left my youngest when
she was only five years old. She was already nine when I saw her again but she
still wanted for me to carry her [weeps]. That hurt me because it showed me that
my children missed out on a lot."

Sometimes the toll it takes on the domestic worker is overwhelming and
suggests that the nanny has not displaced her love onto an employer's child but
rather has continued to long intensely for her own child. As one woman told
Parrenas, "The first two years I felt like I was going crazy. . . . I would catch
myself gazing at nothing, thinking about my child. Every moment, every second
of the day, I felt like I was thinking about my baby. My youngest, you have to
understand, I left when he was only two months old. . . . You know, whenever
I receive a letter from my children, I cannot sleep. I cry. It's good that my job is
more demanding at night."

Despite the anguish these separations clearly cause, Filipina women continue
to leave for jobs abroad. Since the early 1990s, 55 percent of migrants out of the
Philippines have been women; next to electronic manufacturing, their remittances make up the major source of foreign currency in the Philippines. The rate
of female emigration has continued to increase and includes college-educated
teachers, businesswomen, and secretaries. In Parrenas's study, more than half of
the nannies she interviewed had college degrees and most were married mothers
in their 30s.

Where are men in this picture? For the most part, men—especially men at
the top of the class ladder—leave child-rearing to women. Many of the husbands
and fathers of Parrenas's domestic workers had migrated to the Arabian peninsula
and other places in search of better wages, relieving other men of "male work"
as construction workers and tradesmen, while being replaced themselves at
home. Others remained at home, responsible fathers caring or helping to care for
their children. But some of the men tyrannized their wives. Indeed, many of the
women migrants Parrenas interviewed didn't just leave; they fled. As one migrant
maid explained:

> You have to understand that my problems were very heavy before I left
> the Philippines. My husband was abusive. I couldn't even think about my
> children, the only thing I could think about was the opportunity to escape
> my situation. If my husband was not going to kill me, I was probably going
> to kill him. . . . He always beat me up and my parents wanted me to leave

him for a long time. I left my children with my sister. . . . In the plane . . . I felt like a bird whose cage had been locked for many years. . . . I felt free. . . . Deep inside, I felt homesick for my children but I also felt free for being able to escape the most dire problem that was slowly killing me.

Other men abandoned their wives. A former public school teacher back in the Philippines confided to Parrenas: "After three years of marriage, my husband left me for another woman. My husband supported us for just a little over a year. Then the support was stopped. . . . The letters stopped. I have not seen him since." In the absence of government aid, then, migration becomes a way of coping with abandonment.

Sometimes the husband of a female migrant worker is himself a migrant worker who takes turns with his wife migrating. One Filipino man worked in Saudi Arabia for 10 years, coming home for a month each year. When he finally returned home for good, his wife set off to work as a maid in America while he took care of the children. As she explained to Parrenas, "My children were very sad when I left them. My husband told me that when they came back home from the airport, my children could not touch their food and they wanted to cry. My son, whenever he writes me, always draws the head of Fido the dog with tears on the eyes. Whenever he goes to Mass on Sundays, he tells me that he misses me more because he sees his friends with their mothers. Then he comes home and cries."

THE END OF THE CHAIN

Just as global capitalism helps create a third world supply of mothering, it creates a first world demand for it. The past half-century has witnessed a huge rise in the number of women in paid work—from 15 percent of mothers of children aged 6 and under in 1950 to 65 percent today. Indeed, American women now make up 45 percent of the American labor force. Three-quarters of mothers of children 18 and under now work, as do 65 percent of mothers of children 6 and under. In addition, a recent report by the International Labor Organization reveals that the average number of hours of work per week has been rising in this country.

Earlier generations of American working women would rely on grandmothers and other female kin to help look after their children; now the grandmothers and aunts are themselves busy doing paid work outside the home. Statistics show that over the past 30 years a decreasing number of families have relied on relatives to care for their children—and hence are compelled to look for nonfamily care. At the first world end of care chains, working parents are grateful to find a good nanny or child care provider, and they are generally able to pay far more than the nanny could earn in her native country. This is not just a child care problem. Many American families are now relying on immigrant or out-of-home care for their *elderly* relatives. As a Los Angeles elder-care worker, an immigrant, told Parrenas, "Domestics here are able to make a living from the elderly that families abandon." But this often means that nannies cannot take care of their own ailing parents and therefore produce an elder-care version of a child care chain—caring for first world elderly persons while a paid worker cares for their aged mother back in the Philippines.

My own research for two books, *The Second Shift* and *The Time Bind,* sheds some light on the first world end of the chain. Many women have joined the law, academia, medicine, business—but such professions are still organized for men who are free of family responsibilities. The successful career, at least for those who are broadly middle class or above, is still largely built on some key traditional components: doing professional work, competing with fellow professionals, getting credit for work, building a reputation while you're young, hoarding scarce time, and minimizing family obligations by finding someone else to deal with domestic chores. In the past, the professional was a man and the "someone else to deal with [chores]" was a wife. The wife oversaw the family, which—in preindustrial times, anyway—was supposed to absorb the human vicissitudes of birth, sickness, and death that the workplace discarded. Today, men take on much more of the child care and housework at home, but they still base their identity on demanding careers in the context of which children are beloved impediments; hence, men resist sharing care equally at home. So when parents don't have enough "caring time" between them, they feel forced to look for that care further down the global chain.

The ultimate beneficiaries of these various care changes might actually be large multinational companies, usually based in the United States. In my research on a Fortune 500 manufacturing company I call Amerco, I discovered a disproportionate number of women employed in the human side of the company: public relations, marketing, human resources. In all sectors of the company, women often helped others sort out problems—both personal and professional— at work. It was often the welcoming voice and "soft touch" of women workers that made Amerco seem like a family to other workers. In other words, it appears that these working mothers displace some of their emotional labor from their children to their employer, which holds itself out to the worker as a "family." So, the care in the chain may begin with that which a rural third world mother gives (as a nanny) the urban child she cares for, and it may end with the care a working mother gives her employees as the vice president of publicity at your company.

HOW MUCH IS CARE WORTH?

How are we to respond to the growing number of global care chains? Through what perspective should we view them?

I can think of three vantage points from which to see care chains: that of the primordialist, the sunshine modernist, and (my own) the critical modernist. The primordialist believes that our primary responsibility is to our own family, our own community, our own country. According to this view, if we all tend our own primordial plots, everybody will be fine. There is some logic to this point of view. After all, Freud's concept of displacement rests on the premise that some original first object of love has a primary "right" to that love, and second and third comers don't fully share that right. (For the primordialist—as for most all of us—those first objects are members of one's most immediate family.) But the primordialist is an isolationist, an antiglobalist. To such a person, care chains seem wrong—not because they're unfair to the least-cared-for children at the bottom of the chain, but because they are global. Also, because family care has historically been provided by women, primordialists often believe that women should stay home to provide this care.

The sunshine modernist, on the other hand, believes care chains are just fine, an inevitable part of globalization, which is itself uncritically accepted as good. The idea of displacement is hard for the sunshine modernists to grasp because in their equation—seen mainly in economic terms—the global market will sort out who has proper claims on a nanny's love. As long as the global supply of labor meets the global demand for it, the sunshine modernist believes, everything will be okay. If the primordialist thinks care chains are bad because they're global, the sunshine modernist thinks they're good for the very same reason. In either case, the issue of inequality of access to care disappears.

The critical modernist embraces modernity but with a global sense of ethics. When the critical modernist goes out to buy a pair of Nike shoes, she is concerned to learn how low the wage was and how long the hours were for the third world factory worker making the shoes. The critical modernist applies the same moral concern to care chains: The welfare of the Filipino child back home must be seen as some part, however small, of the total picture. The critical modernist sees globalization as a very mixed blessing, bringing with it new opportunities—such as the nanny's access to good wages—but also new problems, including emotional and psychological costs we have hardly begun to understand.

From the critical modernist perspective, globalization may be increasing inequities not simply in access to money—and those inequities are important enough—but in access to care. The poor maid's child may be getting less motherly care than the first world child. (And for that matter, because of longer hours of work, the first world child may not be getting the ideal quantity of parenting attention for healthy development because too much of it is now displaced onto the employees of Fortune 500 companies.) We needn't lapse into primordialism to sense that something may be amiss in this.

I see no easy solutions to the human costs of global care chains. But here are some initial thoughts. We might, for example, reduce the incentive to migrate by addressing the causes of the migrant's economic desperation and fostering economic growth in the third world. Thus one obvious goal would be to develop the Filipino economy.

But it's not so simple. Immigration scholars have demonstrated that development itself can *encourage* migration because development gives rise to new economic uncertainties that families try to mitigate by seeking employment in the first world. If members of a family are laid off at home, a migrant's monthly remittance can see them through, often by making a capital outlay in a small business or paying for a child's education.

Other solutions might focus on individual links in the care chain. Because some women migrate to flee abusive husbands, a partial solution would be to create local refuges from such husbands. Another would be to alter immigration policy so as to encourage nannies to bring their children with them. Alternatively, employers or even government subsidies could help nannies make regular visits home.

The most fundamental approach to the problem is to raise the value of caring work and to ensure that whoever does it gets more credit and money for it. Otherwise, caring work will be what's left over, the work that's continually passed on down the chain. Sadly, the value ascribed to the labor of raising a child has always been low relative to the value of other kinds of labor, and under the impact of globalization, it has sunk lower still. The low value placed on caring work is due neither to an absence of demand for it (which is always high) nor to the simplicity of the work (successful caregiving is not easy) but rather to the cultural politics underlying this global exchange.

The declining value of child care anywhere in the world can be compared to the declining value of basic food crops relative to manufactured goods on the international market. Though clearly more essential to life, crops such as wheat, rice, or cocoa fetch low and declining prices while the prices of manufactured goods (relative to primary goods) continue to soar in the world market. And just as the low market price of primary produce keeps the third world low in the community of nations, the low market value of care keeps low the status of the women who do it.

One way to solve this problem is to get fathers to contribute more to child care. If fathers worldwide shared child care labor more equitably, care would spread laterally instead of being passed down a social-class ladder, diminishing in value along the way. Culturally, Americans have begun to embrace this idea—but they've yet to put it into practice on a truly large scale [see Richard Weissbourd, "Redefining Dad," *TAP,* December 6, 1999]. This is where norms and policies established in the first world can have perhaps the greatest influence on reducing costs along global care chains.

According to the International Labor Organization, half of the world's women between ages 15 and 64 are working in paid jobs. Between 1960 and 1980, 69 out of 88 countries for which data are available showed a growing proportion of women in paid work (and the rate of increase has skyrocketed since the 1950s in the United States, Scandinavia, and the United Kingdom). If we want developed societies with women doctors, political leaders, teachers, bus drivers, and computer programmers, we will need qualified people to help care for children. And there is no reason why every society cannot enjoy such loving paid child care. It may even remain the case that Vicky Diaz is the best person to provide it. But we would be wise to adopt the perspective of the critical modernist and extend our concern to the potential hidden losers in the care chain. These days, the personal is global.

DISCUSSION QUESTIONS

1. What does Hochschild mean by the *globalization of love* and how is this phenomenon linked to the status of women in the United States? In other parts of the world?

2. What different perspectives on the care chain does Hochschild identify? What solutions to the problem does each perspective suggest? What would you recommend?

27

Michael Jordan and the New Global Capitalism

WALTER LAFEBER

LaFeber argues that the international recognition of Michael Jordan exemplifies the global nature of American culture and capitalism. Recent technologies from television to the Internet allow American wealth to be displayed all around the world. The author considers the consequences of this global capitalism.

. . . [T]his [reading] briefly speculates on the meaning of the story in the long history of imperialism, including the rise of what has come to be known as cultural imperialism. It also touches on the question of whether U.S. exploitation of the new information age, so well exemplified by Michael Jordan's successes, will lead to a better twenty-first century or, as some believe, a bloody clash of cultures and civilizations.

The analysis uses the sport of basketball to begin examining these subjects. Why basketball? Because with the help of new media the marketing of basketball has become an important fixture in global as well as American culture. Because some of the pioneering transnational corporations are exploiting American sports so profitably and with far-reaching social consequences. Because basketball has attracted both women and men as players as well as spectators since the game first appeared in 1891. And because basketball produced Michael Jordan.

From: LaFeber, Walter. *Michael Jordan and the New Global Capitalism,* 13–24. New York and London: W. W. Norton & Company Ltd., 2002. Reprinted with permission.

This account assumes that the importance of U.S. foreign relations diminished not at all with the Cold War's end after 1989. The nation's overseas influence and power has only become more fascinating . . . and it has become vastly more important for new generations of Americans to understand this.

Just how far that influence reached became clear to Max Perelman, a young American college student, when he traveled through remote regions of China in January 1997. While stranded by winter weather in west Sichuan, a long fifteen hundred miles from Beijing, he encountered a group of Tibetans bound for their capital, Lhasa. The Tibetans, Perelman recalled, had never strayed far from their native village. They had apparently not seen anything like his camera. As they shared with him bites of meat from the raw, bloody, rib cage of an unspecified animal retrieved from their rucksacks, the group began to discuss things American. Just how, one of the Tibetans asked the young American, was Michael Jordan doing?

How these travelers knew about the Chicago Bulls' star was never made clear. That they knew about him, however, was perhaps not surprising. He was the most famous athlete and one of the most recognizable people in the world. Jordan and his "Red Oxen," as his team was known in much of Asia, had gained renown for their basketball championships. But Jordan was especially famous for another reason: he was the superhuman who flew through the air in television advertisements as he endlessly and effortlessly dunked basketballs and, simultaneously, sold Nike sneakers. These glamorous advertisements flew about the globe thanks to new technologies such as earth satellites and cable. . . .

Much of this post-1970s technology was dominated by U.S. empire-builders, notably the flamboyant Ted Turner. He had sunk the family fortune into the emerging business of cable television and communication satellites, only to go nearly bankrupt in the early 1980s. Not long after, however, he built CNN (Cable News Network) into an international as well as American powerhouse. Indeed, the network became so international that Turner outlawed the use of the word "foreign" in its broadcasts. Nothing was foreign to CNN. In 1997, Turner stunned the world by giving one billion dollars over a ten-year period to the United Nations to help its international humanitarian programs. It stood to reason that he gave this incredible gift to the UN instead of to, say, his home city of Atlanta, Georgia. CNN, like Nike and Michael Jordan, had burst beyond mere city boundaries to become a global institution—and had grown rich by moving far beyond U.S. borders to create a worldwide marketplace.

Jordan's role in this growing Americanization of global media was profound. . . . [H]e became a part of the heated argument over whether African-American athletes increasingly dominated basketball and many other sports because they were physically different from and superior to whites, or because they concentrated on these sports due to racism since other careers were closed to them. And eventually, Jordan became a figure who often transcended race. . . .

. . . [B]asketball has always been a commercial product. In the 1980s its ability to churn out profits reached a new level with the appearance of revolutionary technology and imaginative entrepreneurs . . . who were determined to exploit it. Jordan became a world icon, even in far-off Sichuan, in part because the number of television sets for every hundred people around the globe doubled to 23.4 between 1981 and 1997. . . .

Jordan's career also helps us understand something about the nature of U.S. power in the post–Cold-War era. Phil Knight liked to say that by the 1990s sports had become the world's most important entertainment. No one better exemplified the power of that entertainment than Jordan and Knight. American popular culture (the jazz of Duke Ellington, the musical theatre of George Gershwin, the dance of Fred Astaire and Martha Graham, blue jeans, McDonald's fast food, Coca-Cola), has long been part of U.S. influence and profit overseas. The power of that popular culture, however, multiplied with the technological marvels that appeared in the 1960s and 1970s.

In earlier eras, a culture was transmitted across national boundaries by migration, travel, or reading. Since leisure travel and literacy were often limited to the rich, the understanding—and exploitation—of other cultures was often enjoyed only by elites. Television and the post-1970s media, along with cheaper and more rapid transportation via jet airplanes, changed all that. Culture could move with nearly the speed of sound and reach billions of people, not just the privileged. Jordan and Nike (and McDonald's and Disney), suddenly enjoyed the power to reach vast audiences with an efficiency unimagined several generations earlier.

Jordan and Nike, moreover, exploited yet another kind of power new to the post-1970s media. For centuries, the control over mass dispersal of information was held in the hands of monarchs, the Church, or, more recently, powerful newspaper and radio owners. After the 1970s, however, this power to spread information and culture became more decentralized. Masses of people could pass on information in large globs over computer systems. When only three major U.S. television networks existed, as in the 1950s, the network owners generally controlled what people could see. With 70, 150, or even 500 channels, audiences enjoyed much wider choices. Thus Jordan and Nike could select certain channels (MTV) to target young buyers of sneakers, or use other channels (ESPN) watched by sports fans. . . .

It was an awesome power. Transnational corporations not only played a dominant role in creating and defining American popular culture, but they used that culture's own seductiveness to influence the language, eating habits, clothes, and television watching of peoples around the earth. . . .

Growing resistance to the power of American popular culture led to an intense debate over whether the United States was actually an imperialist spreading its culture so effectively that it was radically changing, if not potentially destroying, other cultures. Some of these observers believed that Americans fooled themselves if they thought other peoples would change their traditional way of living just to enjoy U.S. products. Indeed, some argued that as Americans went abroad to spread their culture and fatten their pocketbooks, they would instead have to change their own culture. That is, they would have to become less nationalistic, less ignorant of and more open to other cultures and religions.

The new global commercial power exemplified by Michael Jordan, Nike, CNN, in other words, is making Americans fear that as they are electronically interspersed into the world community, they are threatened by the loss of their national identity—they are a people becoming too "multicultural" and sympathetic to the global power of groups like the United Nations—just as other peoples begin to eat and dress like Americans. Such fear moved into American political and economic debates during the 1990s. . . . As revolutionary technology thus integrated Americans into the rest of the world, many of them feared the strangeness and challenges that they encountered. Americans had feared the strangeness and challenges of other peoples since the seventeenth century, but never before had such dangers been so instantaneous, so immediate, as they were in the new, tightly wired world. . . .

In this developing battle of capital versus culture, capital will ultimately win. The United States is, and has been since World War I, the world's clearinghouse for capital. By the 1990s, the volume of that capital became overwhelming; some $1.5 trillion moved through New York City financial markets every *day.* This torrent of money developed the new media and powered the new transnational corporations. For good or ill, it wielded the power to bring other governments nearly to their knees during the recurring financial crises of the 1990s. It even forced the world's superpower, the U.S. government, to change social priorities and spending policies. Other nations, such as France and Japan, do not necessarily favor this kind of fundamental change and will certainly resist such power. American culture, if powered by vast sums of capital, will thus change as it becomes more global or else produce conflict that will have explosive results for U.S. politics and security.

The story of how the United States and the world reached this point begins in the 1890s, when the American economy first became the world's greatest, and when basketball was first invented. The history of basketball, especially in the era of Michael Jordan, helps us understand this era known as "the American Century." . . .

It was a clash of capitalisms and cultures of the most dramatic, and important, kind. Then it turned worse. Under tremendous U.S. pressure, 102 nations, led by Japan (the world's second largest economy), agreed to open their financial markets to foreign investors. Suddenly U.S. firms began buying up or controlling Asian firms that had long been protected from foreign influence. The most important American economic official, Alan Greenspan, Chairman of the Federal Reserve system, announced that these other nations were finally seeing the light; they were moving toward "the type of market system which we have in this country." A century of U.S. economic power apparently climaxed with an ultimate triumph. Others, however, were not so sure. Anti-American feelings rose in Japan, Malaysia, Indonesia, and elsewhere as these nations tried to protect themselves from U.S. capitalism's cultural backriders. Otherwise, as one American reporter wrote from Japan, it would only be a matter of time before an Asian family would take cash from their corner U.S. bank, "drive off to Walmart and fill the trunk of their Ford with the likes of Fritos and Snickers," then stop at the American-owned movie theater to see the latest Disney film before returning home to check their U.S. mutual fund accounts and America Online (on their IBM computer with Microsoft software).

Asians see this as nothing less than the U.S. "desire to bury Asian values," and they are not pleased. Nor are many Americans. . . .

At the center of this discussion arises one all-important question. Given Americans' increased dependence on world markets for jobs, given how the new technology is locking Americans into a sometimes violent global community that too easily resorts to terrorism to fight the United States, and given that Americans have no choice but to be participants in that complex, often threatening global community—what kind of participants will Americans be? They wield immense power, and unless that power is accompanied by an understanding of its effects and how it came to be, the twenty-first century will be a continuation of the confrontations and bloodshed of the twentieth century. But with new technologies the clashes will occur in a confined, interlinked global village from which no one can escape to safety.

On January 13, 1999, Jordan announced his retirement from basketball. . . .

The impact was indeed global. JORDAN RETIRES! SHOCK FELT AROUND THE WORLD, a Japanese sportspaper headlined. Basketball was a minor sport in Japan, but thanks to television ads, Air Jordan Nike sneakers had sold for as much as $1,000 a pair and some were collected like jewels. Mexican, Polish, German, Spanish, Chinese, and British headlines, among others, echoed Japanese feelings. Standing at the end of a century in which the United States had come to command global financial power, communications systems, marketing networks, and cutting-edge technologies, Jordan exemplified this imperial control—and also some of the explosively dangerous challenges and high costs Americans now confronted in the newly wired world's new century. . . .

DISCUSSION QUESTIONS

1. Why is Michael Jordan a good example of global capitalism? What is specific about Michael Jordan that makes his international presence so significant?

2. What are the positive consequences of America's global reach? What are the negative consequences of America's presence in nations around the world?

http://infotrac.thomsonlearning.com

InfoTrac College Edition

BONUS READING

Sen, Amartya. "How to Judge Globalism." *The American Prospect* 13 (January 2002): 2–6.

In this reading, Amartya Sen discusses the downside of globalism. While many argue that global westernization brings greater knowledge and higher living standards to poorer nations around the world, Sen argues that the world's poorest people are suffering. How does the international market system influence public policies on issues affecting the poor? What issues or problems are not being addressed by a global market?

SEARCH TERMS

global capitalism
international division of labor
third-world poverty
globalization and sport
globalization and the poor

28

The Souls of Black Folk

W. E. B. DU BOIS

W. E. B. Du Bois, the first African American Ph. D. from Harvard University, is a classic sociological analyst. In this well-known essay, he develops the idea that African Americans have a "double consciousness"—one that they must develop as a protective strategy to understand how Whites see them. Originally written in 1903, Du Bois also reflects on the long struggle for Black freedom.

Between me and the other world there is ever an unasked question: unasked by some through feelings of delicacy; by others through the difficulty of rightly framing it. All, nevertheless, flutter round it. They approach me in a half-hesitant sort of way, eye me curiously or compassionately, and then, instead of saying directly, How does it feel to be a problem? they say, I know an excellent colored man in my town; or, I fought at Mechanicsville; or, Do not these Southern outrages make your blood boil? At these I smile, or am interested, or reduce the boiling to a simmer as the occasion may require. To the real question, How does it feel to be a problem? I answer seldom a word. . . .

After the Egyptian and Indian, the Greek and Roman, the Teuton and Mongolian, the Negro is a sort of seventh son, born with a veil, and gifted with second-sight in this American world,—a world which yields him no true self-consciousness, but only lets him see himself through the revelation of the other world. It is a peculiar sensation, this double-consciousness, this sense of always looking at one's self through the eyes of others, of measuring one's soul by the tape of a world that looks on in amused contempt and pity. One ever feels his twoness,—an American, a Negro; two souls, two thoughts, two unreconciled strivings: two warring ideals in one dark body, whose dogged strength alone keeps it from being torn asunder.

The history of the American Negro is the history of this strife—this longing to attain self-conscious manhood, to merge his double self into a better and truer self. In this merging he wishes neither of the older selves to be lost. He would not Africanize America, for America has too much to teach the world and Africa. He would not bleach his Negro soul in a flood of white Americanism, for he knows that Negro blood has a message for the world. He simply wishes to make it possible for a man to be both a Negro and an American, without being cursed and spit upon by his fellows, without having the doors of Opportunity closed roughly in his face.

This, then, is the end of his striving: to be a co-worker in the kingdom of culture, to escape both death and isolation, to husband and use his best powers and his

From: W. E. B. Du Bois. 1989. *The Souls of Black Folk,* edited and with an introduction by Donald B. Gibson. New York: Penguin, pp. 3–12.

latent genius. These powers of body and mind have in the past been strangely wasted, dispersed, or forgotten. The shadow of a mighty Negro past flits through the tale of Ethiopia the Shadowy and of Egypt the Sphinx. Throughout history, the powers of single black men flash here and there like falling stars, and die sometimes before the world has rightly gauged their brightness. Here in America, in the few days since Emancipation, the black man's turning hither and thither in hesitant and doubtful striving has often made his very strength to lose effectiveness, to seem like absence of power, like weakness. And yet it is not weakness—it is the contradiction of double aims. The double-aimed struggle of the black artisan—on the one hand to escape white contempt for a nation of mere hewers of wood and drawers of water, and on the other hand to plough and nail and dig for a poverty-stricken horde—could only result in making him a poor craftsman, for he had but half a heart in either cause. By the poverty and ignorance of his people, the Negro minister or doctor was tempted toward quackery and demagogy; and by the criticism of the other world, toward ideals that made him ashamed of his lowly tasks. The would-be black *savant* was confronted by the paradox that the knowledge people needed was a twice-told tale to his white neighbors, while the knowledge which would teach the white world was Greek to his own flesh and blood. The innate love of harmony and beauty that set the ruder souls of his people a-dancing and a-singing raised but confusion and doubt in the soul of the black artist; for the beauty revealed to him was the soul-beauty of a race which his larger audience despised, and he could not articulate the message of another people. This waste of double aims, this seeking to satisfy two unreconciled ideals, has wrought sad havoc with the courage and faith and deeds of ten thousand of thousands people,—has sent them often wooing false gods and invoking false means of salvation, and at times has even seemed about to make them ashamed of themselves. . . .

The Nation has not yet found peace from its sins; the freedman has not yet found in freedom his promised land. Whatever of good may have come in these years of change, the shadow of a deep disappointment rests upon the Negro people—a disappointment all the more bitter because the unattained ideal was unbounded save by the simple ignorance of a lowly people. . . .

. . . Merely a concrete test of the underlying principles of the great republic is the Negro Problem, and the spiritual striving of the freedmen's sons is the travail of souls whose burden is almost beyond the measure of their strength, but who bear it in the name of an historic race, in the name of this the land of their fathers' fathers and in the name of human opportunity.

DISCUSSION QUESTIONS

1. What does Du Bois mean by "double consciousness" and how does this affect how African American people see themselves and others?

2. In the contemporary world, what examples do you see that Black people are still defined as "a problem," as Du Bois notes? How does this affect the Black experience?

29

Seeing More than Black and White

ELIZABETH MARTINEZ

Racism in the United States is more than simply a matter of Black and White. It involves intense racism about other groups as well, especially Latinos. Elizabeth Martinez advocates a broader framework involving many groups for studying and combating racism.

The racial and ethnic landscape has changed too much in recent years to view it with the same eyes as before. We are looking at a multi-dimensional reality in which race, ethnicity, nationality, culture and immigrant status come together with breathtakingly new results. We are also seeing global changes that have a massive impact on our domestic situation, especially the economy and labor force. For a group of Korean restaurant entrepreneurs to hire Mexican cooks to prepare Chinese dishes for mainly African-American customers, as happened in Houston, Texas, has ceased to be unusual.

The ever-changing demographic landscape compels those struggling against racism and for a transformed, non-capitalist society to resolve several strategic questions. Among them: doesn't the exclusively Black-white framework discourage the perception of common interests among people of color and thus sustain White Supremacy? Doesn't the view that only African Americans face serious institutionalized racism isolate them from potential allies? Doesn't the Black-white model encourage people of color to spend too much energy understanding our lives in relation to whiteness, obsessing about what white society will think and do?

That tendency is inevitable in some ways: the locus of power over our lives has long been white (although big shifts have recently taken place in the color of capital, as we see in Japan, Singapore and elsewhere). The oppressed have always survived by becoming experts on the oppressor's ways. But that can become a prison of sorts, a trap of compulsive vigilance. Let us liberate ourselves, then, from the tunnel vision of whiteness and behold the many colors around us! Let us summon the courage to reject outdated ideas and stretch our imaginations into the next century.

For a Latina to urge recognizing a variety of racist models is not, and should not be, yet another round in the Oppression Olympics. We don't need more competition among different social groups for the gold medal of "Most Oppressed." We don't need more comparisons of suffering between women and

From: Martinez, Elizabeth. "Seeing More than Black and White." In *De Colores Means All of Us: Latina Views for a Multi-Colored Century*. Cambridge, MA: South End Press, 1998. Reprinted by permission of the South End Press.

Blacks, the disabled and the gay, Latino teenagers and white seniors, or whatever. Pursuing some hierarchy of oppression leads us down dead-end streets where we will never find the linkage between different oppressions and how to overcome them. To criticize the exclusively Black-white framework, then, is not some resentful demand by other people of color for equal sympathy, equal funding, equal clout, equal patronage or other questionable crumbs. Above all, it is not a devious way of minimizing the centrality of the African-American experience in any analysis of racism.

The goal in re-examining the Black-white framework is to find an effective strategy for vanquishing an evil that has expanded rather than diminished. Racism has expanded partly as a result of the worldwide economic recession that followed the end of the post-war boom in the early 1970s, with the resulting capitalist restructuring and changes in the international division of labor. Those developments generated feelings of insecurity and a search for scapegoats. In the United States racism has also escalated as whites increasingly fear becoming a weakened, minority population in the next century. The stage is set for decades of ever more vicious divide-and-conquer tactics.

What has been the response from people of color to this ugly White Supremacist agenda? Instead of uniting, based on common experience and needs, we have often closed our doors in a defensive, isolationist mode, each community on its own. A fire of fear and distrust begins to crackle, threatening to consume us all. Building solidarity among people of color is more necessary than ever—but the exclusively Black-white definition of racism makes such solidarity more difficult than ever.

We urgently need twenty-first-century thinking that will move us beyond the Black-white framework without negating its historical role in the construction of U.S. racism. We need a better understanding of how racism developed both similarly and differently for various peoples, according to whether they experienced genocide, enslavement, colonization or some other structure of oppression. At stake is the building of a united anti-racist force strong enough to resist White Supremacist strategies of divide-and-conquer and move forward toward social justice for all. . . .

. . . African Americans have reason to be uneasy about where they, as a people, will find themselves politically, economically and socially with the rapid numerical growth of other folk of color. The issue is not just possible job loss, a real question that does need to be faced honestly. There is also a feeling that after centuries of fighting for simple recognition as human beings, Blacks will be shoved to the back of history again (like the back of the bus). Whether these fears are real or not, uneasiness exists and can lead to resentment when there's talk about a new model of race relations. So let me repeat: in speaking here of the need to move beyond the bipolar concept, the goal is to clear the way for stronger unity against White Supremacy. The goal is to identify our commonalities of experience and needs so we can build alliances.

The commonalities begin with history, which reveals that again and again peoples of color have had one experience in common: European colonization and/or neo-colonialism with its accompanying exploitation. This is true for all indigenous peoples, including Hawaiians. It is true for all Latino peoples, who were invaded and ruled by Spain or Portugal. It is true for people in Africa, Asia and the Pacific Islands, where European powers became the colonizers. People of color were victimized by colonialism not only externally but also through internalized racism—the "colonized mentality."

Flowing from this shared history are our contemporary commonalities. On the poverty scale, African Americans and Native Americans have always been at the bottom, with Latinos nearby. In 1995 the U.S. Census found that Latinos have the highest poverty rate, 24 percent. Segregation may have been legally abolished in the 1960s, but now the United States is rapidly moving toward resegregation as a result of whites moving to the suburbs. This leaves people of color—especially Blacks and Latinos—with inner cities that lack an adequate tax base and thus have inadequate schools. Not surprisingly, Blacks and Latinos finish college at a far lower rate than whites. In other words, the victims of U.S. social ills come in more than one color. Doesn't that indicate the need for new, inclusive models for fighting racism? Doesn't that speak to the absolutely urgent need for alliances among peoples of color?

With greater solidarity, justice for people of color could be won. And an even bigger prize would be possible: a U.S. society that advances beyond "equality," beyond granting people of color a respect equal to that given to Euro-Americans. Too often "equality" leaves whites still at the center, still embodying the Americanness by which others are judged, still defining the national character. . . .

. . . Innumerable statistics, reports and daily incidents should make it impossible to exclude Latinos and other non-Black populations of color when racism is discussed, but they don't. Police killings, hate crimes by racist individuals and murders with impunity by border officials should make it impossible, but they don't. With chilling regularity, ranch owners compel migrant workers, usually Mexican, to repay the cost of smuggling them into the United States by laboring the rest of their lives for free. The 45 Latino and Thai garment workers locked up in an El Monte, California, factory, working 18 hours a day seven days a week for $299 a month, can also be considered slaves (and one must ask why it took three years for the Immigration and Naturalization Service to act on its own reports about this horror) (*San Francisco Examiner,* August 8, 1995). Abusive treatment of migrant workers can be found all over the United States. In Jackson Hole, Wyoming, for example, police and federal agents rounded up 150 Latino workers in 1997, inked numbers on their arms and hauled them off to jail in patrol cars and a horse trailer full of manure (*Los Angeles Times,* September 6, 1997).

These experiences cannot be attributed to xenophobia, cultural prejudice or some other, less repellent term than racism. Take the case of two small Latino children in San Francisco who were found in 1997 covered from head to toe with flour. They explained they had hoped to make their skin white enough for school. There is no way to understand their action except as the result of fear in the racist climate that accompanied passage of Proposition 187, which denies schooling to the children of undocumented immigrants. Another example: Mexican and Chicana women working at a Nabisco plant in Oxnard, California, were not allowed to take bathroom breaks from the assembly line and were told to wear diapers instead. Can we really imagine white workers being treated that way? (The Nabisco women did file a suit and won, in 1997.)

No "model minority" myth protects Asians and Asian Americans from hate crimes, police brutality, immigrant-bashing, stereotyping and everyday racist prejudice. Scapegoating can even take their lives, as happened with the murder of Vincent Chin in Detroit some years ago. . . .

WHY THE BLACK-WHITE MODEL?

A bipolar model of racism has never been really accurate for the United States. Early in this nation's history, Benjamin Franklin perceived a tri-racial society based on skin color—"the lovely white" (Franklin's words), the Black, and the "tawny," as Ron Takaki tells us in *Iron Cages*. But this concept changed as capital's need for labor intensified in the new nation and came to focus on African slave labor. The "tawny" were decimated or forcibly exiled to distant areas; Mexicans were not yet available to be the main labor force. As enslaved Africans became the crucial labor force for the primitive accumulation of capital, they also served as the foundation for the very idea of whiteness—based on the concept of blackness as inferior.

Three other reasons for the Black-white framework seem obvious: numbers, geography and history. African Americans have long been the largest population of color in the United States; only recently has this begun to change. Also, African Americans have long been found in sizable numbers in most parts of the United States, including major cities, which has not been true of Latinos until recent times. Historically, the Black-white relationship has been entrenched in the nation's collective memory for some 300 years—whereas it is only 150 years since the United States seized half of Mexico and incorporated those lands and their peoples. Slavery and the struggle to end it formed a central theme in this country's only civil war—a prolonged, momentous conflict. Above all, enslaved Africans in the United States and African Americans have created an unmatched heritage of massive, persistent, dramatic and infinitely courageous resistance, with individual leaders of worldwide note.

We also find sociological and psychological explanations of the Black-white model's persistence. From the days of Jefferson onward, Native Americans, Mexicans and later the Asian/Pacific Islanders did not seem as much a threat to racial purity or as capable of arousing white sexual anxieties as did Blacks. A major reason for this must have been Anglo ambiguity about who could be called white. Most of the Mexican *ranchero* elite in California had welcomed the U.S. takeover, and Mexicans were partly European—therefore "semi-civilized"; this allowed Anglos to see them as white, unlike lower-class Mexicans. For years Mexicans were legally white, and even today we hear the ambiguous U.S. Census term "Non-Hispanic Whites."

Like Latinos, Asian Americans have also been officially counted as white in some historical periods. They have been defined as "colored" in others, with

"Chinese" being yet another category. Like Mexicans, they were often seen as not really white but not quite Black either. Such ambiguity tended to put Asian Americans along with Latinos outside the prevailing framework of racism.

Blacks, on the other hand, were not defined as white, could rarely become upper-class and maintained an almost constant rebelliousness. Contemporary Black rebellion has been urban: right in the Man's face, scary. Mexicans, by contrast, have lived primarily in rural areas until a few decades ago and "have no Mau-Mau image," as one Black friend said, even when protesting injustice energetically. Only the nineteenth-century resistance heroes labeled "bandits" stirred white fear, and that was along the border, a limited area. Latino stereotypes are mostly silly: snoozing next to a cactus, eating greasy food, always being late and disorganized, rolling big Carmen Miranda eyes, shrugging with self-deprecation "me no speek good eengleesh." In other words, *not serious.* This view may be altered today by stereotypes of the gangbanger, criminal or dirty immigrant, but the prevailing image of Latinos remains that of a debased white, at best. . . .

Among other important reasons for the exclusively Black–white model, sheer ignorance leaps to mind. The oppression and exploitation of Latinos (like Asians) have historical roots unknown to most Americans. People who learn at least a little about Black slavery remain totally ignorant about how the United States seized half of Mexico or how it has colonized Puerto Rico. . . .

One other important reason for the bipolar model of racism is the stubborn self-centeredness of U.S. political culture. It has meant that the nation lacks any global vision other than relations of domination. In particular, the United States refuses to see itself as one among some 20 countries in a hemisphere whose dominant languages are Spanish and Portuguese, not English. It has only a big yawn of contempt or at best indifference for the people, languages and issues of Latin America. It arrogantly took for itself alone the name of half the western hemisphere, America, as was its "Manifest Destiny," of course.

So Mexico may be nice for a vacation and lots of Yankees like tacos, but the political image of Latin America combines incompetence with absurdity, fat corrupt dictators with endless siestas. Similar attitudes extend to Latinos within the United States. My parents, both Spanish teachers, endured decades of being told that students were better off learning French or German. The mass media complain that "people can't relate to Hispanics (or Asians)." It takes mysterious masked rebels, a beautiful young murdered singer or salsa outselling ketchup for the Anglo world to take notice of Latinos. If there weren't a mushrooming, billion-dollar "Hispanic" market to be wooed, the Anglo world might still not know we exist. No wonder that racial paradigm sees only two poles.

The exclusively Black–white framework is also sustained by the "model minority" myth, because it distances Asian Americans from other victims of racism. Portraying Asian Americans as people who work hard, study hard, obey the established order and therefore prosper, the myth in effect admonishes Blacks and Latinos: "See, anyone can make it in this society if you try hard enough. The poverty and prejudice you face are all *your* fault."

The "model" label has been a wedge separating Asian Americans from others of color by denying their commonalities. It creates a sort of racial bourgeoisie, which White Supremacy uses to keep Asian Americans from joining forces with the poor, the homeless and criminalized youth. People then see Asian Americans as a special class of yuppie: young, single, college-educated, on the white-collar track—and they like to shop for fun. Here is a dandy minority group, ready to be used against others.

The stereotype of Asian Americans as whiz kids is also enraging because it hides so many harsh truths about the impoverishment, oppression and racist treatment they experience. Some do come from middle- or upper-class families in Asia, some do attain middle-class or higher status in the U.S., and their community must deal with the reality of class privilege where it exists. But the hidden truths include the poverty of many Asian/Pacific Islander groups, especially women, who often work under intolerable conditions, as in the sweatshops. . . .

THE DEVILS OF DUALISM

Yet another cause of the persistent Black-white conception of racism is dualism, the philosophy that sees all life as consisting of two irreducible elements. Those elements are usually oppositional, like good and evil, mind and body, civilized and savage. Dualism allowed the invaders, colonizers and enslavers of today's United States to rationalize their actions by stratifying supposed opposites along race, color or gender lines. So mind is European, male and rational; body is colored, female and emotional. Dozens of other such pairs can be found, with their clear implications of superior-inferior. In the arena of race, this society's dualism has long maintained that if a person is not totally white (whatever that can mean biologically), he or she must be considered Black. . . .

Racism evolves; our models must also evolve. Today's challenge is to move beyond the Black-white dualism that has served as the foundation of White Supremacy. In taking up this challenge, we have to proceed with both boldness and infinite care. Talking race in these United States is an intellectual minefield; for every observation, one can find three contradictions and four necessary qualifications from five different racial groups. Making your way through that complexity, you have to think: keep your eyes on the prize.

DISCUSSION QUESTIONS

1. Think of four examples of subtle White-to-Black racism. Now think of four examples of subtle White-to-Latino/Latina racism. What are the similarities and what are the differences?

2. The author advocates going beyond a White-Black paradigm of racism. Describe this paradigm briefly and suggest names for a new racial paradigm.

30

Immigrant America
Who They Are and Why They Come

ALEJANDRO PORTES AND RUBÉN RUMBAUT

In this essay, Alejandro Portes and Rubén Rumbaut convey the varied yet similar experiences of some of the "new" immigrants of color (Mexican, South Vietnamese, Cuban, and East Indian) to the United States. Their experiences after arrival are seen as dependent upon their own socioeconomic successes, occupations, and intense personal ambition. They are ethnically a diverse population, and these new immigrants face discrimination and hardship upon their arrival.

In Guadalajara, Juan Manuel Fernández worked as a mechanic in his uncle's repair shop making the equivalent of $150 per month. At thirty-two and after ten years on the job, he decided it was time to go into business on his own. The family, his uncle included, was willing to help, but capital for the new venture was scarce. Luisa, Juan's wife, owned a small corner grocery shop; when money ran out at the end of the month, she often fed the family off the store's shelves. The store was enough to sustain her and her children but not to capitalize her husband's project. For a while, it looked as if Juan would remain a worker for life.

Today Juan owns his own auto repair shop, where he employs three other mechanics, two Mexicans and a Salvadoran. The shop is not in Guadalajara, however, but in Gary, Indiana. The entire family—Luisa, the two children, and a brother—have resettled there. Luisa does not work any longer because she does not speak English and because income from her husband's business is enough to support the family. The children attend school and already speak better English than their parents. They resist the idea of going back to Mexico.

Juan crossed the border on his own near El Paso in 1979. No one stopped him, and he was able to head north toward a few distant cousins and the prospect of a factory job. To his surprise, he found one easily and at the end of four months was getting double the minimum wage in steady employment. Almost every worker in the plant was Mexican, his foreman was Puerto Rican, and the language of work was uniformly Spanish. Three trips from Gary to Guadalajara during the next two years persuaded him that it made much better sense to move his business project north of the border. Guadalajara was teeming with repair shops of all sorts, and competition was fierce. "In Gary," he said,

From: Alejandro Portes and Rubén Rumbaut. 1996. *Immigrant America,* 2nd ed. Berkeley: University of California Press, pp. 1–9. Reprinted with permission.

"many Mexicans would not get their cars fixed because they did not know how to bargain with an American mechanic." Sensing the opportunity, he cut remittances to Mexico and opened a local savings account instead.

During his last trip, the "migra" (border patrol) stopped him shortly after crossing; that required a costly second attempt two days later with a hired "coyote" (smuggler). The incident put a stop to the commuting. Juan started fixing cars out of a shed in front of his barrio home. Word got around that there was a reliable Spanish-speaking mechanic in the neighborhood. In a few months, he was able to rent an abandoned garage, buy some equipment, and eventually hire others. To stay in business, Juan has had to obtain a municipal permit and pay a fee. He pays his workers in cash, however, and neither deducts taxes from their wages nor contributes to Social Security for them. All transactions are informal and, for the most part, in cash.

Juan and Luisa feel a great deal of nostalgia for Mexico, and both firmly intend to return. "In this country, we've been able to move ahead economically, but it is not our own," she says. "The gringos will always consider us inferior." Their savings are not in the bank, as before the shop was rented, but in land in Guadalajara, a small house for his parents, and the goodwill of many relatives who receive periodic remittances. They figure that in ten years they will be able to return, although they worry about their children, who may be thoroughly Americanized by then. A more pressing problem is their lack of "papers" and the constant threat of deportation. Juan has devised ingenious ways to run the business, despite his illegal status, but it is a constant problem. A good part of his recent earnings is in the hands of an immigration lawyer downtown, who has promised to obtain papers for a resident's visa, so far without results.

At age twenty-six, Nguyen Van Tran was a young lieutenant in the army of the Republic of South Vietnam when a strategic retreat order from the ARVN high command quickly turned into the final rout. Nguyen spent three years in communist reeducation camps, all the while attempting to conceal his past as a skilled electronics technician. He finally got aboard a boat bound for Malaysia and after two more years in a refugee camp arrived in Los Angeles in 1980. He had neither family nor friends in the city, but the government provided some resettlement aid and the opportunity to improve his English. At the end of a year, he had secured a job in a local electronics assembly plant, which brought in enough to support himself and his wife and child.

Seeing this plant double in a single year, Nguyen realized the opportunities opening up in electronics. He enrolled in the local community college at night and graduated with an associate degree in computer science. He pooled his savings with another Vietnamese technician and a Chinese engineer and in 1983 launched his own firm. Two years later Integrated Circuits, Inc. employed approximately three hundred workers; most were not Asians, but undocumented Mexican women. By 1985, the company sold about $20 million worth of semiconductors and other equipment to the local IBM plant and other large firms. ICI has even started its own line of IBM-compatible personal computers, the Trantex, which has sold well so far in the local market.

Nguyen, who is chairman of the company, sports a mustache, a sleek Mercedes, and a brand new name, George Best. Perhaps for fear of the "protection gangs" re-created by former Vietnamese policemen in Los Angeles, he has kept a low profile within the Vietnamese community. The name change is part of this approach. "Mr. Best" is not particularly nationalistic, nor does he dream of returning to Vietnam. He attributes his remarkable five-year ascent to hard work and a willingness to take risks. To underline the point, he has hung a large portrait of himself in his community college graduation gown behind his oversized desk. He and his wife are already U.S. citizens. They vote Republican, and he has recently joined the local chamber of commerce.

Lilia González-Fleites left Cuba at fifteen, sent alone by her formerly wealthy parents, who remained behind. The Catholic Welfare Agency received her in Miami, and she went to live with other refugee children in an orphanage in Kendall, Florida, until released to an aunt. She finished high school promptly and married, without her parents' consent, her boyfriend from Cuba, Tomás. There was little work in Miami, and the young couple accepted an offer from the Cuban Refugee Center to resettle them, along with the rest of Tomás's family, in North Carolina. Everyone found work in the tobacco and clothing factories except Lilia, whom Tomás kept at home. At eighteen, the formerly pampered girl found herself a cook and maid for Tomás's entire family.

By sheer luck, the same order of nuns who ran her private school in Havana had a college nearby. Lilia used her school connections to gain admittance with a small scholarship and found herself a part-time job. Those were hard years, working in one city and attending school in another. Tomás and Lilia rarely saw each other because he also decided to return to school while still working.

At age thirty-nine, Lilia is today a successful Miami architect. Divorced from Tomás, she has not remarried, instead pursuing her professional career with single-minded determination. When Cuban refugees finally abandoned their dreams of return, Lilia entered local politics, affiliating with the Republican party. She ran for state office in 1986 but was defeated. Undaunted, she remained active in the party and became increasingly prominent in south Florida political circles. More than an immigrant success story, she sees herself at the beginning of a public career that will bridge the gap between the Anglo and Cuban communities in south Florida. Her unaccented English, fierce loyalty to her adopted country, and ability to shift easily between languages and cultures bodes well for her political future.

After finishing medical school, Amitar Ray confronted the prospect of working *ad honorem* in one of the few well-equipped hospitals in Bombay or moving to a job in the countryside and to quick obsolescence in his career. He opted instead for preparing and taking the Educational Council for Foreign Medical Graduates (ECFMG) examination, administered at the local branch of the Indo-American Cultural Institute. He passed it on his second attempt. In 1972, there was a shortage of doctors in the United States, and U.S. consulates were directed to facilitate the emigration of qualified physicians from abroad.

Amitar and his wife, also a doctor, had little difficulty obtaining permanent residents' visas under the third preference of the U.S. immigration law, reserved for professionals of exceptional ability. He went on to specialize in anesthesiology and completed his residence at a public hospital in Brooklyn. After four years, nostalgia and the hope that things had improved at home moved the Rays to go back to India with their young daughter, Rita. The trip strengthened their professional and family ties, but it also dispelled any doubts as to where their future was. Medical vacancies were rare and paid a fraction of what he earned as a resident in Brooklyn. More important, there were few opportunities to grow professionally because he would have had to combine several part-time jobs to earn a livelihood, leaving little time for study.

At fifty-one, Amitar is now associate professor of anesthesiology at a midwestern medical school; his wife has a local practice as an internist. Their combined income is in the six figures, affording them a very comfortable life-style. Their daughter is a senior at Bryn Mawr, and she plans to pursue a graduate degree in international relations. There are few Indian immigrants in the mid-sized city where the Rays live: thus, they have had to learn local ways in order to gain entry into American social circles. Their color is sometimes a barrier to close contact with white middle-class families, but they have cultivated many friendships among the local faculty and medical community.

Ties to India persist and are strengthened through periodic trips and through the professional help the Rays are able to provide to colleagues back home. They have already sponsored the immigration of two bright young physicians from their native city. More important, they make sure that information on new medical developments is relayed to a few selected specialists back home. However, there is little chance that they will return, even after retirement. Work and new local ties play a role in this, but the decisive factor is a thoroughly Americanized daughter whose present life and future have very little to do with India. Rita does not plan to marry soon; she is interested in Latin American politics, and her current goal is a career in the foreign service.

After a lapse of half a century, the United States has again become a country of immigration. In 1990, the foreign-born population reached 19.8 million or 7.9 percent of the total. Although a far cry from the situation eighty years earlier, when immigrants accounted for 14.7 percent of the American population, the impact of contemporary immigration is both significant and growing. Numerous books and articles have called attention to this revival and sought its causes—first in a booming American economy and second in the liberalized provisions of the 1965 immigration act. A common exercise is to compare this "new" immigration with the "old" inflow at the turn of the century. Similarities include the predominantly urban destination of most newcomers, their concentration in a few port cities, and their willingness to accept the lowest paid jobs. Differences are more frequently stressed, however, for the "old" immigration was overwhelmingly

European and white; but the present inflow is, to a large extent, nonwhite and comes from countries of the Third World.

The public image of contemporary immigration has been colored to a large extent by the Third World origins of most recent arrivals. Because the sending countries are generally poor, many Americans believe that the immigrants themselves are uniformly poor and uneducated. Their move is commonly portrayed as a one-way escape from hunger, want, and persecution and their arrival on U.S. shores as not too different from that of the tired, "huddled masses" that Emma Lazarus immortalized at the base of the Statue of Liberty. The "quality" of the newcomers and their chances for assimilation are sometimes portrayed as worse because of their non-European past and the precarious legal status of many.

The reality is very different. The four previous cases, each a composite of real life experiences, are certainly not representative of all recent immigrants. Clearly, not all newcomers are doctors or skilled mechanics, and fewer still become politicians or millionaires. Still, these are not isolated instances. Underneath its apparent uniformity, contemporary immigration features a bewildering variety of origins, return patterns, and modes of adaptation to American society. Never before has the United States received immigrants from so many countries, from such different social and economic backgrounds, and for so many reasons. Although pre–World War I European immigration was by no means homogeneous, the differences between successive waves of Irish, Italians, Jews, Greeks, and Poles often pale by comparison with the current diversity. For the same reason, theories coined in the wake of the European's arrival at the turn of the century have been made obsolete by events during the last decades.

Increasingly implausible, for example, is the view of a uniform assimilation process that different groups undergo in the course of several generations as a pre-condition for their social and economic advancement. There are today first-generation millionaires who speak broken English, foreign-born mayors of large cities, and top-flight immigrant engineers and scientists in the nation's research centers; there are also those, at the other extreme, who cannot even take the first step toward assimilation because of the insecurity linked to an uncertain legal status.

. . . Many of the countries from which today's immigrants come have one of their largest cities in the United States. Los Angeles' Mexican population is next in size to those of Mexico City, Monterrey, and Guadalajara. Havana is not much larger than Cuban Miami, and Santo Domingo holds a precarious advantage over Dominician New York. This is not the case for all groups; others, such as Asian Indians, Laotians, Argentines, and Brazilians, are more dispersed throughout the country. Reasons for both these differences and other characteristics of contemporary immigrant groups are not well known—in part because of the recency of their arrival and in part because of the common expectation that their assimilation process would conform to the well-known European pattern. But immigrant America is a different place today from the America that emerged out of Ellis Island and grew up in the tenements of New York and Boston. . . .

DISCUSSION QUESTIONS

1. How do the "new" immigrants (primarily immigrants of color during the second half of the twentieth century) differ from the "old" immigrants of the first half of the twentieth century (who were primarily European Whites)? List some differences.

2. How does the social structure of the United States affect new immigrants and their chances for success?

31

Color-Blind Privilege

The Social and Political Functions of Erasing the Color Line in Post Race America

CHARLES A. GALLAGHER

Charles A. Gallagher discusses the problem of a color-blind approach to race and race relations in this country. By denying race as a structural basis for inequality, we fail to recognize the privilege of Whiteness. With the blurring of racial lines, White college students lack a clear understanding of how the existing social, political, and economic systems advantage or privilege Whites.

INTRODUCTION

An adolescent white male at a bar mitzah wears a FUBU shirt while his white friend preens his tightly set, perfectly braided corn rows. A black model dressed in yachting attire peddles a New England yuppie boating look in Nautica advertisements. It is quite unremarkable to observe white, Asian or African-Americans with dyed purple, blond or red hair. White, black and Asian students decorate their bodies with tattoos of Chinese characters and symbols. In

From: Gallagher, Charles A. "Color-Blind Privilege: The Social and Political Functions of Erasing the Color Line in Post Race America." *Race, Gender and Class* (June 2003). Reprinted with permission of the author.

cities and suburbs young adults across the color line wear hip-hop clothing and listen to white rapper Eminem and black rapper Jay-Z. A north Georgia branch of the NAACP installs a white biology professor as its president. The music of Jimi Hendrix is used to sell Apple Computers. Du-Rag kits, complete with bandana headscarf and elastic headband, are on sale for $2.95 at hip-hop clothing stores and family centered theme parks like Six Flags. Salsa has replaced ketchup as the best selling condiment in the United States. Companies as diverse as Polo, McDonalds, Tommy Hilfiger, Walt Disney World, Master Card, Skechers sneakers, IBM, Giorgio Armani and Neosporin antibiotic ointment have each crafted advertisements that show a balanced, multiracial cast of characters interacting and consuming their products in a post-race, color-blind world. . . .

Americans are constantly bombarded by depictions of race relations in the media which suggest that discriminatory racial barriers have been dismantled. Social and cultural indicators suggest that America is on the verge, or has already become, a truly color-blind nation. National polling data indicate that a majority of whites now believe discrimination against racial minorities no longer exists. A majority of whites believe that blacks have "as good a chance as whites" in procuring housing and employment or achieving middle class status while a 1995 survey of white adults found that a majority of whites (58%) believed that African Americans were better off finding jobs than whites.[1] Much of white America now see a level playing field, while a majority of black Americans see a field which is still quite uneven. . . . The color-blind or race neutral perspective holds that in an environment where institutional racism and discrimination have been replaced by equal opportunity, one's qualifications, not one's color or ethnicity, should be the mechanism by which upward mobility is achieved. Whites and blacks differ significantly, however, in their support for affirmative action, the perceived fairness of the criminal justice system, the ability to acquire the "American Dream," and the extent to which whites have benefited from past discrimination.[2]

This article examines the social and political functions color-blindness serves for whites in the United States. Drawing on interviews and focus groups with whites from around the country I argue that color-blindness maintains white privilege by negating racial inequality. Embracing post-race, color-blind perspective provides whites with a degree of psychological comfort by allowing them to imagine that being white or black or brown has no bearing on an individual's or a group's relative place in the socio-economic hierarchy. My interviews included seventeen focus group and thirty individual interviews with whites around the country. While my sample is not representative of the total white population, I used personal contacts and snowball sampling to purposively locate respondents raised in urban, suburban and rural environments. Twelve of the seventeen focus groups were conducted in a university setting, one in a liberal arts college in the Rocky Mountains and the other at a large urban university in the Northeast. Respondents in these focus groups were selected randomly from the student population. The occupational range for my individual interviews was quite eclectic and included a butcher, construction worker, hair stylist, partner in a prestigious corporate law firm, executive secretary, high school principal, bank president from a small town, retail workers, country lawyer and custodial workers. Twelve of the thirty individual interviews were with respondents who were raised in rural and/or agrarian settings. The remaining respondents lived in suburbs of large cities or in urban areas.

What linked this rather disparate group of white individuals together was their belief race based privilege had ended. As a majority of my respondents saw it, color-blindness was now the norm in the United States. The illusion of racial equality implicit in the myth of color-blindness was, for many whites, a form of comfort. This aspect of pleasure took the form of political empowerment ("what about whites' rights") and moral gratification from being liberated from "oppressor" charges ("we are not responsible for the past"). The rosy picture that color-blindness presumes about race relations and the satisfying sense that one is part of a period in American history that is morally superior to the racist days of the past is, quite simply, a less stressful and more pleasurable social place for whites to inhabit.

THE NORM OF COLOR-BLINDNESS

The perception among a majority of white Americans that the socio-economic playing field is now level, along with whites' belief that they have purged themselves of overt racist attitudes and behaviors, has made color-blindness the dominant lens through which whites understand contemporary race relations. Color-blindness allows whites to believe that segregation and discrimination are no longer an issue because it is now illegal for individuals to be denied access to housing, public accommodations or jobs because of their race. . . . Individuals from any racial background can wear hip-hop clothing, listen to rap music (both purchased at Wal-Mart) and root for their favorite, majority black, professional sports team. Within the context of racial symbols that are bought and sold in the market, color-blindness means that one's race has no bearing on who can . . . live in an exclusive neighborhood, attend private schools or own a Rolex.

The passive interaction whites have with people of color through the media creates the impression that little, if any, socio-economic difference exists between the races. Research has found that whites who are exposed to images of upper-middle class African Americans . . . believe that blacks have the same socio-economic opportunities as whites. Highly visible and successful racial minorities like Secretary of State Colin Powell and National Security Advisor Condelleeza Rice are further proof to white America that the state's efforts to enforce and promote racial equality has been accomplished. Reflecting on the extent to which discrimination is an obstacle to socio-economic advancement and the perception of seeing African-Americans in leadership roles, Tom explained:

> If you look at some prominent black people in society today and I don't really see [racial discrimination] I don't understand how they can keep bringing this problem onto themselves. If they did what society would want them to I don't see that society is making problems for them. I don't see it.

. . . The new color-blind ideology does not, however, ignore race; it acknowledges race while ignoring racial hierarchy by taking racially coded styles and products and reducing these symbols to commodities or experiences which whites and racial minorities can purchase and share. It is through such acts of shared consumption that race becomes nothing more than an innocuous cultural signifier. Large corporations have made American culture more homogeneous through the ubiquitous of fast food, television, and shopping malls but this trend has also created the illusion that we are all the same through consumption. Most adults eat at national fast food chains like McDonalds, shop at mall anchor stores like Sears and J. C. Penney's and watch major league sports, situation comedies or television drama. Defining race only as cultural symbols that are for sale allows whites to experience and view race as nothing more than a benign cultural marker that has been stripped of all forms of institutional, discriminatory or coercive power. The post-race, color-blind perspective allows whites to imagine that depictions of racial minorities working in high status jobs and consuming the same products, or at least appearing in commercials for products whites desire or consume, is the same as living in a society where color is no longer used to allocate resources or shape group outcomes. By constructing a picture of society where racial harmony is the norm, the color-blind perspective functions to make white privilege invisible while removing from public discussion the need to maintain any social programs that are race-based.

. . . Starting with the deeply held belief that America is now a meritocracy, whites are able to imagine that the socio-economic success they enjoy relative to racial minorities is a function of individual hard work, determination, thrift and investments in education. The color-blind perspective removes from personal thought and public discussion any taint or suggestion of white supremacy or white guilt while legitimating the existing social, political and economic arrangements which privilege whites. This perspective insinuates that class and culture, and not institutional racism, are responsible for social inequality. Color-blindness allows whites to define themselves as politically progressive and racially tolerant as they proclaim their adherence to a belief system that does not see or judge individuals by the "color of their skin." This perspective ignores, as Ruth Frankenberg puts it, how whiteness is a "location of structural advantage societies structured in racial dominance."[3] Frankenberg uses the term "color and power evasiveness" rather than color-blindness to convey how the ability to ignore race by members of the dominant group reflects a position of power and privilege. Color-blindness hides white privilege behind a mask of assumed meritocracy while rendering invisible the institutional arrangements that perpetuate racial inequality. The veneer of equality implied in color-blindness allows whites to present their place in the racialized social structure as one that was earned.

Given this norm of color-blindness it was not surprising that respondents in this study believed that using race to promote group interests was a form of racism.

Joe, a student in his early twenties from a working class background, was quite adamant that the opportunity structure in the United States did not favor one racial group over another.

I mean, I think that the black person of our age has as much opportunity as me, maybe he didn't have the same guidance and that might hurt him. But I mean, he's got the same opportunities that I do to go to school, maybe even more, to get more money. I can't get any aid . . . I think that blacks have the same opportunities as whites nowadays and I think it's old hat.

Not only does Joe believe that young blacks and whites have similar educational experiences and opportunity but it is his contention that blacks are more likely or able to receive money for higher education. The idea that race matters in any way, according to Joe, is anachronistic; it is "old hat" in a color-blind society to blame one's shortcomings on something as irrelevant as race.

Believing and acting as if America is now color-blind allows whites to imagine a society where institutional racism no longer exists and racial barriers to upward mobility have been removed. The use of group identity to challenge the existing racial order by making demands for the amelioration of racial inequities is viewed as racist because such claims violate the belief that we are a nation that recognizes the rights of individuals not rights demanded by groups. Sam, an upper middle class respondent in his 20's, draws on a pre- and post-civil rights framework to explain racial opportunity among his peers:

> I guess I can understand in my parents' generation. My parents are older, my dad is almost 60 and my mother is in her mid 50's, ok? But the kids I'm going to school with, the minorities I'm going to school with, I don't think they should use racism as an excuse for not getting a job. Maybe their parents, sure, I mean they were discriminated against. But these kids have every opportunity that I do to do well.

In one generation, as Sam sees it, the color line has been erased. Like Sam's view that opportunity structure is open there is, according to Tara, a reason to celebrate the current state of race relations:

> I mean, like you are not the only people that have been persecuted—I mean, yea, you have been, but so has every group. I mean if there's any time to be black in America it's now.

Seeing society as race-neutral serves to decouple past historical practices and social conditions from present day racial inequality as was the case for a number of respondents who pointed out that job discrimination had ended. Michelle was quite direct in her perception that the labor market is now free of discrimination stating that "I don't think people hire and fire because someone is black and white now." Ken also believed that discrimination in hiring did not occur since racial minorities now have legal recourse if discrimination occurs:

> I think that pretty much we got past that point as far as jobs. I think people realize that you really can't discriminate that way because you will end up losing . . . because you will have a lawsuit against you.

. . . The logic inherent in the color-blind approach is circular; since race no longer shapes life chances in a color-blind world there is no need to take race into account when discussing differences in outcomes between racial groups.

This approach erases America's racial hierarchy by implying that social, economic and political power and mobility are equally shared among all racial groups. Ignoring the extent or ways in which race shapes life chances validates whites' social location in the existing racial hierarchy while legitimating the political and economic arrangements which perpetuate and reproduce racial inequality and privilege. . . .

THE COST OF RACIALIZED PLEASURES

Being able to ignore or being oblivious to the ways in which almost all whites are privileged in a society cleaved on race has a number of implications. Whites derive pleasure in being told that the current system for allocating resources is fair and equitable. Creating and internalizing a color-blind view of race relations reflects how the dominant group is able to use the mass media, immigration stories of upward mobility, rags-to-riches narratives and achievement ideology to make white privilege invisible. Frankenberg argues that whiteness can be "displaced," as is the case with whiteness hiding behind the veil of color-blindness. It can also be made "normative" rather than specifically "racial," as is the case when being white is defined by white respondents as being no different than being black or Asian.[4] Lawrence Bobo and associates have advanced a theory of laissez-faire racism that draws on the color-blind perspective. As whites embrace the equality of opportunity narrative they suggest that

> laissez-faire racism encompasses an ideology that blames blacks themselves for their poorer relative economic standing, seeing it as a function of perceived cultural inferiority. The analysis of the bases of laissez-faire racism underscores two central components: contemporary stereotypes of blacks held by whites, and the denial of societal (structural) responsibility for the conditions in black communities.[5]

As many of my respondents make clear if the opportunity structure is open ("It doesn't matter what color you are"), there must be something inherently wrong with racial minorities or their culture that explains group level differences.

. . . [T]he form color-blindness takes as the nation's hegemonic political discourse is a variant of laissez-faire racism. Historian David Roediger contends that in order for the Irish to be absorbed into the white race in the mid-nineteenth century "the imperative to define themselves as whites came from the particular 'public and psychological wages' whiteness offered" these new immigrants.[6] There is still a "wage" to whiteness, that element of ascribed status whites automatically receive because of their membership in the dominant group. But within the framework of color-blindness the imperative has switched from whites overtly defining themselves or their interests as white, to one where they claim that color is irrelevant; being white is the same as being black, yellow, brown or red. . . .

My interviews with whites around the country suggest that in this post-race era of color-blind ideology Ellison's keen observations about race relations need modification. The question now is what are we to make of a young white man from the suburbs who listens to hip-hop, wears baggy hip-hop pants, a baseball cap turned sideways, unlaced sneakers and a oversized shirt emblazoned with a famous NBA player who, far from shouting racial epithets, lists a number of racial minorities as his heroes? It is now possible to define oneself as not being racist because of the clothes you wear, the celebrities you like or the music you listen to while believing that blacks or Latinos are disproportionately poor or over-represented in low pay, dead end jobs because they are part of a debased, culturally deficient group. Having a narrative that smooths over the cognitive dissonance and oft time schizophrenic dance that whites must do when they navigate race relations is an invaluable source of pleasure.

NOTES

1. The Gallup Organization, "Black/White Relations in the U.S.," *The Gallup Poll Monthly* (June 10, 1997): 1–5; David Shipler, *A Country of Strangers: Blacks and Whites in America* (New York: Vintage, 1998).

2. David Moore, "Americans' Most Important Sources of Information: Local News," *The Gallup Poll Monthly* (September 1995): 2–5; David Moore and Lydia Saad, "No Immediate Signs that Simpson Trial Intensified Racial Animosity," *The Gallup Poll Monthly* (October 1995): 2–5; Kaiser Foundation, *The Four Americas: Government and Social Policy through the Eyes of America's Multi-Racial and Multi-Ethnic Society* (Menlo Park, CA: Kaiser Family Foundation, 1995).

3. O. Ruth Frankenberg, "The Mirage of an Unmarked Whiteness," in *The Making and Unmaking of Whiteness,* ed. Birget Brander Rasmussen, Eric Klineberg, Irene J. Nexica, and Matt Wray (Durham: Duke University Press, 2001).

4. Ibid., 76.

5. Lawrence Bobo and James R. Kluegel, "Status, Ideology, and Dimensions of Whites' Racial Beliefs and Attitudes: Progress and Stagnation," in *Racial Attitudes in the 1990s: Continuity and Change,* ed. Steven A. Tuch and Jack K. Martin, 95 (Westport, CT: Praeger Publishers, 1997).

6. David Roediger, *The Wages of Whiteness: Race and the Making of the American Working Class,* 137 (New York: Verso Press, 1991).

DISCUSSION QUESTIONS

1. Summarize Gallagher's argument for why a color-blind attitude is still a privileged attitude. What does color–blindness not see when viewing race relations in America?

2. What is the problem with a generation of individuals who do not judge others by the "color of their skin"? Can the individualistic ideology of color-blindness coexist with a society of racist practices?

http://infotrac.thomsonlearning.com

InfoTrac College Edition

BONUS READING

Gurin, Patricia Y., Eric L. Dey, Gerald Gurin, and Sylvia Hurtado. "How Does Racial/Ethnic Diversity Promote Education?" *The Western Journal of Black Studies* 27 (Spring 2003): 20–29.

The authors of this reading examine the importance of the recent Supreme Court decision about affirmative action policies for admission into college and graduate school. They argue that students benefit from an educational environment that is diverse in its student body and campus community. How do they counter the criticisms of their work? What evidence do they provide that diversity on campus promotes a more active learning experience and better prepares students to be citizens in a diverse democracy?

SEARCH TERMS

emancipation and slavery
racial paradigm
assimilation and immigrants
color-blind society
White privilege
affirmative action and college admissions

32

The Social Construction
of Gender

MARGARET L. ANDERSEN

In this essay, Margaret Andersen outlines the meaning of the "social construction of gender." She discusses the difference between the terms "sex" and "gender" and defines sexuality as it relates to both. After a brief discussion of the cultural basis of gender, the essay outlines the difference between a gender roles conceptualization of gender and the gendered institutions approach.

To understand what sociologists mean by the phrase *the social construction of gender,* watch people when they are with young children. "Oh, he's such a boy!" someone might say as he or she watches a 2-year-old child run around a room or shoot various kinds of play guns. "She's so sweet," someone might say while watching a little girl play with her toys. You can also see the social construction of gender by listening to children themselves or watching them play with each other. Boys are more likely to brag and insult other boys (often in joking ways) than are girls; when conflicts arise during children's play, girls are more likely than boys to take action to diffuse the conflict (McCloskey and Coleman, 1992; Miller, Danaber, and Forbes, 1986).

To see the social construction of gender, try to buy a gender-neutral present for a child—that is, one not specifically designed with either boys or girls in mind. You may be surprised how hard this is, since the aisles in toy stores are highly stereotyped by concepts of what boys and girls do and like. Even products such as diapers, kids' shampoos, and bicycles are gender stereotyped. Diapers for boys are packaged in blue boxes; girls' diapers are packaged in pink. Boys wear diapers with blue borders and little animals on them; girls wear diapers with pink borders with flowers. You can continue your observations by thinking about how we describe children's toys. Girls are said to play with dolls; boys play with action figures!

When sociologists refer to the social construction of gender, they are referring to the many different processes by which the expectations associated with being a boy (and later a man) or being a girl (later a woman) are passed on through society. This process pervades society, and it begins the minute a child is born. The exclamation "It's a boy!" or "It's a girl!" in the delivery room sets a course that

From: Margaret L. Andersen. 2003. *Thinking About Women: Sociological Perspectives on Sex and Gender.* Boston, MA: Allyn and Bacon, pp. 19–24. Reprinted with permission.

from that moment on influences multiple facets of a person's life. Indeed, with the modern technologies now used during pregnancy, the social construction of gender can begin even before one is born. Parents or grandparents may buy expected children gifts that reflect different images, depending on whether the child will be a boy or a girl. They may choose names that embed gendered meanings or talk about the expected child in ways that are based on different social stereotypes about how boys and girls behave and what they will become. All of these expectations—communicated through parents, peers, the media, schools, religious organizations, and numerous other facets of society—create a concept of what it means to be a "woman" or be a "man." They deeply influence who we become, what others think of us, and the opportunities and choices available to us. The idea of the social construction of gender sees society, not biological sex differences, as the basis for gender identity. To understand this fully, we first need to understand some of the basic concepts associated with the social construction of gender and review some information about biological sex differences.

SEX, GENDER, AND SEXUALITY

The terms *sex, gender,* and *sexuality* have related, but distinct, meanings within the scholarship on women. Sex refers to the biological identity and is meant to signify the fact that one is either male or female. One's biological sex usually establishes a pattern of gendered expectations, although, . . . biological sex identity is not always the same as gender identity; nor is biological identity always as clear as this definition implies.

Gender is a social, not biological, concept, referring to the entire array of social patterns that we associate with women and men in society. Being "female" and "male" are biological facts; being a woman or a man is a social and cultural process—one that is constructed through the whole array of social, political, economic, and cultural experiences in a given society. Like race and class, gender is a social construct that establishes, in large measure, one's life chances and directs social relations with others. Sociologists typically distinguish sex and gender to emphasize the social and cultural basis of gender, although this distinction is not always so clear as one might imagine, since gender can even construct our concepts of biological sex identity.

Making this picture even more complex, sexuality refers to a whole constellation of sexual behaviors, identities, meaning systems, and institutional practices that constitute sexual experience within society. This is not so simple a concept as it might appear, since sexuality is neither fixed nor unidimensional in the social experience of diverse groups. Furthermore, sexuality is deeply linked to gender relations in society. Here, it is important to understand that sexuality, sex, and gender are intricately linked social and cultural processes that overlap in establishing women's and men's experiences in society.

Fundamental to each of these concepts is understanding the significance of culture. Sociologists and anthropologists define culture as "the set of definitions of reality held in common by people who share a distinctive way of life" (Kluckhohn, 1962:52). Culture is, in essence, a pattern of expectations about what are appropriate behaviors and beliefs for the members of the society; thus, culture provides prescriptions for social behavior. Culture tells us what we ought to do, what we ought to think, who we ought to be, and what we ought to expect of others. . . .

The cultural basis of gender is apparent especially when we look at different cultural contexts. In most Western cultures, people think of *man* and *woman* as dichotomous categories—that is, separate and opposite, with no overlap between the two. Looking at gender from different cultural viewpoints challenges this assumption, however. Many cultures consider there to be three genders, or even more. Consider the Navaho Indians. In traditional Navaho society, the *berdaches* were those who were anatomically normal men but who were defined as a third gender and were considered to be intersexed. Berdaches married other men. The men they married were not themselves considered to be berdaches; they were defined as ordinary men. Nor were the berdaches or the men they married considered to be homosexuals, as they would be judged by contemporary Western culture. . . .

Another good example for understanding the cultural basis of gender is the *hijras* of India (Nanda, 1998). Hijras are a religious community of men in India who are born as males, but they come to think of themselves as neither men nor women. Like berdaches, they are considered a third gender. Hijras dress as women and may marry other men; typically, they live within a communal sub-culture. An important thing to note is that hijras are not born so; they choose this way of life. As male adolescents, they have their penises and testicles cut off in an elaborate and prolonged cultural ritual—a rite of passage marking the transition to becoming a hijra. . . .

These examples are good illustrations of the cultural basis of gender. Even within contemporary U.S. society, so-called "gender bending" shows how the dichotomous thinking that defines men and women as "either/or" can be transformed. Cross-dressers, transvestites, and transsexuals illustrate how fluid gender can be and, if one is willing to challenge social convention, how easily gender can be altered. The cultural expectations associated with gender, however, are strong, as one may witness by people's reactions to those who deviate from presumed gender roles. . . .

In different ways and for a variety of reasons, all cultures use gender as a primary category of social relations. The differences we observe between men and women can be attributed largely to these cultural patterns.

THE INSTITUTIONAL BASIS OF GENDER

Understanding the cultural basis for gender requires putting gender into a sociological context. From a sociological perspective, gender is systematically structured in social institutions, meaning that it is deeply embedded in the social structure of society. Gender is created, not just within family or interpersonal relationships (although these are important sources of gender relations), but also within the structure of all major social institutions, including schools, religion, the economy, and the state (i.e., government and other organized systems of authority such as the police and the military). These institutions shape and mold the experiences of us all.

Sociologists define institutions as established patterns of behavior with a particular and recognized purpose; institutions include specific participants who share expectations and act in specific roles, with rights and duties attached to them. Institutions define reality for us insofar as they exist as objective entities in our experience. . . .

Understanding gender in an institutional context means that gender is not just an attribute of individuals; instead, institutions themselves are *gendered*. To say that an institution is gendered means that the whole institution is patterned on specific gendered relationships. That is, gender is "present in the processes, practices, images and ideologies, and distribution of power in the various sectors of social life" (Acker, 1992:567). The concept of a gendered institution was introduced by Joan Acker, a feminist sociologist. Acker uses this concept to explain not just that gender expectations are passed to men and women within institutions, but that the institutions themselves are structured along gendered lines. Gendered institutions are the total pattern of gender relations—stereotypical expectations, interpersonal relationships, and men's and women's different placements in social, economic, and political hierarchies. This is what interests sociologists, and it is what they mean by the social structure of gender relations in society.

Conceptualizing gender in this way is somewhat different from the related concept of gender roles. Sociologists use the concept of social roles to refer to culturally prescribed expectations, duties, and rights that define the relationship between a person in a particular position and the other people with whom she or he interacts. For example, to be a mother is a specific social role with a definable set of expectations, rights, and duties. Persons occupy multiple roles in society; we can think of social roles as linking individuals to social structures. It is through social roles that cultural norms are patterned and learned. Gender roles are the expectations for behavior and attitudes that the culture defines as appropriate for women and men.

The concept of gender is broader than the concept of gender roles. *Gender* refers to the complex social, political, economic, and psychological relations between women and men in society. Gender is part of the social structure—in other words, it is institutionalized in society. *Gender roles* are the patterns through which gender relations are expressed, but our understanding of gender in society cannot be reduced to roles and learned expectations.

The distinction between gender as institutionalized and gender roles is perhaps most clear in thinking about analogous cases—specifically, race and class. Race relations in society are seldom, if ever, thought of in terms of "race roles." Likewise, class inequality is not discussed in terms of "class roles." Doing so would make race and class inequality seem like matters of interpersonal interaction. Although race, class, and gender inequalities are experienced within interpersonal interactions, limiting the analysis of race, class, or gender relations to this level of social interaction individualizes more complex systems of inequality; moreover, restricting the analysis of race, class, or gender to social roles hides the power relations that are embedded in race, class, and gender inequality (Lopata and Thorne, 1978).

Understanding the institutional basis of gender also underscores the interrelationships of gender, race, and class, since all three are part of the institutional framework of society. As a social category, gender intersects with class and race; thus, gender is manifested in different ways, depending on one's location in the race and class system. For example, African American women are more likely than White women to reject gender stereotypes for women, although they are more accepting than White women of stereotypical gender roles for children. Although this seems contradictory, it can be explained by understanding that African American women may reject the dominant culture's view while also hoping their children can attain some of the privileges of the dominant group (Dugger, 1988).

Institutional analyses of gender emphasize that gender, like race and class, is a part of the social experience of us all—not just of women. Gender is just as important in the formation of men's experiences as it is in women's (Messner, 1998). From a sociological perspective, class, race, and gender relations are systemically structured in social institutions, meaning that class, race, and gender relations shape the experiences of all. Sociologists do not see gender simply as a psychological attribute, although that is one dimension of gender relations in society. In addition to the psychological significance of gender, gender relations are part of the institutionalized patterns in society. Understanding gender, as well as class and race, is central to the study of any social institution or situation. Understanding gender in terms of social structure indicates that social change is not just a matter of individual will—that if we changed our minds, gender would disappear. Transformation of gender inequality requires change both in consciousness and in social institutions. . . .

REFERENCES

Acker, Joan. 1992. "Gendered Institutions: From Sex Roles to Gendered Institutions." *Contemporary Sociology* 21 (September): 565–569.

Dugger, Karen. 1988. "The Social Location of Black and White Women's Attitudes." *Gender & Society* 2 (December): 425–448.

Kluckhohn, C. 1962. *Culture and Behavior.* New York: Free Press.

Lopata, Helene Z., and Barrie Thorne. 1978. "On the Term 'Sex Roles.' " *Signs* 3 (Spring): 718–721.

McCloskey, Laura A., and Lerita M. Coleman. 1992. "Difference Without Dominance: Children's Talk in Mixed- and Same-Sex Dyads." *Sex Roles* 27 (September): 241–258.

Messner, Michael A. 1998. "The Limits of 'The Male Sex Role': An Analysis of the Men's Liberation and Men's Rights Movements' Discourse." *Gender & Society* 12 (June): 255–276.

Miller, D., D. Danaber, and D. Forbes. 1986. "Sex-related Strategies for Coping with Interpersonal Conflict in Children Five and Seven." *Developmental Psychology* 22: 543–548.

Nanda, Serena. 1998. *Neither Man Nor Woman: The Hijras of India.* Belmont, CA: Wadsworth.

DISCUSSION QUESTIONS

1. Walk through a baby store. Can you easily identify products for girls and for boys? Could you easily purchase clothing appropriate for either a boy or a girl?

2. Consider an occupation that is traditionally men's work or traditionally women's work. What happens when a member of the opposite sex works in that field? What stereotypes and derogatory assumptions do we make about a woman working in a man's occupation or a man working in a woman's occupation?

33

The Politics of Masculinities

MICHAEL A. MESSNER

Michael Messner uses personal accounts to illustrate three important considerations in the study of masculinity. Specifically, he discusses the institutionalized nature of male privilege that gives men as a group power over women as a group. Next, he argues that the limiting definition of masculinity in U.S. culture comes at some cost for men. Finally, he addresses differences and inequalities among men that give White, heterosexual men more power than others.

Not long ago, I was standing in line behind a woman and a man at the local car wash, waiting to pay the cashier. As the man, a thirtyish white guy wearing a tight tank top, paid his money, the female cashier asked him about the prominent Asian characters that were tattooed on his heavily muscled arm. "Is that Chinese?" she asked him. "No," he replied tersely, "it's Korean." She persisted in her curiosity: "What's it say?" He lifted the arm a bit and flexed: "It says, 'Fear No Man, Trust No Woman.'" An uncomfortable moment of silence passed as he got his change and left. Then, the next woman in line stepped forward to pay her money, and the cashier said to her, "That's pretty scary!" "I don't know," the other woman replied. "I think they *all* should come with warning labels."

From: Michael A. Messner. 1997. *The Politics of Masculinities*, pp. xiii–xv, 1–10. Copyright
© 1997 Rowman and Littlefield Publishers, Inc.

At first, I thought about this scene only in terms of how women today are often so adept at poking fun not only at some men's hyper-masculine posturing but also at the very real danger of violence that all men potentially represent. But as I thought more about it, I began to think and wonder about what might have compelled this man to inscribe—apparently permanently—this depressing message on his arm. "Fear No Man, Trust No Woman." Imagine the isolation, loneliness, and alienation that must underlie such a slogan.

And I began to ponder the "No Fear" slogan that has appeared lately on the baseball caps, T-shirts, and bumper stickers of boys and young men, seemingly everywhere. It seems to me that you don't see hundreds of thousands of people massed in the streets chanting for peace unless you already have war. And you don't have a whole generation of young males publicly proclaiming that they have "No Fear" unless there's something actually scaring the crap out of them.

What are men so afraid of today? Most obviously, they are afraid of other men's violence. Young African American males in particular are falling prey to each other's violence in epidemic proportions, but young males from all social groups feel increasingly vulnerable today. Less obvious, but just as ominous, are young men's worries and fears of an uncertain future. As deindustrialization has eliminated tens of thousands of inner-city jobs, as structural unemployment has risen, and as government has become increasingly unable and unwilling to provide hope, a higher and higher proportion of young males today see that the image of the male family breadwinner is increasingly unattainable for them.

It's actually getting harder and harder for a young male to figure out how to *be* a man. But this is not necessarily a bad thing. Young men's current fears of other men and the continued erosion of the male breadwinner role might offer a historic opportunity for men—individually and collectively—to reject narrow, limiting, and destructive definitions of masculinity and, instead, to create a more humane, peaceful, and egalitarian definition of manhood. . . .

INSTITUTIONALIZED PRIVILEGE

In the early 1970s, when I was an undergraduate, I took a course on social inequality in which I was confronted with research that showed that women in the paid labor force were earning about $.59 to the male's dollar. Even women who were working in the same occupations as men, I learned, were earning substantially less than their male counterparts. These facts radically contradicted what I had been taught about the United States; that this is a country of equal opportunity in which merit is rewarded independent of race, religion, or sex. In a term paper for this course, I explored the reasons for gender inequities in the workforce and stated passionately in my conclusion that it was only fair for women to have equal rights with men. My professor liked the paper, and I felt proud that I had taken a "profeminist" position.

The following summer, I was back in my hometown, working at my regular summer job as a recreation worker in city parks. With the exception of two full-time supervisors, the summer staff consisted of about 15 temporary workers who were, like me, college students. Perhaps a dozen or so of these workers were women, and three of us were men, giving the appearance, perhaps, that this was a female-dominated job. But what had not occurred to me at the time, or struck me as unusual, was that all of the women had been given 20- or 30-hour per week assignments at smaller city parks, whereas each of the men had been assigned 40-hour weeks at the larger parks. What's more, when opportunities for overtime work arose, the supervisors invariably invited the men to do the work. So I regularly chalked up 42, perhaps even 46, hours of work each week. One week, at a staff meeting, a supervisor routinely invited me and another man to come to the recreation center to do some overtime work. Before we had a chance to say yes, we were interrupted by one of the women workers, who firmly stated, "I don't know why the guys always get the extra hours; we women can do that work as well as them. It doesn't really seem fair." I immediately felt threatened and defensive and broke the uncomfortable moment of silence in the room by whispering—far too loudly, as it turned out—to my male coworker, "Who the hell does she think she is, Gloria Steinem?" In response, the woman worker glared and pointed her finger at me: "Don't talk about something you don't know anything about, Mike!"

Immediately, it ran through my mind that I *did*, in fact, know *a lot* about this topic. Why, I had just written this wonderful paper about how women workers are paid less than men and had taken the position that this should change. Why, then, when faced with a concrete situation where I could put that knowledge and those principles to work had I taken a defensive, reactionary position? In retrospect, I can see that I had not yet learned the difference between taking an intellectual position on an issue and actually integrating principles into my life ("the personal is political," my feminist friends would later teach me). But more important, I had not yet come to grips with the reality that men—especially white, heterosexual, middle-class men like myself—tend to take for granted certain *institutional privileges.* In this case, because I was a man, I was "just naturally" afforded greater opportunities than my female coworkers. Yes, I had to work hard, but I worked no harder than did the women. Yet, to receive equal treatment, the women had to stand up for themselves and make public claims based on values of justice and equal opportunity. I just had to show up. What strikes me about this in retrospect is how easy it is for members of a privileged group to remain ignorant of the ways that the social structures of which we are a part grant us privileges, often at the expense of others. . . .

. . . Gender is a system of unequal—but shifting and at times contested—power relations between women and men (Connell, 1987). In the current historical moment, men's institutional privileges still persist, by and large, but they no longer can be entirely taken for granted. For the past three decades, women have organized to actively challenge unequal and unfair gender arrangements. This reality was brought home to me over 20 years ago quite dramatically by my female workmate.

THE COSTS OF MASCULINITY

In 1977, a major support undergirding my world suddenly gave way when my father died. It just didn't seem fair: A nonsmoker, moderate social drinker, and former athlete only a few pounds overweight, he appeared to have lived a fairly healthy life. Seemingly vibrant at the age of 56, he was just too young to die. But within a few short months, cancer quickly dropped the man who, to our family, had always seemed like the Rock of Gibraltar. And maybe, I can now see, that was part of the problem. As a high school and college football player in the 1930s and 1940s, then in the navy in World War II, he had been taught that a real man ignores his own pain and pays whatever price is necessary for the good of the team or country. Throughout his adulthood, this lesson was buttressed by his conservative Lutheranism, which taught him that a man's first responsibility was as a family breadwinner, which meant to work hard and sacrifice himself, day in and day out, for the good of his family.

Indeed, as my mother cared for me and my two sisters, my father worked very hard during the school year as a high school teacher and a coach and on summer "tours" in the navy reserve. He prided himself that he had never let a little cold or flu or a sore back keep him from work. He'd been taught to "play through the pain," to keep his complaints to himself, never to show his own hurt, pain, or fears. . . .

He died with nearly a year's worth of accumulated "sick leave" at the high school.

I've come to see my father's story as paradigmatic of the story of men in general. The promise of public status and masculine privilege comes with a price tag: Often, men pay with poor health, shorter lives, emotionally shallow relationships, and less time spent with loved ones. Indeed, the current gap between women's and men's life expectancy is about 7 years; men tend to consume tobacco and alcohol at higher rates than do women, resulting in higher rates of heart disease, cirrhosis of the liver, and lung cancer; men tend to be slower than women to ask for professional medical help; men tend to engage in violence and high-risk behavior at much higher rates than do women; and men are taught to downplay or ignore their own pain (Harrison, Chin, & Ficarratto, 1995; Sabo, 1994; Sabo & Gordon, 1995; Stillion, 1995; Waldron, 1995). In short, conformity with narrow definitions of masculinity can be lethal for men. . . .

DIFFERENCES AND INEQUALITIES AMONG MEN

In the early 1980s, at one of the first National Conferences on Men and Masculinity, I sat with several hundred men and listened to a radical feminist male exhort all of us to "renounce masculinity" and "give up all of our male privileges" as we unite with women to work for a just and egalitarian world. Shortly after this moving speech, a black man stood up and angrily shouted, "When you ask me to give up my privileges as a man, you are asking me to give up something that white America has never allowed me in the first place! I've never been allowed to *be* a man in this racist society." After a smattering of applause and confused chatter, another man stood and said, "Yeah—I feel the same way as a gay man. My struggle is not to learn how to cry and hug other men. That's what you straight guys are all hung up on. I am oppressed in this homophobic society and need to empower myself to fight that oppression. I can't relate to your guilt-tripping us all into *giving up* our power. What power?"

This meeting illustrated one of the major issues faced by feminists, especially beginning in the 1980s: Women and men of color, gay men and lesbians, and differently abled people have all challenged the simplistic assumption that we can neatly discuss "women" and "men" as discrete categories within which members are assumed to share certain life experiences, life chances, and worldviews (Baca Zinn, Cannon, Higginbotham, & Dill, 1986; Collins, 1990; Wittig, 1992). In fact, although it may be true that "men, as a group, enjoy institutional privileges at the expense of women, as a group," men share very unequally in the fruits of these privileges. Indeed, one can make a good case that the economic, political, and legal constraints facing poor African American, Latino, or Native American men, institutionally disenfranchised disabled men, illegal immigrant men, and some gay men more than overshadow whatever privileges these people might have as men in this society (Anderson, 1990; Gerschick & Miller, 1995; Hondagneu-Sotelo & Messner, 1994; Nonn, 1995; Staples, 1995). And the "costs of masculinity," such as poor health and shorter life expectancy are paid out disproportionately by socially and economically marginalized men (Sabo, 1995; Staples, 1995).

When we examine gender relations along with race and ethnicity, social class, sexuality, and age as crosscutting, interrelated systems of power and inequality, it becomes clear that studying men and women is far more complicated than it might first have seemed. In fact, as R. W. Connell (1987) has argued, it makes little sense to talk of a singular masculinity (or femininity, for that matter) as did much of the 1970s "sex role" literature. Instead, Connell observes that at any given historical moment there are various and competing masculinities. Hegemonic masculinity, the form of masculinity that is dominant, expresses (for the moment) a successful strategy for the domination of women, and it is also constructed in relation to various marginalized and subordinated masculinities (e.g., gay, black, and working-class masculinities). . . .

REFERENCES

Anderson, E. (1990). *Streetwise: Race, class, and change in an urban community*. Chicago: University of Chicago Press.

Baca Zinn, M., Cannon, L. W., Higginbotham, E., & Dill, B. T. (1986). The costs of exclusionary practices in women's studies. *Signs: Journal of Women in Culture and Society 11*, 290–303.

Collins, P. H. (1990). *Black feminist thought: Knowledge, consciousness, and the politics of empowerment*. Boston: Unwin Hyman.

Connell, R. W. (1987). *Gender and power*. Stanford, CA: Stanford University Press.

Gerschick, T. J., and Miller, A. S. (1995). Coming to terms: Masculinity and physical disability. In M. S. Kimmel & M. A. Messner (Eds.), *Mens's Lives* (3rd ed., pp. 262–276). Boston: Allyn & Bacon.

Harrison, J., Chin, J. & Ficarratto T. (1995). Warning: Masculinity may be dangerous to your health. In M. S. Kimmel & M. A. Messner (Eds.), *Men's Lives* (3rd ed., pp. 237–249). Boston: Allyn & Bacon.

Hondagneu-Sotelo, P., and Messner, M. A. (1994). Gender displays and men's power: The "new man" and the Mexican immigrant man. In H. Brod and M. Kaufman (Eds.), *Theorizing masculinities* (pp. 200–218). Thousand Oaks, CA: Sage.

Nonn, T. (1995). Hitting bottom: Homelessness, poverty, and masculinity. In M. S. Kimmel and M. A. Messner (Eds.), *Men's lives* (3rd ed., pp. 225–234). Boston: Allyn & Bacon.

Sabo, D. F. (1994b). Pigskin, patriarchy, and pain. In M. A. Messner and D. F. Sabo (Eds.), *Sex, violence, and power in sports: Rethinking masculinity* (pp. 82–88). Freedom, CA: Crossing Press.

Sabo, D. F. (1995). Caring for men. In J. M. Cookfair (Ed.) *Nursing care in the community* (2nd ed., pp. 346–365). St. Louis: C. V. Mosby.

Sabo, D., and Gordon, D. F. (Eds). (1995). *Men's health and illness: Gender, power, and the body*. Thousand Oaks, CA: Sage.

Staples, R. (1995a). Health among Afro-American males. In D. Sabo and D. F. Gordon (Eds.), *Men's health and illness: Gender, power, and the body* (pp. 212–238). Thousand Oaks, CA: Sage.

Stillion, J. M. (1995). Premature death among males: Extending the bottom line of men's health. In D. Sabo and D. F. Gordon (Eds.), *Men's health and illness: Gender, power, and the body* (pp. 46–67). Thousand Oaks, CA: Sage.

Waldron, I. (1995). Contributions of changing gender differences in behavior and social roles to changing gender differences in mortality. In D. Sabo and D. F. Gordon (Eds.), *Men's health and illness: Gender, power, and the body* (pp. 22–45). Thousand Oaks, CA: Sage.

Wittig, M. (1992). *The straight mind and other essays*. Boston: Beacon.

DISCUSSION QUESTIONS

1. What was your reaction to hearing about the "Million Man March" or another recent men's movement gathering? How do race and sexuality fit into men's movements? Are their any issues addressed by these groups that are universally a concern for all men?

2. In what ways can men's lives improve if we alter our conceptualization of masculinity? What can men learn from women's roles? What privilege or power would men need to give up for this to work?

34

Catching Sense

Learning from Our Mothers
to Be Black and Female

SUZANNE C. CAROTHERS

In this research, Suzanne Carothers examines Black mother-daughter relationships and the everyday, commonplace activities involved in mothering. The daughters learn important lessons for negotiating both gender and race in society. The women in this study develop identities as women and as Black Americans through interaction with other family members.

Black parents are required to prepare their children to understand and live in two cultures—Black American culture and standard American culture. To confront the bicultural nature of their world, these parents must respond in distinctive ways. In the following essay, I show how this can be seen in the practices and beliefs of several generations of Black women through their descriptions of seemingly ordinary and commonplace activities.

The first setting in which people usually experience role negotiations is the home. Boys and girls will draw from important lessons learned at home during childhood to negotiate their future roles as viable members of society. Distant though the lessons may seem from the perspective of an adult, they were taught directly and indirectly in the context of day-to-day family life. Of the many dyads occurring within families, the interactions between mothers and daughters are a critical source of information on how women perceive what it means to be female. I have been particularly interested in these perceptions among Black mothers and daughters because of the unique socio-economic, political circumstances in which these women find themselves in American culture. During the fall, winter, and spring of 1980–81, I returned to my home town of Hemington, fictitiously named, to collect data for my study. It is the contradiction that emerged between my experience of having grown up in the Black community of Hemington and my graduate school reading of the social science literature on Black family life and mother-daughter relationships that led me to engage in this research. . . .

From: Faye Ginsburg and Anna Lowenhaupt Tsing, eds. 1990. *Uncertain Times: Negotiating Gender in American Culture.* Boston: Beacon Press, pp. 232–247. Reprinted with permission of Beacon Press, Boston.

BACKGROUND

The implications of Black women's strength have not been explored fully in the literature on American mother-daughter relationships. Black women have traditionally combined mothering and working roles, while white middle-class women in the United States until recently have not. In Western cultures, mothering is regarded as a role that directly conflicts with women's other societal roles. In response to this condition, many theorists of female status consider the mothering role to be the root cause of female dependence on and subordination to men.[1] Yet, this has not been the experience of Black mothers. As others have argued, "Women have been making culture, political decisions, and babies simultaneously and without structural conflicts in all parts of the world."[2]

During the 1970s, a recurring theme in the United States literature on mother-daughter relationships was the ambivalence and conflict existing between mothers and daughters. The literature describes competition and rivalry and suggests a negative cycle of influences passed from mothers to their daughters. For example, Judith Arcana suggests,

> The oppression of women created a breach among us, especially between mothers and daughters. Women cannot respect their mothers in a society which degrades them; women cannot respect themselves. Mothers socialize their daughters into the narrow role of wife-mother; in frustration and guilt, daughters reject their mothers for their duplicity and incapacity—so the alienation grows in the turning of the generations.[3]

The above quote is generally inapplicable to the relationship between Black mothers and their daughters. The Black cultural tradition assumes women to be working mothers, models of community strength, and skilled women whose competence moves beyond emotional sensitivity. It is through this tradition of a dual role that Black women acquire their identity, develop support systems (networks), and are surrounded by examples of female initiative, support, and mutual respect.

Black mothers do not raise their children in isolation. In contrast, Nancy Chodorow argues,

> The household with children has become an exclusively parent and child realm; infant and child care has become the exclusive domain of biological mothers who are increasingly isolated from other kin, with fewer social contacts and little routine assistance during their parenting time. . . .[4]

The families to whom Chodorow refers above are child-centered. In the arrangement she describes, the needs of the domestic unit are shaped and determined primarily by those of the children. Scholars of Black family life offer evidence of other arrangements.[5] According to them, Black women raise their children in the context of extended families in which social and domestic relations, as well as kinship and residence structures offer a great deal of social interaction among adults that includes children. In addition, these researchers have shown that child rearing is only one of many obligations to be performed within Black family households. They agree that child rearing cannot be evaluated in the singular context of an individual but rather in the plural context of the household. The process through which daughters learn from their mothers in Black families, therefore, contradicts the wave of literature on mother-daughter relationships.

In order to appreciate the contrast in orientation of Black mothers, it is necessary to consider the wider social context of Black parenting. Black parents in American society have a unique responsibility. They must prepare their children to understand and live in two cultures—Black American culture and standard American culture.[6] Or as Wade W. Nobles[7] has suggested, Black families must prepare their children to live near and be among white people without becoming white. This phenomenon has been referred to as *biculturality* by Ulf Hannerz and by Charles Valentine, an idea derived from W. E. B. Du Bois who wrote in the early 1900s about the idea of double consciousness: "that Blacks have to guard their sense of blackness while accepting the rules of the games and cultural consciousness of the dominant white culture."[8] Because Black parents recognize that their children must learn to deal with institutional racism and personal discrimination, Black children are encouraged to test absolute rules and absolute authority.[9] It is therefore critical to the socialization of Black children that their parents provide them with ample experiences dealing with procedures of interpersonal interaction rather than rules of conduct. The children are socialized to be part of a Black community rather than just Black families or "a fixed set of consanguinal and affixed members."[10] Beginning early in childhood, the wider social context in which Black children are raised usually involves not only their mothers, but also many adults—all performing a variety of roles in relation to the child, the domestic unit, and the larger community. Furthermore, the transmission of knowledge and skills in Black family life is not limited to domestic life but occurs in public life arenas in which Black women are expected to participate. Working outside their homes to contribute to the economic resources of the family has been only one of the many roles of the majority of Black women. As members of the labor market, Black mothers simultaneously manage their personal lives, raise their children, organize their households, participate in community and civic organizations, and create networks to help each other cope with seemingly insurmountable adversities.

Participation in work and community activities broadens the concept and practice of mothering for Black women. How do the women learn these roles? What must they pass on to their female children if they are to one day perform these roles? An exploration of the social interactions between Black working mothers and their daughters, as well as the cultural context and content can extend our knowledge of the cultural variation in mothering roles and mother-daughter relationships and the processes by which mothers shape female identity.

THE STUDY

Several reasons prompted my decision to return to Hemington for this research. Typically, studies of Black family life have been conducted in urban ghetto communities in the north and mid-west, some in the deep rural South and most with lower class or poor people.[11] In developing my research, I wanted to avoid choosing a location where there were vast differences between the Black and white standard of living that so often characterize the research settings in which Blacks are studied. In these settings, the usual distinguishing features for the Black population are poverty, unemployment, under employment, low wages, and inadequate housing. The white communities are more varied economically, ranging from relatively wealthy managerial elites to welfare recipients living in housing projects. Yet, there are many middle-sized and large southern cities with large, varied and stable Black communities that researchers have not adequately explored. Hemington is such an example.[12]

Prior to 1960 and urban redevelopment, the Black community of Hemington was primarily concentrated in the area closest to the main business district of the city. At the time of the study, most Blacks still lived in predominantly all Black neighborhoods located on the west side of town, where a broad range of housing suggests an economically varied Black community. Blacks in Hemington live in public housing projects, low income housing (both private homes and apartments), modern apartment complexes, and privately owned homes, ranging from modest to lavish.

Although I had not lived in Hemington for more than ten years, my kinship ties to the community meant that I had access to people, situations, and information that an outsider might not have or would need a considerably longer time to acquire. My experiences of growing up there and then studying and working in educational institutions in northern white society made me sensitive to differences in cultural patterns and more eager to analyze them.

Forty-two women and nine girls between the ages of 11 and 86 from twenty families agreed to participate in the study. I asked them to help me understand the meaning of mothering and working in their lives and how this meaning was passed on from mother to daughter—generation to generation. The stories that the women and girls shared with me about the very ordinary day-to-day activities of their lives became a rich source for understanding how women found and created meaning in the less-than-perfect world in which they lived.[13]

The study was of women whom Alice Walker would call the anonymous Black mothers whose art goes unsigned and whose names are known only by their families.[14] Many of the women in the study have known me all my life. They are great-grandmothers, who have lived to see their grandchildren's children born and grandmothers who have raised their children. I grew up with some of the mothers who are raising their children. Still others, the young girls, I remember from the time they were born.[15]

Seventy-five years separate the births of the oldest from the youngest participants. . . .

These women perceive themselves as middle-class. All are or have been employed. They are people who share a common system of values, attitudes, sentiments, and beliefs which indicate that an important measure of "class" for these Black Americans is the range of resources available to the extended family unit. This system, then—based on extensive inter-household sharing—is not synonymous with traditional criteria for social class structure which includes wealth, prestige, and power.[16]

WHAT BLACK MOTHERS TEACH: CONCRETE
LEARNINGS AND CRITICAL UNDERSTANDINGS

In order to understand the teaching and learning process taking place between Black mothers and daughters, I observed the women and asked them questions about their seemingly ordinary and commonplace daily activities. When the participants in the study were asked from whom they learned, their answers included their mothers, fathers, stepfathers, stepmothers, grandfathers, grandmothers, great-grandmothers, aunts, the lady next door, an older sister, a brother, a teacher—in short, their community. When asked *what* they had learned from these people, their responses touched a range of possibilities, which can be grouped into two broad categories. Cooking, sewing, cleaning, and ironing are examples of activities that are associated with the daily routines of households that I call "concrete learnings." The regular performance of these leads to what I refer to as "critical understandings," which include such things as achieving independence, taking on responsibility, feeling confident, getting along with others, or being trustworthy. The acquisition of these is not easy to pinpoint and define. They are not taught as directly as the concrete learnings, but they are consistently expected. Their outcomes are not as immediately measurable because they usually take a longer time to develop.

What do mothers do to pass on to their daughters the understandings that are considered critical to a daughter's well being, and the skills, the learned power of doing a thing competently? Mothers teach their daughters what to take into account in order to figure out how to perform various tasks, recognizing that the individual tasks that they and their daughters are required to perform change over time. Therefore, the preparation that mothers provide includes familiarity with the task itself, as well as a total comprehension of the working of the home or other situations within which the task is being done. The women's interviews reveal that mothers teach by the way that they live their own lives ("example"), by pointing out critical understandings they feel their daughters need ("showing"), and by instructing their daughters how to do a task competently. Their teaching is both verbal and nonverbal, direct and indirect. Daughters learn not only from their mothers, but also from other members of family and the community.

The data indicate that concrete learnings teach—in ways that verbal expression alone does not—a route toward mastery and pride that integrates the child into the family and community. Daughters learn competency through a sense of aesthetics, an appreciation for work done beautifully. The women described this notion as follows: "You don't see pretty clothes hanging on the line like you used to"; "Mama could do a beautiful piece of ironing"; "You always iron the back of the collar first. The wrinkles get on the back and it makes the front of the collar smooth"; or "Now if I got in the kitchen and say I saw these pretty biscuits, I might say Mama how did you get your biscuits to look this pretty . . ." This aesthetic quality becomes one of the measures of competently done work as judged by the women themselves and by other members of their community.

As each generation encountered technological changes in household work, mothers became less rigid about teaching their daughters concrete learnings. However, like previous generations, mothers still teach their daughters responsibility through chores, which gives them opportunities to practice and get better at doing them, both alone and with others. These activities are not contrived but rather they constitute real work and contribute to the daily needs of their households. Participation in these activities encourages mastery of them. . . .

Each woman described a certain kind of pride in herself for having learned and accomplished a task well. Such mastery reinforced and established the woman's confidence in her ability to perform well. Having chores to do was the important link bridging concrete learnings to critical understandings germane to a daughter's well-being.

DEMANDS OF DOUBLE CONSCIOUSNESS
OR LESSONS OF RACISM

While mothers teach critical understandings through example, maxims, and practical lessons, they use what I am calling "dramatic enactments" as a powerful tool to teach their daughters ways to deal with white people in a racist society. Thus, daughters learn critical understandings that are specific to the Black experience.

When mothers teach by example, they enact before their daughters the particular skills necessary to achieve the task at hand. By contrast, dramatic enactments expose children to conflict or crisis and are often reserved for complex learning situations. "Learning to deal with white people," for example, was viewed by some women in the study as important to their survival, and dramatic enactment was identified as being a powerful technique for acquiring this skill. One thirty-one-year-old woman . . . explained how she learned this critical understanding through dramatic enactments from her grandmother, who did domestic work.

> My sister and I were somewhat awed of white people because when we were growing up, we did not have to deal with them in our little environment. I mean you just didn't have to because we went to an all-Black school, an all-Black church, and lived in an all-Black neighborhood. We just didn't deal with them. If you did, it was a clerk in a store.
>
> Grandmother was dealing with them. And little by little she showed us how. First, [she taught us that] you do not fear them. I'll always remember that. Just because their color may be different and they may think differently, they are just people.
>
> The way she did it was by taking us back and forth downtown with her. Here she is, a lady who cleans up peoples' kitchens. She comes into a store to spend her money. She could cause complete havoc if she felt she wasn't being treated properly. She'd say things like, "If you don't have it in the store, order it." It was like she had $500,000 to spend. We'd just be standing there and

watching. But what she was trying to say [to us] was, they will ignore you if you let them. If you walk in there to spend your 15 cents, and you're not getting proper service, raise hell, carry on, call the manager but don't let them ignore you.

Preparing their daughters to deal with encounters in the world beyond home was a persistent theme in the stories offered by the women in this study. By introducing their daughters and granddaughters to such potentially explosive situations and showing the growing girls how older women could handle the problems spurred by racism, mothers and grandmothers taught the lessons needed for survival, culturally defined as coping with the wider world. . . .

THE COMMUNITY CONTEXT

The concept of Black family units working in concert to achieve the common goals they value, arises out of the inherent expectations of helping and assuming responsibility for each other as part of a conscious model of social exchange.[17] Thus, giving and receiving are the understood premises for participating in community life. Different from guilt, this system has been fueled by the racial and economic oppressions that have plagued Black families since their introduction into American society. Women in these family units traditionally assume a critical role in meeting these responsibilities. This does not end when children reach maturity, nor is it hierarchical, from mother to child; it is part of the larger community value that the women believe in and sustain. . . .

Given the difficult conditions that racism and economic discrimination have imposed on the Black community (including, of course, the middle-class Black community), it is important for children to know that their parents can survive the difficult situations they encounter. Children need to trust that the world is sufficiently stable to give meaning to what they are learning. *Dependability*—based on elements of character such as hard work, faith, and the belief that their children can live a better life—provides the context for that trust and the daughters' sense of their mothers' competence to deal with the world.

Despite the conflicts that sometimes arose, daughters generally acknowledged the ongoing lessons their mothers had to teach them and that the process of *lifelong learning* was central to their relationship, as Mrs. Washington's quote illustrates,

Kitty and James, either one haven't got to the place today where I couldn't tell them if they were doing something wrong. And people gets on me for that. And I say, well you never get too old to learn. I say, if I know it's right why can't I correct them? They say, "When a child gets up on his own, you ought to let 'em alone." Then I say, well I'm going to bother mine as long as I live if I see 'em doing something wrong. I'm going to speak to them and if they don't do what I say, at least they don't tell me that they aren't. They just go someplace else and do it.

This sense of teaching and learning as an ongoing part of the mother-child relationship adds an impetus to the daughters' frequently expressed belief that they fulfilled their mothers' dreams and in a sense justified their mothers' lives through gaining an education.

> This is what Mama was working toward [my going to college]. This was her ambition. It was just like she was going to college herself. The first summer I finished high school and every summer after that, Mama took sleep-in jobs up in the mountains to earn extra money for my college education . . . She very much wanted us to take advantage of all the things she never had. This is what she worked for. . . .

Black daughters learn their mothers' histories by seeing their mothers in the roles of mamas who nurture as friends, who become confidantes and companions; as teachers, who facilitate and encourage their learning about the world; and as advisors who counsel. For these women, the role of mother is not seen as "a person without further identity, one who can find her chief gratification in being all day with small children, living at a pace tuned to theirs."[18] From early on, the women in the present study see their mothers as complex beings. Knowing her mother's history intensifies the bond between mother and daughter, and helps daughters understand more about the limits under which their mothers have operated.

Getting along however, is not necessarily dependent on the women always reaching agreement, or daughters following the advice of their mothers. It would not be unusual for a mother and daughter to fall out about an issue one day and speak to each other the next. Such interactions provide continuing opportunities for daughters to practice developing and defending their own points of view, a skill useful in the outside world. These interactions insure the back and forth between Black mothers and daughters, which promotes the teaching and learning process and increases Black daughters' respect for what mothers have done and who they are.

Although respect remains a key value in mother daughter relationships and helps to foster the teaching and learning process, generational differences between mothers and daughters lead to tensions that threaten the teaching and learning context. Daughters need to *balance* their loyalty to their mothers with their own needs to grow up in accord with the terms of their own generation: realities. Mothers, on the other hand, need to balance their need for their daughters' allegiance with the knowledge that the daughters require a high degree of independence to survive and achieve in the world.

Loyalty is the unspoken but clear message in the words spoken by the mothers and daughters in this study. They describe it in terms of faithfulness and continuing emotional attachment. As loyalty defines what Black mothers and daughters expect from each other, it also is part of the conflict between them—when Black women are unable to separate the tangled threads that bind them so closely. Their obligation to each other and the deep understanding of the plight they share sometimes nurtures a desire to protect, rather than commit what would be seen as an act of desertion. . . .

CONCLUSION

Women in this study routinely have confronted very early on the contradictions between the world in which they were born and raised and the one away from their homes. The result is that the women are not thrown by that which is different or contradictory to their home practices; rather, they can accept, understand, negotiate and deal with the differences in reasonable ways.

A high degree of mutual respect and camaraderie characterize the teaching and learning processes taking place between Black mothers and daughters. The community value of mutual responsibility makes this possible. Because of the multiple roles that these mothers play, the interactions between Black mothers and daughters require that mothers balance these roles and determine which one is appropriate in different situations. It also requires that daughters actively consider the context and purpose of the interaction and the mood of her mother to determine the appropriate response. Thus, the issue of authority that is often a major concern and obstacle in school learning for both teacher and students, shifts to mutuality in the teaching and learning process occurring between Black working mothers and their daughters.

The daughters have learned from their mothers by being exposed to the complications, complexities, and contradictions that as working women, their mothers faced in a society which has traditionally viewed working and mothering as incompatible roles. The recognition of this difference requires that Black women, as a condition of their daily existence, constantly negotiate an alternative understanding of female identity that challenges the dominant gender paradigm in American culture.

NOTES

1. Nancy Chodorow, *The Reproduction of Mothering: Psychoanalysis and the Sociology of Gender* (Berkeley: University of California Press, 1978).

2. Karen Sacks, *Sisters and Wives: The Past and Future of Sexual Equity* (Westport, CT: Greenwood Press, 1979).

3. Judith Arcana, *Our Mothers' Daughters* (Berkeley: Shameless Hussy Press, 1979).

4. Chodorow, p. 5.

5. Joyce Aschenbrenner, *Lifelines: Black Families in Chicago* (New York: Holt, Rinehart & Winston, 1975). Cynthia Epstein, "Positive Effects of the Multiple Negatives: Explaining the Success of Black Professional Women," in *Changing Women in a Changing Society,* ed. J. Huber (Chicago: University of Chicago Press, 1973). T. R. Kennedy, *You Gotta Deal With It: Black Family Relations in a Southern Community* (New York: Oxford University Press, 1980). E. P. Martin & J. M. Martin, *The Black Extended Family* (Chicago: University Press, 1978). Karen Sacks, *Sisters and Wives* (Westport, CT: Greenwood Press, 1979). V. H. Young, "Family and Childhood in a Southern Negro Community," in *American Anthropologist* 72 (1970), 269–88.

6. T. Morgan, "The World Ahead: Black Parents Prepare Their Children for Pride and Prejudice," in *The New York Times Magazine* (1985, October 27), 32. V. H. Young, "A Black American Socialization Pattern," in *American Ethnologist* 1 (1974), 405–513.

7. See Nobles in J. E. Hale, *Black Children: Their Roots, Culture and Learning Styles* (Provo, UT: Brigham Young University Press, 1982).

8. Du Bois, W. E. B. *The Gift of Black Folk: The Negroes in the Making of America,* (New York: Washington Square Press, 1970), xii. For "biculturality" see Ulf Hannerz, *Soulside: Inquiries Into Ghetto Culture a Community* (New York: Columbia University Press, 1969), and Charles Valentine, "Deficit, Difference, and Bicultural Models of Afro American Behavior," in *Harvard Educational Review* 41, no. 2 (1971).

9. Young (1974), 405–513.

10. Kennedy, 223.

11. Kennedy, 226.

12. According to the 1980 census approximately one-third of Hemington's population was Black. The population of Hemington County was 404,270. Blacks were 27% of this population. In the City of Hemington Blacks were 32% of the population.

13. The research consisted of 51 taped interviews of two, three and four generations of mothers and their daughters. In addition, a questionnaire was given to each of the participants on a day other than the interview.

14. Alice Walker, *In Search of our Mothers' Gardens,* (New York: Harcourt Brace and Jovanovich, 1983), 231–243.

15. They represent five different sets of consanguineous generations including: 1) seven grandmothers and mothers; 2) four great grandmothers, grandmothers, and mothers; 3) two great-grandmothers, grandmothers, mothers, and unmarried daughters; 4) three grandmothers, mothers, and unmarried daughters; and 5) four mothers and teenage daughters. The method of selecting the participants was primarily through a snowball sample technique using personal contacts of women in my mother's network of friends, neighbors, and co-workers. The initial source of participants was an older subdivision called Fenbrook Park. Names of participants, when used, have been changed.

16. This study employs the definition of social class as discussed by John F. Cuber and William F. Kenkel in *Social Stratification in the United States* (New York: Appleton-Century-Crofts, 1954). They suggest that "*Social class* has been defined in so many different ways that a systematic treatment would be both time consuming and of doubtful utility. One central core of meaning, however, runs throughout the varied usages, namely, the notion that the hierarchies of differential statuses and of privilege and disprivilege fall into certain clearly distinguishable categories set off from one another. Historically, this conception seems to have much better factual justification than it does in contemporary America . . . Radical differences, to be sure, do exist in wealth, privilege, and possessions; but the differences *seem to range along a continuum with imperceptible gradation from one person to another,* so that no one can objectively draw 'the line' between the 'haves' and 'the have nots,' the 'privileged' and 'underprivileged,' or for that matter, say who is in the 'working class,' who is 'the common man,' or who is a 'capitalist.' The differences are not categorical, but continuous" (p. 12). For an in-depth discussion of class see Rayna Rapp's article, "Family and Class in Contemporary America: Notes Toward an Understanding of Ideology," *Science and Society* 42(3): 278–300.

17. See I. G. Joseph and J. Lewis, *Common Differences: Conflicts in Black and White Feminist Perspectives* (Garden City, NY: Anchor Books/Doubleday, 1981), 76–126.

18. A. Rich, *Of Woman Born* (New York: W. W. Norton, 1976), 3.

DISCUSSION QUESTIONS

1. Think of the things you learned from your parents/guardians. What are some examples of how you were socialized to behave in ways consistant with your parents' expectations? Do your values and beliefs reflect or challenge your parents/guardians?

2. The women in Carothers' article talk about learning how to interact with white people. How has your interaction with members of a different racicial–ethinic group been influenced by your parents/guardians? Can you think of times when someone from a different racial–ethnic group reacted to you in a particular way because of your race?

35

Challenges for Middle Eastern Women

ELIZABETH FERNEA

Elizabeth Fernea presents recent evidence of the changing role of women in Middle Eastern countries. She argues that, despite the continued absence of women in political positions of authority, the grass roots movements of change are led and carried out by many women. Women's changing roles, both in and outside the home, will lead to great change in the Middle East.

The Middle East, during the twentieth century, has been transformed from a territory, loosely governed at local, even "grass roots," level by a distant Ottoman Emperor, into a collection of some two dozen separate nation states, headed by different kinds of leaders than in the past. The Ottoman Sultan is gone, but so are most of the Kings. The monarchy exists today only in the Gulf, in Jordan, and in Morocco, and it is certainly a very different kind of institution than the absolute ruler of medieval times. It has been the followers, the grass roots movements, combined with some elites, who have challenged the common enemy, the foreign colonial rulers, and successfully achieved nationhood and independence. . . .

I would argue that one category of followers has in recent decades become more important and more participatory. At the dawn of a new century, this group is actively competing at different levels, for positions of power and/or leadership in the Middle East. I refer to women. Women constitute at least half of the population of the Middle East today. They are no longer passive accepters of the status quo, of the ideology that men are in charge of women. They are

From: Fernea, Elizabeth. "Challenges for Middle Eastern Women." *Middle East Journal* 54, no.2 (Spring 2000): 185–93.

participating and struggling at every level for jobs, promotions, improvements in standard of living, and political clout. How and why? Why should women, as a group, be considered competitors for power and leadership in the next decades and centuries? Are they not, ideologically speaking, under the control of men, their fathers, husbands and sons, the patriarchy? The answer, quite simply, is that the stated ideology of men dominating women is being contested by social practice. This has occurred as a result of a number of processes.

To begin with, we are no longer talking in terms of old paradigms like the public/private split (men in the public sphere, women in the private sphere). Though this may have had some validity as a working concept in the long-distant past, my personal belief . . . is that this paradigm was constructed by male social scientists who, banned from studying the private or family sphere, simply discounted the private sphere as unimportant to the world of politics, commerce and religion, and therefore not worthy of study. A further implicit assumption based on the public/private paradigm was that women were kept in the home and their decisions came from there. All that has changed. First of all, women are no longer in the home, which has created problems as well as achievements. This is perhaps the biggest change in the Middle East, a change which has far-reaching implications for politics as well as for the labor market and for family cohesiveness. Incomplete statistics from Tunisia, Morocco, and the Gulf suggest that nearly a third of Middle East women from these countries work full-time outside the home, and that most of the rest work part-time in or out of their homes.[1]

Further, in many countries, men are absent from their old positions of control. Millions of migrant laborers drawn from Morocco, Tunisia, Algeria, Yemen, Turkey, and Egypt have been away from home off and on for the past two generations. Their labor has provided hard currency for the coffers of their nations as well as a means of improvement in life styles (brick houses, modern appliances, television sets) for the families involved. But the migrant labor has also created women-centered households and absent fathers. Men may still be, ideologically, the head of the household, but if the men are not present, children as well as women folk may, in reality, have to go their own way. . . .

A new vacuum thus exists in the family, the heart of Middle East society, a vacuum created by the absence of adult males and the subsequent diminution in the dimensions of male power. This vacuum is being filled by women—in the labor force, in religious practice, in politics, in religious organizations, and, in that most crucial area of life, economics. . . . Organizations headed by women abound, in the plethora of Non-Governmental Organizations (NGOs) now operating in every Middle East country. . . .

The labor pool? In "traditional" occupations, such as teaching and social work, women are in the majority, filling the familiar role of "shaping the next generation." They continue as domestic laborers, cooks and nannies as well as part-time agricultural workers during harvest, and piece-workers at home, though these efforts are generally not counted in official labor statistics. Women are also filling the ranks in occupations hitherto closed to them: medicine, law, engineering. "Of course, they tend to women," is the way women's medical participation is dismissed; the same cannot be said of engineering, where the percentage of women engineers is supposedly proportionately higher in Egypt than in Germany. Women lawyers are found in almost every country in the region. A prohibition against women judges persists, yet women judges do sit on the bench in countries like Tunisia, where they constitute 25 percent of the total number of judges in the nation, and in Morocco, where 20 percent of the judges are women. This is a higher proportion of women on the bench than in the United States.[2] . . .

How did it happen? A feminist revolution? Active women's movements? Western influence? And are there not powerful forces operating in reaction to these new practices, such as the Taliban in Afghanistan, and religious "fundamentalist" movements in the area? Is there not an increase in honor crimes? Let us consider these questions. First, how did it happen that women's power has increased? Western attitudes toward women's rights constituted an important element in the first missionary schools for girls, and this was certainly an influence in early years. There were women's movements in the 1920's and 30's, formed by upper class elite women like Hoda Sha'rawi, who banded together and who agitated for and gained support for women's education, for raising the legal level of marriage for girls. These early fighters deserve recognition for seeing the key to change in women's worlds as well as men's. As in early Islam, education would be the new merit basis for leadership and power, outdistancing blood relationship and class. The new nationalist governments which emerged after the end of colonialism promised free compulsory education for men and women. After forty years, governments have more or less achieved the goal of a majority of literate citizens, and have been able to mobilize manpower as well as womanpower for the important task of nation-building. Class lines are blurring in many countries. Not all statistical sources agree on literacy rates in Middle Eastern countries. Nevertheless; some trends are clear. UNESCO has documented the situation in Morocco, where in 1959, only 2,500 students in the entire country were enrolled in university programs; by 1997 that number had risen to 250,000 students in 13 universities. Half of this total were women.[3] Literacy figures for Tunisia are disappointing overall, but for those aged 10, literacy stands at 63% for both men and women.[4] Saudi Arabia's investment in education has produced a startling rise, from the 1970 literacy rate of 2 percent for women, and 15 percent for men, to the 1990 figures of 73 percent for men, and 48 percent for women! The Saudi education of women takes place in women's universities, of course. The same is true of the new all-women's medical college in Qom, Iran.[5]

Education has been, and continues to be, the spur to women's activism and participation in the public sphere, combined with the economic need for new kinds of skilled labor. For the second great change in Middle Eastern society has been the shift from country to city, from agriculture to industrial production. Forty years ago, two thirds of Middle Eastern peoples lived in rural areas; at the end of the twentieth century, more than half the population lives in cities. . . .

Women's movements have been important in the Middle Eastern countries since the turn of the 20th century, but because of their class base, they have not touched the majority of women whose needs are economic, and who live in poverty. Yet new developments indicate other changes. . . .

But it is in the area of religious revival that one sees for the first time a woman's movement that, from its outset, has cut across class lines and been concerned with wider political and social issues. The Western media-based hoopla about the spreading evils of Islamic fundamentalism seldom mention how the movement has affected national and local politics, to the advantage, rather than the disadvantage of large numbers of ordinary people. . . .

One could argue that, though this new women's participation is important in education, the labor force, and religion, it is still not affecting political life. Granted, here the movement appears slower. But as of the year 2000, women will vote in almost every country except Saudi Arabia and Afghanistan. Kuwait's women had been promised the vote, but lost out in a 1999 Parliamentary vote on the issue. . . .

Perhaps the most disappointing news is from the Palestinian Authority, where the original Charter's promise of equality in terms of gender has been watered down. A hundred years ago, Palestinian women might have been content to accept the new male directive: no longer. When I said, in some concern, to a Palestinian woman that I didn't know what to write about this despairing turn of events, she said sharply, "Write about how we're fighting off these new strictures! We're not like we were before. We were never passive as you in the West liked to think. We're even less so now. We're ready to struggle—and we'll continue until we succeed. If we don't, our daughters will!"

Who then will be the leaders in the next century? Those men or women who dare to take on the still formidable problems of the area—poverty, unemployment, corruption in high places, unequal access to education, and health care. Young men as well as young women fault their male elders for giving in too easily, for not accepting the responsibilities that come with power, which in the past was synonymous with patriarchy. Women are emerging, and they are also meeting growing resistance, as ideologues mount the barricades for a concerted stand. The glass ceiling, so well known in the West, has come crashing down on women who were welcomed into professional ranks when need was greatest and who now, in a crowded labor market, must compete for promotion with men. . . .

In the past, women who were seen as "getting ahead of themselves" were divided and castigated, like early Western feminists, for placing individual desire over group good. The familiar epithets were hurled at those women: "home breakers," "unworthy mothers," "Western-influenced tarts" and "strumpets." But today such epithets are hard to hurl at modestly-attired women who do not seem interested in breaking up the home, but instead are attempting to rebuild it—reforming divorce laws, on a more egalitarian rather than a patriarchal base. The strident tone of some men's voices is instructive, for it suggests that women are indeed a force to be reckoned with, now that social practices have, indeed, far outstripped the stated ideology, that a "fundamental" change in the power base has already taken place. . . .

Of course, great diversity and variability in the speed and range of these developments exists across the region. Class and wealth still remain crucial in the rise of any leader. But the pattern I have been describing can be observed in greater or lesser fashion in almost every country.

What then is a leader? More than a manager, more than a diplomat, more than a controller by force. To lead anywhere or anything, one must have followers, and such followers must respect the incipient leader. Respect is what is emerging in the new competition for power between men and women. In the past, women were seen by men as either sexual objects or maternal icons. Now for the first time, women outside the home are also perceived as fellow Muslims and hence worthy of respect. This is a big change in women's identity, as seen by themselves as well as by men.

Finally, one asks again, how can we even conceive of women as leaders if they do not hold political office? But has political leadership by women brought change or improvement in social relations between gender and classes? Not necessarily. Think of Margaret Thatcher, Benazir Bhutto and even the United States. Has American politics changed much because of women in political life? Will more elected women mean big changes? I am not at all sure. Male patriarchy is still a subtle force in the majority of countries. In the U.S. particularly, we have failed, both men and women, to trace and identify the links in our thought and practice to patriarchal ideology. However, one must say that the ideology of patriarchy is slowly changing in America, as belief in inherent male superiority is being questioned in many parts of our post-modern world. It is not necessarily the election of women to public office, but the shifting and changing of a whole universe of long-honored assumptions about the male and his power that is important. This, I suggest, is already beginning to take place in the Middle East, as practice erodes ideology and in many countries has overtaken it. And Middle Eastern women and men have an advantage over the United States, for they have never pretended that patriarchal ideology did not exist. We, on the other hand, have been telling ourselves for generations that the patriarchy is nearly gone. Indeed, when and if women vote in Kuwait, we may be seeing the first real change of heart by a male majority.

Does all this suggest that future leadership patterns everywhere will be formed less on the basis of gender, class or wealth and more on knowledge, aptitude, merit? Do women qualify? Do men? How will this quiet power struggle between Middle Eastern men and women be resolved? Will old assumptions about patriarchal control eventually be subsumed by the realities of new identities and eventually disappear in the face of social practice in everyday life? In America we have paid very little attention to these new developments, which is our loss. The outcome of this struggle, I believe, will determine the leadership patterns of the next century in the Middle East.

NOTES

1. "Official Statistics." *Europa World Year Book* (London: Europa Publications, Ltd., 1999) for Kuwait, Egypt, and Tunisia; *Middle East Executive Reports* 21, no. II: 8, for Lebanon.

2. Tunisia figures from Akila Jarraya, "Women's Rights Achievements in Tunisian Legislation and Juresprudence," Lecture at the U.S. Supreme Court, 27 September 1999;

Moroccan figures from E. Fernea personal interviews with women judges in Rabat, January 1995.

3. UNESCO, Moroccan Country Survey, 1985, and recent Moroccan statistical surveys; see also Douglas Ashford, "Second and Third Generation Elites in the Maghreb," in *Man, State and Society in the Contemporary Maghreb,* ed. I. William Zartman, 93–108 (London: Pall Mall Press, 1973); also Salloum Habeeb, "King Hassan II—Creator of Modern Morocco," *Contemporary Review* 263, no. 1534 (November 1993): 253–55.

4. Keith Walters, *Without Tunisian Arabic We're Not Tunisian: Identity and the Changing Political Economy of Language in Tunisia* (book in preparation).

5. Saudi Arabian figures from interviews conducted in the Kingdom by E. Fernea, December 1996. Footage of the new women's medical school in Qom, Iran, and interviews with students and faculty were recorded in summer 1999 by Maysoon Pachachi, London.

DISCUSSION QUESTIONS

1. What do the American media tell us about Middle Eastern women? How are the images portrayed in the news media different from the evidence provided by Fernea?

2. What examples of women's involvement in grass roots movements do we have in America? What can we learn from the successes or failures of those movements about the future of Middle Eastern countries?

http://infotrac.thomsonlearning.com

InfoTrac College Edition

BONUS READING

Williams, Christine L., Patti A. Giuffre, and Kirsten Dellinger. "Sexuality in the Workplace: Organizational Control, Sexual Harassment, and the Pursuit of Pleasure." *Annual Review of Sociology* 25 (1999): 73–93.

This reading reviews existing literature about sexual harassment compared with sexual freedom in the workplace. Sexuality of any sort in the workplace has traditionally been associated with gender inequality and women's disadvantage. Williams and her colleagues discuss the possibility that sexual expression within the workplace is empowering for women. Can we make a distinction between sexual harassment that is harmful and sexual freedom that is not? What issues need to be considered in an effort to definitively define sexual harassment and its negative consequences?

SEARCH TERMS

gendered institutions
hegemonic masculinity
Middle Eastern women
sexual harassment in the workplace and gender
social construction of gender

36

Pluralistic Ignorance
and Hooking Up

TRACY A. LAMBERT, ARNOLD S. KAHN, AND KEVIN J. APPLE

Sexual norms are powerful influences on social behavior, including our most intimate forms of interaction. Here the authors conduct a research study examining the influence of perceived norms about sex on college campuses.

Although one-night stands and uncommitted sexual behaviors are not a recent phenomenon, past research has focused on personality traits, attitudes, and individual differences in willingness to engage in such behaviors (e.g., Gerrard, 1980; Gerrard & Gibbons, 1982; Simpson & Gangestad, 1991; Snyder, Simpson, & Gangestad, 1986). The tacit assumption in this past research was that sexual behaviors within a committed and loving relationship were unproblematic, but that unloving, uncommitted sexual relations had to be explained. However, today on college campuses across the United States what was once viewed as problematic has now become normative, and students refer to this process as "hooking up."

Hooking up occurs when two people who are casual acquaintances or who have just met that evening at a bar or party agree to engage in some forms of sexual behavior for which there will likely be no future commitment (Boswell & Spade, 1996; Kahn et al., 2000; Paul, McManus, & Hayes, 2000). The couple typically does not communicate what sexual behaviors they will or will not engage in, and frequently both parties have been drinking alcohol (Kahn et al., 2000; Paul et al., 2000). Paul et al. (2000) found that 78% of women and men on the campus being studied had engaged in hooking up at least once. In the Kahn et al. (2000) sample of college students, 86% of the women and 88% of the men indicated they had hooked up. Almost one half (47%) of the men and one third of the women in the Paul et al. sample engaged in sexual intercourse during the hookup, and Kahn et al. found that their sample believed petting below the waist, oral sex, and sexual intercourse occurred with some regularity in the process of hooking up.

Pluralistic ignorance, a concept first coined by Floyd Allport (1924, 1933), exists when, within a group of individuals, each person believes his or her private attitudes, beliefs, or judgments are discrepant from the norm displayed by the public behavior of others. Therefore, each group member, wishing to be seen as a desirable member of the group, publicly conforms to the norm, each believing he or she is the only one in the group experiencing conflict between his or her private attitude and his or her public behavior. Group members believe that most others in their group, especially those who are popular and opinion leaders (Katz & Lazarsfeld, 1955), actually endorse the norm and want to behave that way, while they themselves privately feel they are going along with the norm because of a desire to fit in with the group and exemplify the norm (Prentice & Miller, 1993, 1996). In this study we examined the extent to which pluralistic ignorance might be related to U.S. college students' comfort levels with sexual behaviors involved in hooking up. Consistent with the premise of pluralistic ignorance, we hypothesized that college students would perceive others as having a greater comfort level engaging in a variety of sexual behaviors than they themselves would have. . . .

Pluralistic ignorance might have consequences when beliefs about the norm condone intimate sexual behaviors. In the process of hooking up, pluralistic ignorance may lead one or both sexual partners to act according to the perceived norm rather than to their own convictions. There is a large literature showing that men have more liberal attitudes towards sexual behaviors and expect sexual intercourse sooner in a relationship than do women (Cohen & Shotland, 1996; Knox & Wilson, 1981; Oliver & Hyde, 1993) and that men are much more receptive than are women to offers of sexual intercourse (Clark & Hatfield, 1989). Byers and Lewis (1988) found that disagreements among dating partners on the desired level of sexual behavior were almost always in the direction of the male partner wanting a higher level of sexual intimacy than that desired by the female partner. Thus, it is possible that many men go into hooking-up situations hoping to engage in more intimate sexual behaviors than are desired by their female partners. Because men are expected to initiate sexual activity (DeLamater, 1987; Peplau & Gordon, 1985), it is possible that in the process of hooking up, some women will experience unwanted sexual advances and possibly even sexual assault or rape.

In their research on hooking up, Kahn et al. (2000) asked 92 female and 50 male college students if they had ever had a "really terrible hooking up experience." Nearly one half of the women (42%) and the men (46%) indicated they had had such an experience. A "terrible experience" for the men was usually due to the women wanting a relationship or to the use of too much alcohol or drugs; none mentioned pressure to go further than they desired. However, nearly one half of the women (48.3%) who reported having a terrible hooking-up experience indicated that they were pressured to go further than they had wanted to go. They gave responses such as "I hooked up with a guy who didn't understand the meaning of 'no'" and "I didn't want to—he did—he wouldn't back off." These women may have experienced sexual assault during a hook up but did not label their experiences as such because they believed the behaviors to be normative. In addition, 10.3% of the women and 11.1% of the men in this sample said the hook up was terrible because they had gone too far without mentioning pressure from partner. Going too far might have been the consequence of pluralistic ignorance, conforming to a presumed norm.

. . . Further, we wanted to examine whether pluralistic ignorance occurred with other sexual behaviors besides sexual intercourse. Based on the research on pluralistic ignorance and gender differences in expected sexual behaviors, we hypothesized that both male and female students would see other students as more comfortable with various hooking-up behaviors than they were them-selves. Although we expected individuals would vary in their own comfort levels with various hooking-up behaviors, we expected they would believe other stu-dents to be uniformly more comfortable engaging in those behaviors than they were themselves. Furthermore, consistent with previous literature, we hypothe-sized that men would be significantly more comfortable than women with engaging in all hooking-up behaviors. Finally, we hypothesized that due to plu-ralistic ignorance, both women and men would overestimate the other gender's comfort with all hooking-up behaviors.

METHOD

Participants

One hundred seventy-five female and 152 male undergraduate students from a mid-sized residential southeastern public university that has few nontraditional students served as participants for the study. The convenience sample represented a moderately even distribution of year in school: for first years, n = 79 (41 females, 38 males); for sophomores, n = 70 (37 females, 33 males); for juniors, n = 84 (45 females, 39 males); and for seniors, n = 93 (52 females, 41 males). A female experimenter approached students as they entered the university library and asked them to volunteer to answer some questions about hooking up and sexual behaviors as part of her senior honors project. She approached other students in their residence halls. No differences appeared between these two samples for any of the dependent measures. Analyses concerning pluralistic igno-rance and comfort with hooking up are based on the data from 136 women (77.7%) and 128 men (84.2%) who indicated that they had hooked up.

RESULTS

. . . We found that both women and men reported less comfort with their per-ceived norm of hooking up than they believed was experienced by their same-sex peers, with men showing a greater difference between self- and peer-ratings than women. In addition, both men and women believed members of the other gender experienced greater comfort with hooking-up behaviors than members of the other gender actually reported. Men were less comfortable with engaging in hooking-up behaviors than women believed them to be, and women were less comfortable with engaging in hooking-up behaviors than men believed them to be. These findings appear to be due to pluralistic ignorance: Hooking up has become the norm for heterosexual sexual relationships on this campus, and since the great majority of students do in fact hook up, it appears that most stu-dents believe that others are comfortable—more comfortable than they are themselves—with engaging in a variety of uncommitted sexual behaviors. It is likely that most students believe others engage in these hooking-up behaviors primarily because they enjoy doing so, while they see themselves engaging in these behaviors primarily due to peer pressure.

Consistent with other pluralistic ignorance research (e.g., Prentice & Miller, 1993), this study showed evidence of an illusion of universality. The students failed to appreciate the extent to which others have different comfort levels with hooking up behaviors. That is, students wrongly assumed that the attitudes of others about hooking up were more homogenous than they actually were.

Similar to other researchers (Cohen & Shotland, 1996; Knox & Wilson, 1981; Oliver & Hyde, 1993), we found that men expressed greater comfort than did women with sexually intimate hooking-up behaviors. In the context of hooking up, this could lead to serious consequences. Our study suggests that men believe women are more comfortable engaging in these behaviors than in fact they are, and also that women believe other women are more comfortable engaging in these behaviors than they are themselves. As a consequence, some men may pressure women to engage in intimate sexual behaviors, and some women may engage in these behaviors or resist only weakly because they believe they are unique in feeling discomfort about engaging in them. In this context it is possible for a woman to experience sexual assault but not interpret the behavior as such, believing it to be normative behavior with which her peers are comfortable.

"Most of Us" is a campaign implemented on many college campuses in an attempt to reveal pluralistic ignorance about alcohol consumption among college students (DeJong & Langford, 2002; Haines, 1998). The campaign is based on providing students with statistical evidence about actual student attitudes and behaviors regarding alcohol consumption. The goal of the campaign is to show that pluralistic ignorance exists regarding college students' heavy alcohol consumption, and that most students prefer to drink less than what is commonly perceived to be the norm. Considering the results of this study, we propose that a similar campaign highlighting students' beliefs about and comfort levels with sexual behaviors while hooking up might help reduce pluralistic ignorance about hooking up.

REFERENCES

Allport, F. H. 1924. *Social psychology*. Boston: Houghton Mifflin.

————. 1933. *Institutional behavior*. Chapel Hill: University of North Carolina Press.

Boswell, A., and J. Spade. 1996. Fraternities and collegiate rape culture. *Gender & Society* 10: 133–47.

Byers, S., and K. Lewis. 1988. Dating couples' disagreements over the desired level of sexual intimacy. *The Journal of Sex Research* 24: 15–29.

Clark, R. D., and E. Hatfield. 1989. Gender differences in receptivity to sexual offers. *Journal of Psychology and Human Sexuality* 2: 39–55.

Cohen, L. L., and R. L. Shotland. 1996. Timing of first sexual intercourse in a relationship: Expectation, experiences, and perceptions of others. *The Journal of Sex Research* 33: 291–99.

DeJong, W., and L. A. Langford. 2002. Typology for campus-based alcohol prevention: Moving toward environmental management strategies. *Journal of Studies on Alcohol Supplement* 14: 140–47.

DeLamater, J. 1987. Gender differences in sexual scenarios. In *Females, males, and sexuality,* ed. K. Kelley, 127–39. Albany, NY: SUNY Press.

Gerrard, M. 1980. Sex guilt and attitudes towards sex in sexually active and inactive female college students. *Journal of Personality Assessment* 44: 258–61.

Gerrard, M., and F. X. Gibbons. 1982. Sexual experience, sex guilt, and sexual moral reasoning. *Journal of Personality* 50: 345–59.

Haines, M. 1998. Social norms: A wellness model for health promotion in higher education. *Wellness Management* 14(4): 1–8.

Kahn, A. S., K. Fricker, J. Hoffman, T. Lambert, M. Tripp, K. Childress, et al. 2000, August. Hooking up: Dangerous new dating methods? In A. S. Kahn (chair), *Sex, unwanted sex, and sexual assault on college campuses.* Symposium conducted at the annual meeting of the American Psychological Association, Washington, DC.

Katz, E., and P. E. Lazarsfeld. 1955. *Personal influence: The part played by people in the flow of mass communication.* Glencoe, IL: Free Press.

Knox, D., and K. Wilson. 1981. Dating behaviors of university students. *Family Relations* 30: 255–58.

Oliver, M. B., and J. S. Hyde. 1993. Gender differences in sexuality: A meta-analysis. *Psychological Bulletin* 114: 129–51.

Paul, E. L., B. McManus, and A. Hayes. 2000. "Hookups": Characteristics and correlates of college students' spontaneous and anonymous sexual experiences. *The Journal of Sex Research* 37: 76–88.

Peplau, L. A., and S. L. Gordon. 1985. Women and men in love: Gender differences in close heterosexual relationships. In *Women, gender, and social psychology,* ed. V. E. O'Leary, R. K. Unger, and B. S. Wallston, 257–92. Hillsdale, NJ: Lawrence Erlbaum Associates.

Prentice, D. A., and D. T. Miller. 1993. Pluralistic ignorance and alcohol use on campus: Some consequences of misperceiving the social norm. *Journal of Personality and Social Psychology* 64: 243–56.

Prentice, D. A., and D. T. Miller. 1996. Pluralistic ignorance and the perpetuation of social norms by unwitting actors. In *Advances in experimental socialpsychology,* ed. M. P. Zanna, vol. 28, 161–209. San Diego, CA: Academic Press.

Simpson, J. A., and S. W. Gangestad. 1991. Individual differences socio-sexuality: Evidence for convergent and discriminant validity. *Journal of Personality and Social Psychology* 60: 870–83.

Snyder, M., J. A. Simpson, and S. Gangestad. 1986. Personality and sexual relations. *Journal of Personality and Social Psychology* 51: 181–90.

DISCUSSION QUESTIONS

1. What are areas of student behavior that might be influenced by pluralistic ignorance?

2. To what extent does hooking up define the sexual norms on your campus? How might the age, gender, race, and social class of students at your school affect this?

37

Masculinity as Homophobia

MICHAEL S. KIMMEL

Michael Kimmel argues that American men are socialized into a very rigid and limiting definition of masculinity. He states that men fear being ridiculed as too feminine by other men and this fear perpetuates homophobic and exclusionary masculinity. He calls for politics of inclusion or broadened definition of manhood to end gender struggle.

The great secret of American manhood is: *We are afraid of other men.* Homophobia is a central organizing principle of our cultural definition of manhood. Homophobia is more than the irrational fear of gay men, more than the fear that we might be perceived as gay. "The word 'faggot' has nothing to do with homosexual experience or even with fears of homosexuals," writes David Leverenz (1986). "It comes out of the depths of manhood: a label of ultimate contempt for anyone who seems sissy, untough, uncool" (p. 455). Homophobia is the fear that other men will unmask us, emasculate us, reveal to us and the world that we do not measure up, that we are not real men. We are afraid to let other men see that fear. Fear makes us ashamed, because the recognition of fear in ourselves is proof to ourselves that we are not as manly as we pretend, that we are, like the young man in a poem by Yeats, "one that ruffles in a manly pose for all his timid heart." Our fear is the fear of humiliation. We are ashamed to be afraid . . .

The fear of being seen as a sissy dominates the cultural definitions of manhood. It starts so early. "Boys among boys are ashamed to be unmanly," wrote one educator in 1871 (cited in Rotundo, 1993, p. 264). I have a standing bet with a friend that I can walk onto any playground in America where 6-year-old boys are happily playing and by asking one question, I can provoke a fight. That question is simple: "Who's a sissy around here?" Once posed, the challenge is made. One of two things is likely to happen. One boy will accuse another of being a sissy, to which that boy will respond that he is not a sissy, that the first boy is. They may have to fight it out to see who's lying. Or a whole group of boys will surround one boy and all shout "He is! He is!" That boy will either burst into tears and run home crying, disgraced, or he will have to take on several boys at once, to prove that he's not a sissy. (And what will his father or older brothers tell him if he chooses to run home crying?) It will be some time before he regains any sense of self-respect.

Violence is often the single most evident marker of manhood. Rather it is the willingness to fight, the desire to fight. The origin of our expression that one has a chip on one's shoulder lies in the practice of an adolescent boy in the country or small town at the turn of the century, who would literally walk around with a chip of wood balanced on his shoulder—a signal of his readiness to fight with anyone who would take the initiative of knocking the chip off (see Gorer, 1964, p. 38; Mead, 1965).

As adolescents, we learn that our peers are a kind of gender police, constantly threatening to unmask us as feminine, as sissies. One of the favorite tricks when I was an adolescent was to ask a boy to look at his fingernails. If he held his palm toward his face and curled his fingers back to see them, he passed the test. He'd looked at his nails "like a man." But if he held the back of his hand away from his face, and looked at his fingernails with arm outstretched, he was immediately ridiculed as a sissy.

As young men we are constantly riding those gender boundaries, checking the fences we have constructed on the perimeter, making sure that nothing even remotely feminine might show through. The possibilities of being unmasked are everywhere. . . . Even the most seemingly insignificant thing can pose a threat or activate that haunting terror. On the day the students in my course "Sociology of Men and Masculinities" were scheduled to discuss homophobia and male-male friendships, one student provided a touching illustration. Noting that it was a beautiful day, the first day of spring after the brutal northeast winter, he decided to wear shorts to class. "I had this really nice pair of new Madras shorts," he commented. "But then I thought to myself, these shorts have lavender and pink in them. Today's class topic is homophobia. Maybe today is not the best day to wear these shorts."

Our efforts to maintain a manly front cover everything we do. What we wear. How we talk. How we walk. What we eat. Every mannerism, every movement contains a coded gender language. Think, for example, of how you would answer the question: How do you "know" if a man is homosexual? When I ask this question in classes or workshops, respondents invariably provide a pretty standard list of stereotypically effeminate behaviors. He walks a certain way, talks a certain way, acts a certain way. He's very emotional; he shows his feelings. One woman commented that she "knows" a man is gay if he really cares about her; another said she knows he's gay if he shows no interest in her, if he leaves her alone.

Now alter the question and imagine what heterosexual men do to make sure no one could possibly get the "wrong idea" about them. Responses typically refer to the original stereotypes, this time as a set of negative rules about behavior. Never dress that way. Never talk or walk that way. Never show your feelings or get emotional. Always be prepared to demonstrate sexual interest in women that you meet, so it is impossible for any woman to get the wrong idea about you. In this sense, homophobia, the fear of being perceived as gay, as not a real man, keeps men exaggerating all the traditional rules of masculinity, including sexual predation with women. Homophobia and sexism go hand in hand. . . .

POWER AND POWERLESSNESS IN THE LIVES OF MEN

. . . Manhood is equated with power—over women, over other men. Everywhere we look, we see the institutional expression of that power—in state and national legislatures, on the boards of directors of every major U.S. corporation or law firm, and in every school and hospital administration. Women have long understood this, and feminist women have spent the past three decades challenging both the public and the private expressions of men's power and acknowledging their fear of men. Feminism as a set of theories both explains women's fear of men and empowers women to confront it both publicly and privately. Feminist women have theorized that masculinity is about the drive for domination, the drive for power, for conquest.

This feminist definition of masculinity as the drive for power is theorized from women's point of view. It is how women experience masculinity. But it assumes a symmetry between the public and the private that does not conform to men's experiences. Feminists observe that women, as a group, do not hold power in our society. They also observe that individually, they, as women, do not feel powerful. They feel afraid, vulnerable. Their observation of the social reality and their individual experiences are therefore symmetrical. Feminism also observes that men, as a group, are in power. Thus, with the same symmetry, feminism has tended to assume that individually men must feel powerful.

This is why the feminist critique of masculinity often falls on deaf ears with men. When confronted with the analysis that men have all the power, many men react incredulously. "What do you mean, men have all the power?" they ask. "What are you talking about? My wife bosses me around. My kids boss me around. My boss bosses me around. I have no power at all! I'm completely powerless!"

Men's feelings are not the feelings of the powerful, but of those who see themselves as powerless. These are the feelings that come inevitably from the discontinuity between the social and the psychological, between the aggregate analysis that reveals how men are in power as a group and the psychological fact that they do not feel powerful as individuals. They are the feelings of men who were raised to believe themselves entitled to feel that power, but do not feel it. No wonder many men are frustrated and angry. . . .

Often the purveyors of the mythopoetic men's movement, that broad umbrella that encompasses all the groups helping men to retrieve this mythic deep manhood, use the image of the chauffeur to describe modern man's position. The chauffeur appears to have the power—he's wearing the uniform, he's in the driver's seat, and he knows where he's going. So, to the observer, the chauffeur looks as though he is in command. But to the chauffeur himself, they note, he is merely taking orders. He is not at all in charge.

Despite the reality that everyone knows chauffeurs do not have the power, this image remains appealing to the men who hear it at these weekend workshops. But there is a missing piece to the image, a piece concealed by the framing of the image in terms of the individual man's experience. That missing piece is that the person who is giving the orders is also a man. Now we have a relationship *between* men—between men giving orders and other men taking those orders. The man who identifies with the chauffeur is entitled to be the man giving the orders, but he is not. ("They," it turns out, are other men.)

The dimension of power is now reinserted into men's experience not only as the product of individual experience but also as the product of relations with other men. In this sense, men's experience of powerlessness is *real*—the men actually feel it and certainly act on it—but it is not *true,* that is, it does not accurately describe their condition. In contrast to women's lives, men's lives are structured around relationships of power and men's differential access to power, as well as the differential access to that power of men as a group. Our imperfect analysis of our own situation leads us to believe that we men need more power, rather than leading us to support feminists' efforts to rearrange power relationships along more equitable lines. . . .

Why, then, do American men feel so powerless? Part of the answer is because we've constructed the rules of manhood so that only the tiniest fraction of men come to believe that they are the biggest of wheels, the sturdiest of oaks, the most virulent repudiators of femininity, the most daring and aggressive. We've managed to disempower the overwhelming majority of American men by other means— such as discriminating on the basis of race, class, ethnicity, age, or sexual preference.

Masculinist retreats to retrieve deep, wounded masculinity are but one of the ways in which American men currently struggle with their fears and their shame. Unfortunately, at the very moment that they work to break down the isolation that governs men's lives, as they enable men to express those fears and that shame, they ignore the social power that men continue to exert over women and the privileges from which they (as the middle-aged, middle-class white men who largely make up these retreats) continue to benefit—regardless of their experiences as wounded victims of oppressive male socialization.

Others still rehearse the politics of exclusion, as if by clearing away the playing field of secure gender identity of any that we deem less than manly—women, gay men, nonnative-born men, men of color—middle-class, straight, white men can reground their sense of themselves without those haunting fears and that deep shame that they are unmanly and will be exposed by other men. This is the manhood of racism, of sexism, of homophobia. It is the manhood that is so chronically insecure that it trembles at the idea of lifting the ban on gays in the military, that is so threatened by women in the workplace that women become the targets of sexual harassment, that is so deeply frightened of equality that it must ensure that the playing field of male competition remains stacked against all newcomers to the game.

Exclusion and escape have been the dominant methods American men have used to keep their fears of humiliation at bay. The fear of emasculation by other men, of being humiliated, of being seen as a sissy, is the leitmotif in my reading of the history of American manhood. Masculinity has become a relentless test by which we prove to other men, to women, and ultimately to ourselves, that we have successfully mastered the part. The restlessness that men feel today is nothing new in American history; we have been anxious and restless for almost two centuries. Neither exclusion nor escape has ever brought us the relief we've sought, and there is no reason to think that either will solve our problems now. Peace of mind, relief from gender struggle, will come only from a politics of inclusion, not exclusion, from standing up for equality and justice, and not by running away.

REFERENCES

Gorer, G. (1964). *The American people: A study in national character.* New York: Norton.

Leverenz, D. (1986, Fall). Manhood, humiliation and public life: Some stories. *Southwest Review,* 71.

Mead, M. (1965). *And keep your powder dry.* New York: William Morrow.

Rotundo, E. A. (1993). *American manhood: Transformations in masculinity from the revolution to the modern era.* New York: Basic Books.

DISCUSSION QUESTIONS

1. Kimmel discusses men's fear of being called a "sissy." Can you think of other examples where men criticize each other's manhood? What are some other terms used to denote femininity as a negative attribute in men?
2. How is manhood defined in other cultures? Is the U.S. ideal of manhood more or less rigid than other examples you identify?

38

The Long Goodbye

DIANE VAUGHAN

Sociologists typically explain how social relationships are formed in society, but here sociologist Diane Vaughan twists the sociological imagination to ask how relationships end. As she shows, there are discernible patterns in social behavior that are the steps to "breaking up."

One week before the wedding, the bride-to-be had dreamed that for the ceremony she would be dressed all in white—except for black shoes. She interpreted this as a warning, and on the second day of the honeymoon her worst fears began to come true. A disagreement turned into a heated argument during which the husband threw his wedding ring across the room and then punched his new bride in the nose. Before leaving for the hospital, the woman searched for the ring and put it back on her husband's finger. This was the first of a long string of similar episodes. Six years later, while they were walking on a snowy

From: Vaughan, Diane. "The Long Goodbye." *Psychology Today* 21 (July 1987): 36–39.

night in Japan, the same thing happened. The woman spent three hours crawling in the snow looking for the ring.

The woman's search for the ring in the snow symbolizes the importance we place on our relationships. They can become so important that we hang on even when they no longer make us happy. We even come up with plausible reasons for doing so: We believe in commitment, we feel bound by the law, we don't want to hurt the other person. We are afraid there may be no one better out there, we believe in our ability to fix things, we are not quitters. We do it for the children, for our parents, for God.

Given all these constraints, it's a wonder couples ever manage to uncouple, but they do. The question is how.

In an attempt to find out, I collected case histories, interviewed counselors and sat in on group sessions for separated and divorced people. For each person, the experience was unique, but I found that no matter what type of couple was involved—married or living together, straight or gay, young or old—there was a discernible pattern in each breakup.

In most cases, one person—the initiator—wants out while the partner wants the relationship to continue. Typically, by the time the partner realizes the relationship is going down the drain, the initiator is already so far gone that efforts to save the relationship are futile.

Partners often report that they were unaware or only remotely aware, even at the time of separation, that the relationship was deteriorating. Initiators tell a different story, reporting months or even years of trying to get the other person's attention.

This apparent breakdown in communication occurs because the closeness of relationships sometimes encourages secrecy rather than openness. In the beginning, each person is so intent on discovering everything about the other person that they develop a sensitivity that allows them to pick up on the smallest cues. They are tuned in to each other and constantly explore and test the nature of the relationship. But once the relationship seems secure, couples often replace the intense energy-consuming monitoring of the early days with a style of communication that suppresses rather than reveals information. Instead of attentively probing, couples confirm that the relationship is healthy by using familiar cues and signals: the words "I love you," holidays with the family, good sex, good times with friends, the routine exchange of "How was your day?"

When a close relationship begins to fall apart, this shorthand method of testing the solidity of the union obscures the changes that are taking place and obstructs the sending and receiving of new information. People may even use these familiar cues and signals to avoid confrontation and to hide their true feelings. Taking this easy way out, however, can have a devastating effect on the relationship. Problems can't be resolved until both parties admit that the relationship is in trouble. And the longer one person's unhappiness remains unacknowledged, the more time he or she has to back off from the relationship.

Dissatisfied partners tend to keep their unhappiness secret at first as they contemplate, brood on and quietly assess the situation. This allows them to continue participating in routine aspects of life with the other person. As unhappiness grows, they begin to reveal it but often in a vague and ill-defined manner.

These early attempts to communicate dissatisfaction usually take the form of complaints aimed at changing the other person or the relationship to better suit the needs of the initiator. He or she may complain, for example, about how the other person spends leisure time. "Why do you watch TV all the time? Why don't you turn it off and do something else?" But television isn't usually the issue. The initiator may be questioning the partner's level of commitment or the appropriateness of a relationship with a person who has nothing better to do than watch the tube.

Such complaints fail to communicate that the relationship is the real problem. At this stage, the initiator may not be able to articulate the deeper reasons for his or her unhappiness. The partner, trying to be cooperative, may even turn off the television, thus eliminating the symptom but not the problem.

Instead of complaining, initiators may use sullenness, anger and a decrease in intimacy to suggest unhappiness. Having always remembered to buy a birthday present, they may bring an inappropriate or a thoughtless gift. The reporting of daily events, usually exchanged at dinner, may get briefer and briefer. When such subtle signals fail, initiators may then turn to interests outside the relationship—a hobby, graduate school, the children, friends. In a healthy relationship, outside interests can add to the excitement and well-being of the situation because they are shared in ways that strengthen ties. In a deteriorating relationship, these activities exclude the partner and widen the already-open gap.

When initiators turn away from the relationship and seek solace in others, they often choose people they believe will be sympathetic to their complaints: a relative, a close friend or work associate, a therapist, a lover. By sharing their unhappiness with others, they create bonds and begin (still unintentionally) to prepare for life without the partner.

As outside bonds grow stronger and more important, the thought of breaking up becomes stronger. Initiators often choose confidants who have gone through a breakup or are in the midst of one—people who can provide relevant information and insights. As initiators gather information, they apply it to their own situation, weighing the costs and benefits of leaving the relationship: Can they make it on their own, will the partner get violent, how will friends react, will the children suffer, is there anyone better out there, what about loneliness?

Even at this advanced stage, initiators are likely to communicate their unhappiness indirectly. They may spend less time with the partner, complain about mundane failings, engage in perfunctory sex. They withhold signals that would be sure to get the partner's attention. Their behavior suggests, "I'm unhappy," but not, "I'm unhappy, and I'm seeing someone else."

Initiators tend to avoid direct confrontation for several reasons. They may be uncertain about their ability to leave the relationship. Many cannot bear to hurt their partner. Some hide their unhappiness to protect themselves from arguments and possible retaliation. Others want out of the relationship but don't want to lose the partner completely.

The signals become increasingly bold, however, as initiators gradually come to see the relationship as unsavable. As outside interests grow more important, they devote less and less time to restoring the relationship. They express discontent to convince the partner that the relationship is not working, rather than to change or improve things.

But even these intensified signals fail to get attention because they are buried in daily routine. A serious airing of feelings, for example, may be interpreted as a sign of some deeper problem at first but as time passes, be seen as only a momentary upset. Or the signals themselves may become routine: What begins as a break in the pattern becomes the pattern. Arguments, working late at the office or even threats to leave lose their impact with repetition.

Even strong, repeated signals of unhappiness may be so inconsistent with the partner's conception of the relationship that they are denied or reinterpreted rather than confronted. Partners may say, for example, "All relationships have trouble. Ours wouldn't be normal if we didn't." "After a while, all couples lose interest in sex." Or partners may interpret negative signals as nothing more than a temporary aberration. When the pressure of work lets up, when the midlife crisis is over, when the children are all in school, when the parents are told that the liaison is gay—then the couple will be happy again.

Ironically, though they are separated in many ways, both members of the relationship remain full-fledged partners in one endeavor: suppressing the true status of the relationship. Even when signals grow so bold that the partner begins to question the initiator, the questions are asked in gentle ways. Nobody wants to discover that a valued way of life is ending. So the partner, too, avoids direct confrontation and gets to hang on a while longer. But by helping the initiator keep the status of the relationship a secret, the partner is missing a chance to do something to solve the problem.

Only when initiators have created what seems to be a secure niche elsewhere and when the costs of staying in the relationship outweigh the benefits will they reveal all to their partners. Until then, they continue to tell without telling. The partners who are about to be left behind will acknowledge the negative signals and confront the problem directly only when the situation becomes so costly in terms of emotional energy, tension and human dignity that they have more to lose by suppression than by revelation. Until then they will continue to know without knowing. It's this conspiracy of secrecy that makes breaking up so very, very hard to do.

DISCUSSION QUESTIONS

1. How might social factors like gender, the amount of money each partner makes, the presence of children, and so forth influence the process of breaking up that Vaughan describes?

2. People usually think of relationships as ending because of problems between individuals, but what external social forces exert pressure on couples that might influence the stability of relationships? How have such social forces changed in recent years and does this increase or decrease the stability of coupled relationships?

http://infotrac.thomsonlearning.com

InfoTrac College Edition

BONUS READING

Morrow, Ronald G., and Diane L. Gill. "Perceptions of Homophobia and Heterosexism in Physical Education." *Research Quarterly for Exercise and Sport* 74 (June 2003): 205–15.

Based on interviews with physical education teachers and young adults, this research documents the hostile enviroment that gay and lesbian youth find in public schools, especially physical education. But it also identifies how teachers can transform homophobic environments.

SEARCH TERMS

homophobia
hooking up
intimate relationships
gender and sexuality
sexual orientation
social construction of sexuality

39

Weaving Work and Motherhood

ANITA GAREY

This article examines the connection between family and work, as experienced in the lives of mothers. Garey argues that an "orientation model" of balancing work and family permeates representations of working mothers in popular culture and academic studies. She suggests instead that work and motherhood are no longer separate forces but are woven together in mothers' lives.

I grew up in the 1950s, but my 1950s was not the one I read about, many years later, in sociology texts, where families were nuclear, fathers were the sole breadwinners, and mothers stayed out of the labor force and were called housewives. And because children assume that the world they know represents the way the world is, my 1950s family was my norm. And in my 1950s, mothers were employed.

My grandmother, my mother, my aunts, and the woman down the street all held jobs. They weren't always full-time jobs, or year-round jobs, or day jobs (although my grandmother sometimes held two jobs), and they certainly weren't "careers," but the women I saw around me were employed. Everywhere I went I saw employed women: my elementary-school teachers, the receptionist and the assistant in my dentist's office, the nurse at the clinic, the beautician who did my grandmother's hair, the grocery clerks at the A&P, the "cafeteria ladies" at school, bank tellers, the school secretary, the salesclerks at Woolworth's and at the candy store, the ticket-seller at the movie theater, the saleswomen in the clothing department of Sears, the "Avon ladies" who came to the door, and the voices of the telephone operators. I had no doubt that women, including mothers, *worked;* I would have been surprised to find that they didn't.

What I've discovered since then is that not all work counts *as work,* in discussions of "working mothers." As a child, I had not yet learned that what counts as *real* work is full-time (forty hours or more), day-shift, year-round employment in a defined occupation. I had not yet learned how not to see the employment of large numbers of women, many of them mothers. This disjuncture between my experience of women and employment and what I read about the 1950s family sensitized me to the missing stories in generalizations about families, mothers, and employment. . . .

Since the 1950s, there has been a steady increase in the proportion of women in the formal labor force. The major part of this increase has been the result of married women entering the labor force, and this has been true across racial-ethnic groups. . . . Two dramatic changes have drawn significant attention to the topic

From: Garey, Anita. *Weaving Work and Motherhood.* Temple University Press, Philadelphia, 1999. Reprinted with permission.

of "working mothers": the increase in the percentage who are employed of mothers with children under six years of age . . . and the increase in the percentage of employed mothers not working in home-based employment. . . .

Although more than two-thirds (70 percent) of all married mothers with children under the age of eighteen are in the labor force, scholarly and popular attention to the topic of working mothers has been for the most part narrowly focused on the small percentage of those mothers employed in managerial or professional positions. While there are a number of excellent studies of women employed in blue-collar or pink-collar jobs, sociological studies of work and family use primarily a dual-career family model that focuses on elite and restrictive careers, despite the fact that this does not reflect the experiences of most women, particularly those of most married women with children. . . . The general image of the "working mother" in popular magazines or in most scholarly discussions of work and family is of the professional or corporate woman, briefcase in hand. The women I saw around me when I was growing up do not fit this image. . . .

While inroads have been made by women, including women with children, into male-dominated, professional occupations, most employed women are concentrated in particular categories of work. In 1990, more women were employed as secretaries than as any other single occupation. . . . Fifty-two percent of employed women in 1995 were in nonsupervisory positions in sales, service, and secretarial occupations. . . . Of all employed women in 1995, for example, only 0.3 percent were physicians, 0.4 percent were lawyers and 0.7 percent were college and university professors. . . . There is a perception that mothers are employed in managerial and professional positions in far larger numbers than is the case, because the majority of employed mothers are missing from cultural images and social analyses of working mothers. . . .

THE ORIENTATION MODEL OF WORK AND FAMILY

. . . Although there is great variation by race-ethnicity and class in the work experiences of women in the United States, most discussions of women, work, and family are embedded in a conceptual frame that I have termed the orientation model of work and family.* For women in the United States, employment and family have been portrayed dichotomously—and women are described as being either "work oriented" or "family oriented." These concepts are not similarly linked for men. To be a "family man" not only includes but necessitates, providing economically for one's family . . . , and while some men may be

---

*Patricia Hill Collins argues that, "in contrast to the cult of true womanhood where work is defined as being in opposition to and incompatible with motherhood, work for Black women has been an important and valued dimension of Afrocentric definitions of Black motherhood" (Collins 1987:5). However, the dominant culture in the United States continues to define employment and motherhood as incompatible. So while that combination may be valued within the African-American community, it is likely to be defined as a problem from outside that community. I am talking here about the conceptual frameworks that shape the images we see in popular culture and about the frameworks used in scholarship and policy-making.

referred to as workaholics, this term is reserved for those perceived to be extreme in the time they give to their employment. For men, employment and family are not portrayed as inevitably detracting from one another. For women, work and family are represented as oppositional arenas that have a zero-sum relationship. In this representation, the more a woman is said to be oriented to her work (employment), the less she is seen as oriented to her family. Regardless of the experience of work and family for individual women or within particular groups of women, *the dominant cultural portrayal of work and family for women in the United States classifies women as either work oriented or family oriented.* . . .

The orientation model of work and family, which has framed most sociological analyses of work and family, fits into a larger ideological framework of separate spheres. The ideology of separate spheres divides the social world into two mutually exclusive areas: the public realm of economic and civic life and the private realm of domestic life. This ideology relegates women to the domestic sphere and men to the public arena. . . . It is not surprising that this ideology emerged in England and the United States in the wake of the industrial revolution, when production moved out of households and into factories. With respect to family life, this meant that men, as fathers, were expected to go outside the home to work in the public world, while women, as mothers, were expected to stay home and be in charge of the domestic world of home and children. . . . Phrases like "breadwinner father" and "stay-at-home mom" epitomize the ideology of separate spheres. But when mothers are employed, the model cannot accommodate them very well; thus employed mothers are described in terms of divided relationships to arenas that are seen as separate and oppositional.

In the same way, conceptualizing women's involvement with employment and family in terms of an orientation obscures the integration and connectedness of that involvement. For example, women who postpone having children until they complete their education, or are established in their occupation, or have saved a certain amount of money are acting in ways connected to *both* family and employment considerations, but they would be categorized as work oriented in an orientation model. Women who work non-day shifts or part-time schedules are also attempting to mesh work and family, but anyone using the orientation model would categorize these women as family oriented. The orientation model of work and family is primarily a behaviorist model that categorizes observable behavior, including what people say, without considering the meaning and context of the behavior to the actors. . . .

"WORKING MOTHERS": MAKING SENSE OF OPPOSITIONAL IMAGES

. . . What does it mean to be a working mother, given that work and family are portrayed as conflicting with each other? How do employed women with children think about their lives and about the ways in which work and motherhood fit into those lives? What does it mean to be a worker with children—and a mother who works?

Basically, a working mother is an employed woman who is also responsible for and in a parental relationship with one or more children. But clearly more is contained in the term "working mother" than this simple definition. We don't, for example, have an equivalent term, "working father," for men—even though most men are employed and most men have or will have children. "Working mother" is a conceptual category meant to encompass the relationship for women between being employed and being a parent, but since that relationship is presented as oppositional, it is a conceptual category that often creates more confusion than clarity.

When I interviewed Danielle, a thirty-six-year-old, Euro-American, full-time clerical worker and the mother of a five-year-old child, I told her that I was interested in what being a working mother meant to women who were employed and who were mothers. Danielle's response illustrates the confusion that can occur in the attempt to get at the meaning of "working mother":

> I don't think that means anything. Not to me. I have a real problem with that phrase, because it implies that women who are at home don't work; it also— men who are parents are never called working fathers [exasperated laugh]. And I guess, from that, men who are parents work and a lot of women who are parents work and that's just, that's what *is*. And it's just this nice phrase that can be attached to one group of those parents—but anyway, I'm trying to think if I would describe myself as a working mother. I don't think I would! I think I'd have to take a few sentences to describe myself. I mean that's just too easy a phrase . . . and it doesn't have a lot of meaning.

This employed mother finds that the term "working mother" doesn't fit her experience; it is a gendered and asymmetrical conceptual category. She wonders what it says about her relationship to work if men are not called working fathers and what it says about the work of mothering if the term "mothers" is qualified by "working" when mothers are employed. She does not, however, find available alternative terms or concepts that simultaneously capture her identity as a mother and an employed person; as she says, "I think I'd have to take a few sentences to describe myself."

Why does Danielle have so much trouble using the term "working mother" or finding terms that would articulate her identity as both a worker and a mother? I suggest that the difficulty stems from the fact that the term "working mother" juxtaposes two words with antithetical cultural images: worker/mother; provider/homemaker; public/private. . . .

. . . The very construction of the term "working mother" points our attention in particular directions: "mother" is a noun, and "working" modifies "mother." We do not say "mothering worker," which conjures up an image of a nurturant employee; nor do we say "mother worker," which sounds more like a job category for nannies. Clearly, when one goes from being a "working woman" to being a "working mother," it is "mother" that, linguistically, stands for the essential self. It is the mother who works, not the worker who has children. It is the mother who must fit into the workplace, not the workplace that must adjust to the needs of workers with children. When workers do adjust their work to their family responsibilities, their actions are interpreted as the actions of their selves *as mothers,* not of their selves *as workers.* It is "mother" that is an identity, "working" that is an activity; it is "mother" that is *being,* "working" that is *doing.*

On the one hand, when we conceptualize mothers as being rather than doing, the work of mothering is hidden. I mean not only the work of maintaining children, although that is certainly a part of the work mothers do, but also all of the ways of acting in relation to one's children that constitute not only the expected norm, but also the actual practice of a great many people. On the other hand, by directing our attention to working as *an activity,* we divert our attention away from what it means to women, including mothers, to *be* workers. Everett Hughes . . . notes that "a man's work is one of the more important parts of his social identity, of his self, indeed of his fate, in the one life he has to live" . . . . Whether Hughes meant men or humankind, it has been shown that work is also an important part of the social identity or being of women, including mothers. . . . Therefore, instead of the orientation model of work and family, we need a framework that makes sense of both the experiences of employed women with children and the experience of *being* employed woman with children.

WEAVING: AN ALTERNATIVE FRAMEWORK

. . . I suggest that we use the metaphor of weaving as a way to look at the lives of employed women with children. Weaving is both a process (an activity—to weave something) and a product (an object—a weaving, something constituted from available materials), and I use the image in both senses. As a process, weaving is a conscious, creative act. It requires not only vision and planning, but also the ability to improvise when materials are scarce, to vary color and texture in response to available resources, to change direction in design, and to splice new yarn. As a product, a weaving reveals both grand patterns and minor designs; it reveals the connections between pattern changes and how what has come before is linked to what follows; and it reveals the richness or thinness of the materials used. . . .

. . . I argue that the metaphor of "weaving" better represents the actions and intentions of employed women with children than the current dominant model of individual orientation that pervades discussions of work and family for women. The conceptual framework of weaving allows us to step back and view the whole, to think of the fabric of a life, the strength of the weave, and the intricacy of design. It reminds us not to get lost in the close examination of one moment or one strand, and to remember that moments and strands are parts of the weave but not the weave itself. Work, family, friendships, reflection, vocation, and recreation are parts of a person's life. They are not, separately and on their own, the life or the person.

NOT A ZERO-SUM GAME

. . . The orientation model of work and family that underlies much of the discussion and thinking about working mothers misrepresents the intersection of employment and motherhood in the lives of women. Motherhood and employment are not incompatible activities in a zero-sum game. And "mothers" and "paid workers" are not opposed categories. The way that we conceptualize the relationship of employment and motherhood is important because the way that we think about an issue shapes what we do about it—or what we think we should do about it. Sociologists stress the importance of understanding the individual's or the group's "definition of the situation," because those definitions—those answers to the question "What's going on here?"—have concrete implications. . . .

There are consequences to defining the situation of employment and motherhood for women in terms of opposition and orientation—and these consequences are not good for women, men, children, and families. By describing as "family oriented" women who are employed on the night shift or who are employed less than forty hours a week (or who put in less than the sixty or seventy hours a week expected in some professions), the orientation model reinforces standard myths about women's marginal relationship to employment—myths that are belied by women's work histories and their expressed feelings about the place of employment in their lives. Defining women's relationship to employment as marginal has had negative consequences for wages, benefits, job security, and national employment and childcare policies. . . .

By describing as "work oriented" mothers who are in male-dominated professions, mothers who are employed forty hours or more per week, mothers who travel as part of their jobs, or even mothers who admit to liking their jobs, the orientation model reproduces cultural ideas that mothers who care about their families do not work full time, or like their jobs, or want to be employed. The expectation that mothers should immerse themselves, to the exclusion of other activities, in the care and nurturance of their children has consequences. For mothers who are employed, the results are often exhaustion from trying to do everything and guilt from feeling they are never doing enough. For nonemployed mothers, the results include economic vulnerability and feelings of resentment toward "working mothers." For fathers, the assumption that children are primarily the responsibility of mothers often results in assumptions that support their nonparticipation in their children's lives. And for children, the result is the lack of societal responsibility for their care and the associated child-unfriendly environment. . . .

Not only does the categorization of these mothers as work oriented reinforce certain cultural constructions of motherhood, it also reproduces cultural understandings of career or commitment to one's job as all-encompassing. This definition of work has consequences for men's and women's relationship to employment and for the organization of the workplace. If work commitment is understood as inherently conflicting with family responsibilities, then men, as breadwinner-fathers, will not be socially expected to share in family work as fathers, sons, and brothers. And women, as mothers, daughters, and sisters, will not be treated seriously in the workplace. Furthermore, employers can use this definition and standard of commitment to rationalize ever-increasing work loads, speed-ups, and demands on workers. If a commitment to job or career is conceptualized as being in conflict with participation in family life, then there is little reason to expect changes in the workplace that will accommodate *both* work *and* family needs. . . .

DISCUSSION QUESTIONS

1. Why does Garey think that the orientation model misrepresents women's experiences of work and motherhood? Why is "weaving" a more appropriate metaphor?

2. Given Garey's argument, what social policies are needed to help women and men weave work and parenthood?

40

Family Rituals and the Construction of Reality

SCOTT COLTRANE

Sociologists see rituals as patterned activities that help groups define their beliefs and social roles. As such, various kinds of rituals are an important element of the culture of a society. Here, Scott Coltrane analyzes a common family ritual—a Thanksgiving dinner—to reveal the family relationships and gender roles that this ritual reaffirms.

HOME FOR THE HOLIDAYS

In the Jodie Foster film *Home for the Holidays,* Claudia (played by Holly Hunter) is a single mother who takes a trip to her parents' house for Thanksgiving. After getting fired from her job at a Chicago museum, Claudia gets a ride to the airport from her teenage daughter. As Claudia grabs her bags and leans over to say good-bye, her daughter tells her that she is planning to have sex with her boyfriend over the weekend. Thus begins a humorous look at one of the most popular family rituals in America—the Thanksgiving holiday feast.

This movie, like television sitcoms, shows what is expected of a "typical" American family at holiday time and simultaneously illustrates some departures

From: Coltrane, Scott. "Family Rituals and the Construction of Reality." In *Gender and Families,* 13–22. Thousand Oaks, CA: Pine Forge Press, 1997. Reprinted with permission.

from the "normal." Suffering from a cold, lamenting her job loss, worrying about her daughter, and fretting about the upcoming visit, Claudia picks up the airplane telephone and calls her brother in Boston. When he doesn't answer, she frantically recounts her tales of woe for the answering machine and pleads with him to join her at their parents' house for the holiday. (This little performance leaves her feeling even more stupid and foolish than she felt before.)

Arriving at her parents' house, Claudia is forced to answer probing questions from her mother (played by Anne Bancroft) about why her daughter did not come with her. Her mother immediately intuits that Claudia was fired from her job, despite Claudia's saying only that she is thinking about a change, whereon her mother closes the door and admonishes her not to tell her father. When Claudia's gay brother, Tommy (played by Robert Downey Jr.), arrives with a handsome friend in tow, Claudia mistakenly assumes that he is her brother's new lover. Later in the movie, we learn that Tommy and his longtime lover, Jack, had a marriage ceremony on a Massachusetts beach some months before. Among the immediate family, only Claudia seems ready to accept Tommy's being gay, and in the beginning, only she accepts Jack as an important part of Tommy's life.

The plot thickens as Claudia and Tommy's high-strung sister shows up with her uptight husband and two bratty children. For some unstated reason, and with much tension, the sister brings in her own turkey and several other Thanksgiving dishes. The women (and daughter) migrate to the kitchen to finish the meal preparations, while the men (and son) sit in the living room. As the feast begins, the women recruit men to carve the turkeys, with obnoxious Tommy reluctantly hacking away at his mother's bird. Unintentionally, but with little remorse, Tommy ends up dumping the entire greasy turkey onto the lap of his uptight sister, who freaks out and flies into a rage. Aunt Gladie, eccentric and senile, has too much to drink and tells everyone how wonderful it was to kiss her brother-in-law some 30 years before. With the tension high and decorum broken, Claudia asks Tommy to field a question about how *her* life is going. To her surprise and amusement, he reveals that she was fired from her job, kissed her 60-year-old boss, and is expecting her daughter to have sex with her boyfriend that evening (which doesn't actually happen after all). Obviously flustered and angry, the mother storms out of the room, retreating to the pantry for a cigarette. Claudia catches up to her and tries to console her, followed by Tommy, who ends up eating with Claudia in the kitchen while the others finish the feast in the dining room.

After the meal, the women go to the kitchen to clean up, and the men go outside to play football. Tommy gets into a fistfight with his sister's husband, and the father, who is washing the sister's car, turns the hose on the "boys," creating even more havoc and leading to a swift departure of the daughter's family. Claudia cannot patch things up with the sister, has both harsh and tender words with her mother, shares a beer watching football with her father, resists having sex with Tommy's amorous friend, and in the end is surprised and delighted to find him joining her on the airplane for a romantic flight back to Chicago.

MAKING SENSE OF THE THANKSGIVING RITUAL

This movie version of a Thanksgiving holiday ritual is a good place to begin exploring how families and gender are socially constructed. Many of today's families look different from this one, and holidays are celebrated differently, depending on family composition, geographical location, ethnicity, income, and family traditions. But we can use this Eastern, white, suburban middle-class family Thanksgiving, with all its quirks and comic relief, to see how family rituals work and why they are important.

Family rituals such as Thanksgiving appear timeless to many people and invoke nostalgic images of "the good old days" and "old-fashioned family values." As historians point out, however, most family rituals looked different in the past, and most holiday rituals such as Thanksgiving and Christmas were not the isolated family-centered emotional events that they have become during the 20th century. Prior to the 1900s, civic festivals and Fourth of July parades in America were much more important occasions for celebration and strong emotion than family holidays. Only in the 20th century did the family come to be the center of festive attention and emotional intensity (Coontz, 1992, p. 17; Skolnick, 1991).

Celebrating family holidays such as Thanksgiving has become more common during the past century, and holidays have increasingly taken on special emotional significance. Why do many American families go through some version of this feast year after year, and what purpose does it serve? Sociologists and anthropologists suggest that celebrating holidays such as Thanksgiving and sharing special meals is one way that families create and reaffirm a sense of themselves. As noted above, families do not have an automatic meaning or definition. That is, they tend to have fuzzy boundaries and need some shared activities to give them shape (Gubrium & Holstein, 1990). Although our culture provides us with an overarching sense of what a family should be, we need to learn what it means to be in a particular family through direct experience and learning, and we need to re-create a sense of belonging over and over again. As described below, periodic family rituals such as birthdays and holiday feasts, along with other activities, actually construct family boundaries and teach us who we are as family members. Who is in, who is out, and what it means to be part of a particular family are literally created and re-created through these routine ceremonial events (see, for example Berger & Kellner, 1964; Gubrium & Holstein, 1990; Imber-Black, Roberts, & Whiting, 1988). Routine rituals help create family scripts—mental representations of ordered events that guide people's actions within and across family settings (Byng-Hall, 1988; Stack & Burton, 1994).

. . . These ceremonial occasions—along with countless other routine family practices—combine to create a sense of ourselves as gendered beings: as mothers or fathers, wives or husbands, daughters or sons, women or men, boys or girls. Ideals from the larger society contribute to these gender ideals, but routine family practices give personal meaning to the larger social definitions and provide us with interpersonal scripts to follow. Ritual family events such as Thanksgiving reinforce conventional expectations about what it is to be a man or a woman, and these messages are then generalized to other social settings.

In Claudia's family's Thanksgiving, as in most American families, the women orchestrated the ritual event. They prepared and served the meal, as well as cleaned up afterward, and we can assume that they also planned the menus, bought the food, and began preparing it well in advance of Thanksgiving Day. The holiday was organized around the feast, and it was the women who made it happen. This was a common pattern in the past and is still typical in many contemporary American families, although things are changing in some families (Coltrane, 1996; DeVault, 1991; diLeonardo, 1987; Fenstermaker-Berk, 1985; Luxton, 1980; Thompson & Walker, 1989). Why aren't men more involved in meal preparation? Why don't boys get asked to set the table or help out with cooking as much as girls? This gender-based division of labor has important implications for the development of different feelings of competence and entitlement in girls and boys, men and women (Hochschild, 1989; Pyke & Coltrane, 1996).

Throughout history, family work—such as cooking and preparing food—and productive work—such as growing and harvesting food—have been closely tied together. Only relatively recently have housework and paid work seemed like separate things, although they are still closely intertwined. Without being fed, clothed, and housed, workers could not stay on the job. Without the resources that come from jobs, people could not maintain households. Recent research shows how paid and unpaid work continue to be linked in modern American families. Studies by economists and sociologists demonstrate that men's ability to get higher-paying jobs allows them to avoid housework (Becker, 1981; Coltrane, 1996; Delphy, 1984; Hartmann, 1981; Shelton, 1992). Because most employers have assumed that wives will perform family work, women's wages and chances for job promotion have been limited (Baxter, 1993; Hochschild, 1989; Luxton, 1980). Opportunities in the job market thus affect decisions about family work at the same time that assumptions about family work affect the structure of the job market (Reskin & Padavic, 1994). What happens in individual families has profound effects on our assumptions about which type of work is appropriate for men and which is appropriate for women (Coltrane, 1989; West & Fenstermaker, 1993).

What makes something women's work or men's work? . . . The tasks performed by one gender or the other are subject to change as historical circumstances change and people's needs are evaluated differently. It seems as if men are supposed to carve turkeys only because people have been exposed to it year after year in family rituals such as Thanksgiving. In *Home for the Holidays,* the father is physically unable to carve the turkey (or at least his wife insists that he is), so Tommy is recruited to perform the task, even though he makes a mockery of it and drops the greasy bird in his horrified sister's lap. Why didn't Claudia or one of the other women assume the task of carving the turkey?

A look back across different civilizations shows that divisions of labor are not the same everywhere. In one society, men might be the only ones to set up the dwellings, whereas in another, this might be a woman's job. In the United States, men are rarely asked to cook but are expected to barbecue hamburgers and carve turkeys. Why should it be "feminine" to stuff and bake the turkey but "masculine" to cut it up? Why are the top professional bakers and chefs virtually always men, although women do most of these tasks in the vast majority of American homes? These questions cannot be answered without looking at issues of power and control in families, topics discussed throughout this book (see also Blumberg & Coleman, 1989; Ferree, 1987; Komter, 1989; Thompson & Walker, 1989; Thorne, 1992; Vannoy-Hiller, 1984).

As shown in *Home for the Holidays,* watching or playing football is also a Thanksgiving tradition in some American families—or more accurately, it is a tradition for some men. While women cook or clean up, men congregate in the living room to watch football on television or go outside to throw the ball around. Although not often acknowledged, this ritual teaches boys that they are entitled to special privileges. They get to play games while the girls are expected to help with the meal preparation, cleanup, and child care. Of course, this does not happen in every family, but for many American men, Thanksgiving is a relaxing event. For most women, it entails work. This does not mean that women do not also find such events to be fun and rewarding. It is mostly women who initiate, plan, and conduct them year after year. But it is clear that ceremonial meals are primarily downtime for men, whereas for women they include work as well as pleasure. This is another example of what sociologist Arlie Hochschild (1989) calls the "leisure gap"—men's greater opportunities for relaxation at home. The leisure gap is narrowing in some families, but it is still likely that at Thanksgiving gatherings in homes across America, the women are waiting on the men.

Another gender division in many families revolves around caring for others and talking about emotions. In *Home for the Holidays,* Claudia and her mother check in with other family members to see how they are doing and initiate conversations about their feelings. They show their concern by talking. Tommy, in contrast, is joking most of the time, and although he is expressive and playful, his teasing is often at others' expense. His father shows his affection for his wife by saying "come 'ere, gorgeous" and dancing with her, but when things get tense around the dinner table, he gets quiet. He demonstrates his concern for his daughter by going outside and washing her car.

In most American families, women consider it their duty to worry about family members and to take care of their everyday needs. They derive great satisfaction from doing so. Often, mothers (and sometimes fathers) focus on making sure that everyone is well fed. Preparing and serving food are thus more than just work because they represent love and care that are given to family members. Women also stereotypically show their love by talking about feelings and trying to help everyone get along. In *Home for the Holidays,* as the women cooked and served, they chatted about the food but also about people and relationships. When the conversation turned argumentative and offensive at the dinner table, Claudia's mother tried, in vain, to make it "nicer," and Claudia eventually took responsibility for letting things get out of hand. The men, in contrast, talked less and focused their conversation on things: work, investments, and football. The tone and style of talk differed as well. The men argued and teased one another, talked louder, changed topics more frequently, and interrupted the women and each other. These conversational patterns reflect gender differences that are common in many American families (Pearson, West, & Turner, 1995; West & Zimmerman, 1987; Wood, 1996).

RITUALS REAFFIRM FAMILY TIES

Home for the Holidays showed an atypical Thanksgiving feast, insofar as the usual tensions and disagreements were contentiously and comically revealed, rather than staying submerged. More typically, people at family gatherings get along a little better, downplay their disagreements and hostilities, and pretend that everything is OK. Acting as if everything is OK (even when it isn't) is one way to normalize the situation and maintain a sense of family unity and continuity. Ritual celebrations such as Thanksgiving allow families to do this on a regular basis.

In a more general way, rituals help construct group identity and create a shared sense of reality. Historically, most rituals started out as community affairs, whereby people got together on special occasions to reaffirm their commitment to some common purpose—an alliance with another clan, a shared religion, a new community settlement, or allegiance to a king or other ruler. These ceremonial gatherings brought people together in face-to-face interaction; focused their attention on some common symbols; heightened their emotions through a group activity such as singing, chanting, and dancing; and linked those emotions to the symbols and to the common purposes of the group (Collins, 1988; Durkheim, 1915/1957; Goffman, 1967). These rituals served many purposes but, most important, gave people a sense of belonging to the group and reaffirmed everyone's commitment to its purposes and symbols.

Although times have changed, we still have many rituals, and they still give people a sense of belonging. Increasingly, modern rituals—at least the face-to-face ones—are centered on family activities. In the past, many public rituals and celebrations were also family based because wealthy families sponsored them, and they tended to solidify alliances between families. Today, family alliances are still important, but they are no longer the central basis of marriages, politics, business, and warfare as they once were. As public displays of family alliances have become less important, family rituals have become more focused on the personal relationships within them. Like earlier rituals, modern family celebrations and activities continue to provide family members with an important sense of place and reinforce feelings of joint membership. These rituals are sometimes linked to national holidays, such as Thanksgiving, but everyday routines, such as eating meals together and watching television together, can serve the ritual function of solidifying family bonds.

Various types of rituals create families, insofar as the family has no fixed definition, and ritualized practices are needed to define and reinforce the meaning of family. If we did not have family rituals such as weddings, anniversaries, and birthdays and holiday get-togethers, we would have a weaker sense of what it means to be a family member. If we did not have more mundane family rituals such as eating together, going on outings, and sharing inside jokes, family would seem less real. Periodic holidays and everyday rituals thus combine to create a shared sense of the family and reinforce our connections to other family members.

In the past, families had many more connections that gave a strong sense of belonging. Most people lived on farms, and family members and relatives contributed on a regular basis to the needs of everyday life. Young adults were dependent on parents and other relatives to learn skills and get jobs. Most people had frequent contact with their parents and grandparents (if they were still alive) because they relied on them for a place to live and for their means of survival. In those days, changes in family membership shaped one's access to resources, and a family member's marriage or death could significantly change the allocation of wealth in the family. The ritual of a wedding symbolically joined the fates of two families, and the ritual of a funeral reaffirmed the living family members' commitments to each other and to the larger family. . . .

REFERENCES

Baxter, J. (1993). *Work at home: The domestic division of labor*. St. Lucia, Australia: Queensland University Press.

Becker, G. (1981). *A treatise on the family*. Cambridge, MA: Harvard University Press.

Berger, P., & Kellner, H. (1964). "Marriage and the construction of reality." *Diogenes, 46,* 1–23.

Blumberg, R. L. (1984). "A general theory of gender stratification." In R. Collins (Ed.), *Sociological Theory 1984*. San Francisco: Jossey-Bass.

Blumberg, R. L., & Coleman, M. T. (1989). "A theoretical look at the gender balance of power in the American couple." *Journal of Family Issues 10*, 225–250.

Byng-Hall, J. (1988). "Scripts and legends in families and family therapy." *Family Process, 27,* 167–179.

Collins, R. (1988). *Theoretical sociology*. San Diego, CA: Harcourt Brace Jovanovich.

Coltrane, Scott. (1989). "Household labor and the routine production of gender." *Social Problems 36*, 473–490.

Coltrane, S. (1996). *Family man: Fatherhood, housework, and gender equity*. New York: Oxford University Press.

Coontz, S. (1992). *The way we never were*. New York: Basic Books.

Delphy, C. (1984). *Close to home: A materialist analysis of women's oppression* (D. Leonard, Trans.). London: Hutchison.

DeVault, M. (1991). *Feeding the family: The social construction of caring as gendered work*. Chicago: University of Chicago Press.

diLeonardo, M. (1987). "The female world of cards and holidays: Women, families, and the work of kinship." *Signs, 12,* 440–453.

Durkheim, E. (1957). *The elementary forms of the religious life*. New York: Free Press (original work published 1915).

Fenstermaker-Berk, S. (1985). *The gender factory*. New York: Plenum.

Ferree, M. (1987). "She works hard for a living." In B. Hess & M. Ferree (Eds.), *Analyzing gender* (pp. 322–347). Newbury Park, CA: Sage.

Goffman, E. (1967). *Interaction ritual*. New York: Doubleday.

Gubrium, J., & Holstein, J. (1990). *What is family?* Mountain View, CA: Mayfield.

Hartmann, H. (1981). "The family as the locus of gender, class, and political struggle: The example of housework." *Signs, 6,* 366–394.

Hochschild, A. (with Manning, A.). (1989). *The second shift.* New York: Viking.

Imber-Black, E., Roberts, J., & Whiting, R. (Eds.). (1988). *Rituals in families and family therapy.* New York: Norton.

Komter, A. (1989). "Hidden power in marriage." *Gender & Society, 3,* 187–216.

Luxton, M. (1980). *More than a labor of love: Three generations of women's work in the home.* Toronto, Ontario, Canada: Women's Press.

Pearson, J., West, R., & Turner, L. (1995). *Gender and communications.* Dubuque, IA: Brown & Benchmark.

Pyke, K., & Coltrane, S. (1996). "Entitlement, gratitude, and obligation in family work." *Journal of Family Issues, 17,* 60–82.

Reskin, B. F., & Padavic, I. (1994). *Women and men at work.* Thousand Oaks, CA: Pine Forge.

Shelton, B. A. (1992). *Women, men, time.* New York: Greenwood.

Skolnick, A. (1991). *Embattled paradise: The American family in an age of uncertainty.* New York: Basic Books.

Stack, C. B., & Burton, L. M. (1994). "Kinscripts: Reflections of family, generation, and culture." In E. N. Glenn, G. Chang, & L. R. Forcey (Eds.), *Mothering: Ideology, experience, and agency* (pp. 33–44). New York: Routledge.

Thompson, L., & Walker, A. J. (1989). "Gender in families: Women and men in marriage, work, and parenthood." *Journal of Marriage and the Family, 51,* 845–871.

Thorne, B. (with Yalom, M., Eds.). (1992). *Rethinking the family: some feminist questions* (Rev. ed.). Boston: Northeastern University Press.

Vannoy-Hiller, D. (1984). "Power dependence and division of family work." *Sex Roles,* 1003–1019.

West, C., & Fenstermaker, S. (1993). "Power and the accomplishment of gender: An ethnomethological perspective." In P. England (Ed.), *Theory on gender/feminism on theory* Social institutions and social change (pp. 151–174). New York: Aldine de Gruyter.

West, C., & Zimmerman, D. (1987). "Doing gender." *Gender & Society, 1,* 125–151.

Wood, J. T. (1996). "She says/he says: Communication, caring, and conflict in heterosexual relationships." In J. T. Wood (Ed.), *Gendered relationships* (pp. 149–162). Mountain View, CA: Mayfield.

DISCUSSION QUESTIONS

1. Coltrane sees family rituals as constructing and reaffirming people's social roles in families. How does gender shape social roles within families and what evidence do you see of this in various family occasions?

2. Families organize household labor through a division of labor. What social factors shape the division of labor in families? Can you explain why?

41

Divorce and Remarriage

TERRY ARENDELL

Increasingly common in family experience, divorce and remarriage are producing new family patterns and family experiences. Terry Arendell reviews patterns in divorce and remarriage, including the impact on children, custody arrangements, and economic consequences of divorce. In addition, she reviews some of the social dynamics found in stepfamilies.

DEMOGRAPHIC PATTERNS IN MARITAL DISSOLUTION AND REMARRIAGE

The divorce rate more than doubled between the early 1960s and mid-1970s. Despite some fluctuations in the annual divorce rates, more than 1 million marriages still are dissolved each year. If trends continue as anticipated, as many as three in five first marriages will end in legal dissolution, as they have since 1980. Second marriages have a somewhat higher termination rate (Gottman, 1994; Kitson & Holmes, 1992; Martin & Bumpass, 1989).

Most likely to divorce are younger adults in shorter term marriages with dependent children. Indeed, children are involved in approximately two thirds of all divorces (U.S. Bureau of the Census, 1995a), and more than half of all children experience their parents' divorces before they reach 18 years of age (Cherlin & Furstenberg, 1994; Furstenberg & Cherlin, 1991; Martin & Bumpass, 1989). Nearly twice as many black children as white children born to married parents will experience parental divorce if trends persist as expected (Amato & Keith, 1991). Marital separation and dissolution rates among parents in other racial and ethnic groups, which generally have been lower than those among whites and blacks, also are increasing (U.S. Bureau of the Census, 1995b). Children who experience divorce spend an average of 5 years in single-parent homes (Glick & Lin, 1986); even among those whose custodial mothers remarry, about half spend 5 years with their mothers alone (Furstenberg, 1990).

Separation and divorce are not the only transitions in parents' marital status and household arrangements experienced by children. Even though remarriage rates are declining, with only about two-thirds of separated or divorced women

From: Arendell, Terry. "Divorce and Remarriage." In *Contemporary Parenting: Challenges and Issues,* edited by Terry Arendell, 154–95. Thousand Oaks, CA: Sage, 1997. Reprinted with permission.

and about three-fourths of men likely to remarry compared to three-fourths and four-fifths, respectively, in the 1960s, more than one third of adults currently in first marriages will divorce and remarry before their youngest children reach age 18. Thus a high proportion of children will experience the remarriage of the parent, if not both, and the formation of a stepfamily or stepfamilies. Moreover, many children will experience the dissolution of a stepfamily when a parent and stepparent divorce. About one in six children will experience two divorces of the custodial parent before the child reaches age 18 (Furstenberg & Cherlin, 1991). Additionally, increasing numbers of adults, including those who are custodial parents of minor children, are cohabitating. Whether they eventually will marry remains to be seen (see Cherlin & Furstenberg, 1994).

Approximately 1 in 10 children in 1992 lived with a biological parent and a stepparent, and this proportion is expected to increase. About 15% of all children lived in blended families—homes in which children lived with at least one stepparent, stepsibling, or half-sibling. More children lived with at least one half-brother or half-sister than with a stepparent or with at least one stepsibling (Furukawa, 1994). Because the practice of cohabitation, or sharing domestic life and intimacy without legal marriage, is increasing steadily, the number of children who reside with a custodial parent and her or his adult partner, who presumably functions, at least to some extent, as a stepparent—*a quasi-stepparent*— probably is much higher than the official numbers indicate (Cherlin & Furstenberg, 1994, pp. 363–365). Because the large majority of children whose parents divorce live with their mothers, most residential stepparents and quasi-stepparents are men. Census data for 1991 show that among children in single-mother families (which includes never-married as well as divorced), 20% also lived with an adult male (related or unrelated) present in the household. About 37% of children living with a single father also lived with an adult female (related or unrelated) (Furukawa, 1994, pp. 1–2). . . .

DIVORCE

Postdivorce Parenting

With respect to family functioning, marital dissolution can be a lengthy process, often underway years before the actual spousal separation occurs. Children show the effects of marital dissension and discontent long before divorce (e.g., Amato & Booth, 1996; Block, Block, & Gjerde, 1986, 1988; Shaw, Emery, & Tuer, 1993). Adjustment to the changes wrought by divorce itself can be a gradual and lengthy process, and many parents and children enter a "crisis" period after the marital separation that can last for several years (e.g., Chase-Lansdale & Hetherington, 1990; Hetherington, 1987, 1988; Morrison & Cherlin, 1995). Maccoby and Mnookin (1992), in the Stanford Custody Project, concluded that

> divorcing parents find it difficult to take the time and trouble required to negotiate with children over task assignments and joint plans. Under these

conditions of diminished parenting, children tend to become bored, moody, and restless and to feel misunderstood; these reactions lead to an increase in behaviors that irritate their parents, and mutually coercive cycles ensue. (pp. 204–205)

A related phenomenon is single parents' lesser ability to make control demands on their children. Examining data from the National Survey of Families and Households, Thomson, McLanahan, and Curtin (1992) found that single parents of both sexes seem to be "structurally limited" in their ability to control and make demands on a child without the presence of another adult.

The extent and duration of uneven parenting, however, varies by families, with some family units adapting fairly rapidly to their altered circumstances and arrangements, achieving stable and healthy family functioning rather soon after divorce. Some units take much longer to find an equilibrium. Others have a delayed reaction, functioning well initially and then encountering adjustment difficulties (e.g., Kitson & Holmes, 1992). In addition, "some show intense and enduring deleterious outcomes" (Hetherington, 1993, p. 40). Whatever the pattern, parental functioning usually recovers over time, returning nearly to the level found in intact families (Hetherington, 1988; Hetherington, Cox, & Cox, 1982). That is, most family units formed by divorce establish workable and functional interactional processes (Maccoby & Mnookin, 1992; Wallerstein & Blakeslee, 1989).

One of the first major tasks facing parents in divorce is that of determining children's living arrangements as family members separate into two households. Most custody decisions occur with little discussion between the parents, and relatively few custody allocations are actually litigated. Yet the working out of parenting and parental relationships after divorce, including children's access to and involvement with the nonresidential parent if parenting is not shared, can be complicated and difficult, involving various changes and intraparental conflicts. Of the four relationships between married persons that must be altered in divorce—parental, economic, spousal, and legal (Maccoby & Mnookin, 1992)—the parental divorce is perhaps the most difficult to achieve (Ahrons & Wallisch, 1987, p. 228; see also Bohannon, 1970).

Custody Arrangements for Minor Children

Three residential patterns are available for children in divorce: maternal, paternal, and dual. Primary physical custody, maternal or paternal, is the situation in which children spend more than 10 overnights in a 2-week period with a particular parent (Mnookin, Maccoby, Albiston, & Depner, 1990, pp. 40–41). Dual or shared custody is defined as the situation in which "the children spend at least a third of their time in each household" (Maccoby & Mnookin, 1992, p. 203). Shared custody is unusual. Even in California, where dual custody probably is more common than anywhere else, only about one in six children actually lives in a shared custody situation. And in these circumstances, "more often than not" mothers handle the bulk of the managerial aspects of child rearing (Maccoby & Mnookin, 1992, p. 269).

As has been the case for most of this century, maternal custody is predominant; more than 85% of children whose parents are divorced are in the custody of their mothers (U.S. Bureau of the Census, 1995a). A somewhat higher proportion of offspring actually reside with their mothers because, in legally mandated dual-custody situations, children often spend relatively little time with their fathers (e.g., Maccoby & Mnookin, 1992; Seltzer, 1991; Seltzer & Bianchi, 1988). Overwhelmingly, then, it is mothers who become the primary parents in divorce. . . .

Economic Support of Children

Although the preseparation parenting division of labor persists after divorce with mothers doing most of the parenting, what does change is the economic providing for minor children. Whereas men's earned incomes provide the larger share of the economic resources available to intact married families, divorced custodial women assume most financial responsibilities for their offspring. The overwhelming body of scholarly research and governmental and other policy studies shows that fathers' contributions to the economic support of their children are much reduced after marital dissolution (e.g., Kellan, 1995; Maccoby & Mnookin, 1992) despite many men's claims to the contrary (e.g., Arendell, 1995). Approximately three fourths of divorced mothers have child support agreements, but only about half of those women receive the full amounts ordered in the agreements (Holden & Smock, 1991; Scoon-Rogers & Lester, 1995). One fourth receive no payment whatsoever, and the other one fourth receive irregular payments in amounts less than those ordered. According to the Congressional Research Service, only about $13 billion of the $34 billion in outstanding support orders was collected in 1993 (Kellan, 1995, p. 27). Moreover, child support payments amounted to only about 16% of the incomes of divorced mothers and their children in 1991. The average monthly child support paid by divorced fathers contributing economic support in 1991 was $302, amounting to $3,623 for the year (Scoon-Rogers & Lester, 1995). Fathers' limited or lack of financial contributions to the support of their children not residing with them is not offset by other kinds of assistance (Teachman, 1991, p. 360).

As a group, women's incomes drop more than 30% following divorce. About 40% of divorcing women lose more than half of their family incomes, whereas fewer than 17% of men experience this large a drop (Hoffman & Duncan, 1988). Men, in general, experience an increase in their incomes—an average of 15%—partially because they share less of their incomes with their children (Furstenberg & Cherlin, 1991; Kitson & Holmes, 1992; Maccoby & Mnookin, 1992). For many women, the financial hardships accompanying divorce become the overriding experience, affecting psychological well-being and parenting as well as dictating decisions such as where to live, what type of child care to use, and whether or not to obtain health care (Arendell, 1986; Kurz, 1995).

Children's economic well-being after divorce is directly related to their mothers' economic situations. Those living with single mothers are far more likely to be poor than are children in other living arrangements; families headed by single mothers are nearly six times as likely to be impoverished as are families having both parents present (U.S. Bureau of the Census, 1995a). This is not the experience of children being raised by single fathers because men's wages are higher than women's (Holden & Smock, 1991; Scoon-Rogers & Lester, 1995; Seltzer & Garfinkel, 1990); about one eighth of custodial fathers, compared to nearly two fifths of divorced custodial mothers, are poor (Scoon-Rogers & Lester, 1995). Divorced women and their children do not regain their predivorce standards of living until 5 years after the marital breakups. Women's decisions to remarry often involve economic considerations; the surest route to financial well-being for many women is remarriage, not their employment, even when it is full-time (Furstenberg & Cherlin, 1991; Kitson & Morgan, 1990). . . .

CHILD OUTCOMES IN DIVORCE

How children fare with divorce is a crucial question, one intimately related to issues of parenting. The arguments vary, with assertions ranging from children being irreparably damaged to children adapting successfully to divorce. Most research evidence suggests that a large majority of children adjust reasonably well to their parents' marital dissolutions. . . .

Some argue that the research findings on the effects of divorce on children are not so clear-cut (see, for review, Bolgar, Sweig-Frank, & Paris, 1995). But even those arguments are tempered when large data sets are the bases of analysis, especially those involving longitudinal studies and not just small, nonrepresentative samples (Amato & Booth, 1996). That a majority of children seem to cope with and adapt well to the change in their parents' marital status is particularly salient because many children enter the divorce phase already disadvantaged by exposure, often of long duration, to parental strife and conflict (Block et al., 1986, 1988; Chase-Lansdale & Hetherington, 1990). Furthermore, as numerous scholars point out, children who experience the dissolution of their parents' marriages may well have to cope with multiple adverse circumstances including family events prior to divorce (e.g., Furstenberg & Teitler, 1994). Allen (1993, p. 47), for example, argued that when scholars (and others) uncritically compare divorced families to nondivorced ones, they imply that two-parent intact families inevitably result in positive parenting outcomes. Other events that might be more detrimental than divorce itself, as she notes, are father abandonment; failure to pay child support; neglect; intersection of class, race, and gender with poverty; and women's inequality in traditional families.

Some earlier findings suggested that a child's sex and age mattered in postdivorce adjustment. Hetherington (1993, pp. 48–49), drawing from recent work, concluded that these variables—sex and age—are not pivotal factors in children's divorce responses and adjustments (see also Furstenberg & Teitler, 1994; Garasky, 1995). Sex differences in adverse responses, previously attributed to boys, disappeared in Hetherington's (1993) longitudinal study as children moved into adolescence. Where age mattered, it was for adolescents, all of whom showed somewhat increased problem behaviors. Children with divorced and remarried parents did show more such problem behaviors than did those whose parents remained married. "Adolescence often triggered problems in children from divorced and remarried families who had previously seemed to be coping well" (p. 49). Fursten-berg and Teitler (1994) summarized findings pertaining to adolescents:

> The findings indicate that certain effects of divorce are quite persistent even when we consider a wide range of predivorce conditions. Early timing of sexual activity, nonmarital cohabitation, and high school dropout do appear to be more frequent for children from divorced families. (p. 188).

The researchers note that these outcomes may be a result of growing up in single-parent homes or witnessing parents' marital transitions, among other things, not just divorce itself (p. 188).

Also, in contrast to earlier arguments, being reared by same-sex parents appears not to be inherently beneficial to children (Powell & Downey, 1995). And, although it may seem counterintuitive, children's overall well-being in divorce does not seem related to the extent of involvement or quality of parent-child relationship with the noncustodial father (e.g., Amato & Keith, 1991; Bolgar et al., 1995; Furstenberg & Cherlin, 1991). . . .

REMARRIAGE

Research attention to stepparenting has increased dramatically in the past 15 years as the divorce and remarriage rates have escalated and remain high. For instance, Coleman and Ganong (1990, p. 925) noted that there were only a handful of studies published prior to 1980 but more than 200 during the decade of the 1980s. The increased attention has continued into the 1990s (e.g., see Booth & Dunn, 1994).

The circumstances leading to the formation of stepfamilies vary. They espe-cially include the marriages of formerly unmarried teen mothers, widowed parents, and divorced ones. Prior to the early 1970s, the death of a spouse was the principal prior circumstance leading a parent to remarry, not divorce as is now the case. Even just among those formed by divorced parents, stepparent families are diverse in composition. For instance, Dunn and Booth (1994, p. 220) noted that two scholars, Burgoyne and Clark (1982), had identified 26 different types. Chil-dren may reside with either a stepmother or a stepfather, although the latter is far more common given the preponderance of mother custody. Or, children may have a nonresidential stepparent. Additionally, children may have stepsiblings and half-siblings with whom they may or may not share residences. More specifically,

> somewhere between two-fifths and half of these children [whose parents remarry] will have a stepsibling, although most will not typically live with him or her. And for more than a quarter, a half-sibling will be born within four years. Thus, about two-thirds of children living in stepfamilies will have either half-siblings or stepsiblings. (Furstenberg, 1990, p. 154)

Depending on their cognitive developmental stage, children construct family relatedness with stepsiblings and half-siblings in various ways, adding to the com-plexity in understanding family relationships (Bernstein, 1988). . . .

The amount of domestic life and parenting shared with nonresidential family members can range greatly between family units and across time for particular children. Variations among stepfamilies occur, moreover, not only in their configurations but also in their functioning.

The remarriage of a divorced parent and creation of a stepfamily entail numerous disruptions and transitions. Altered by the entry of another adult into the family is the family system established by the custodial parent and children following divorce.

. . . Children, and sometimes the custodial parent, often resist a newcomer's efforts to exert authority and alter the existing family dynamics (Hetherington, 1993; Hetherington et al., 1992). Disruption is not limited to the relationship between the stepparent and stepchildren; it can involve the relationship between the custodial parent and children as well. Conflict within the original unit often increases (Brooks-Gunn, 1994, p. 179). Other problems may include a decline in parental supervision and responsibility as the parent divides her time between a new spouse and her children, shifting alliances between family members, and open tension and disputes between children and stepparent and between children of the original unit (e.g., Brooks-Gunn, 1994, p. 170; Hetherington, 1993; Hetherington et al., 1992). Nor are interpersonal tensions and difficulties limited to the residential unit. They may involve the noncustodial parent, his spouse, or other relatives, such as grandparents, aunts, or uncles. Dealing with the larger family context is an ongoing, lengthy, and demanding process (Mills, 1988; more generally, see Beer, 1988).

Some stepparents respond to children's resistance by becoming more authoritarian and dogmatic. Others, on the other hand, withdraw emotionally and cease their attempts to forge intimate relationships. They move to "exhibiting little warmth, control, or monitoring. [These] stepparents are not necessarily negative, they are just distant" (Brooks-Gunn, 1994, p. 179; Hetherington, 1993). Whatever the strategy assumed by stepparents, it has direct impacts on the home ambiance and parent-child relationships. In turn, these all affect the interactional dynamics between spouses; the effects become circular and interactive.

As with other kinds of family transitions, restabilization often follows the initial disequilibrium experienced by the newly formed stepfamily (Ahrons & Wallisch, 1987; Hetherington, 1993; Hetherington et al., 1992). The successful integration of a stepparent into a family is a gradual process, sometimes taking years (e.g., Papernow, 1988, p. 60). Not all families reach such a level; indeed, a large number of stepfamilies dissolve through divorce long before they ever approach the place of becoming smoothly functioning households. . . .

CONCLUSION

In conclusion, a sizable proportion of American children will experience their parents' divorce. The majority of these children will be parented predominantly by one parent, not by both parents. Many of these children also will experience the formation of a stepparent family when a parent remarries. In many families, both parents will remarry, resulting in situations where children have both a live-in and a live-out stepparent. And numerous children will experience another parental divorce. Children, then, are experiencing multiple transitions in the composition and arrangements of their families. Current evidence indicates that the vast majority of children adjust to these changes successfully. What is most crucial in children's well-being and positive outcomes, according to a growing body of research, is the quality and constancy of the parenting by the primary parent. Experiencing relatively low intraparental and other family conflict is crucial for children's adjustment to changing circumstances and positive development.

REFERENCES

Ahrons, C. R., & Wallisch, L. (1987). Parenting in the binuclear family relationships between biological and stepparents. In K. Pasley & M. Ihinger-Tallman (Eds.), *Remarriage and stepparenting: Current research and theory* (pp. 225–256). New York: Guilford.

Allen, K. R. (1993). The dispassionate discourse of children's adjustment to divorce. *Journal of Marriage and the Family, 55,* 46–50.

Amato, P. R., & Booth, A. (1996). A prospective study of divorce and parent-child relationships. *Journal of Marriage and the Family, 58,* 356–365.

Amato, P. R., & Keith, B. (1991). Parental divorce and the well-being of children: A meta-analysis. *Psychological Bulletin, 11,* 26–46.

Arendell, T. (1986). Mothers and divorce: Legal, economic, and social dilemmas. Berkley: University of California Press.

Arendell, T. (1995). *Fathers and divorce.* Thousand Oaks, CA: Sage.

Beer, W. R. (Ed.). (1988). *Relative strangers: Studies of stepfamilies processes.* Totowa, NJ: Rowman & Littlefield.

Bernstein, A. C. (1988). Unraveling the tangles: Children's understanding of stepfamily kinship. In W. Beer (Ed.), *Relative Strangers* (pp. 83–111). Totowa, NJ: Rowman & Littlefield.

Block, J. H., Block, J., & Gjerde, P. F. (1986). Personality of children prior to divorce: A prospective study. *Child Development, 57,* 827–840.

Block, J. H., Block, J., & Gjerde, P. F. (1988). Parental functioning and the home environment in families of divorce: Prospective and concurrent analyses. *Journal of American Academy of Child and Adolescent Psychiatry, 27,* 207–213.

Bohannon, P. (1970). Divorce and after. Garden City, NY: Doubleday.

Bolgar, R., Sweig-Frank, H., & Paris, J. (1995). Childhood antecedents of interpersonal problems in young adult children of divorce. *Journal of the American Academy of Child and Adolescent Psychiatry, 34*(2), 143–150.

Booth, A., & Dunn, J. (Eds.). (1994). *Stepfamilies: Who benefits? Who does not?* Mahwah, NJ: Lawrence Erlbaum.

Brooks-Gunn, J. (1994). Research on stepparenting families: Integrating disciplinary approaches and informing policy. In A. Booth & J. Dunn (Eds.), *Stepfamilies: Who benefits? Who does not?* (pp. 167–204). Mahwah, NJ: Lawrence Erlbaum.

Burgoyne, J., & Clark, D. (1982). Parenting in stepfamilies. In R. Chester, P. Diggory, & M. Sutherland (Eds.), *Changing patterns of child-bearing and child-rearing* (pp. 133–147). London: Academic Press.

Chase-Lansdale, P. L., & Hetherington, E. M. (1990). The impact of divorce on life-span development: Short and long term effects. In D. Featherman & R. Lerner (Eds.), *Life span development and behavior* (Vol. 10, pp. 105–150). Hillsdale, NJ: Lawrence Erlbaum.

Cherlin, A. J., & Furstenberg, F. F., Jr. (1994). Stepfamilies in the United States: A reconsideration. *Annual Review of Sociology, 20,* 359–381.

Coleman, M. & Ganong, L. H. (1990). Remarriage and stepfamily research in the 1980s: Increased interest in an old family form. *Journal of Marriage and the Family, 52,* 925–940.

Dunn, J., & Booth, A. (1994). Stepfamilies: An overview. In A. Booth & J. Dunn (Eds.), *Stepfamilies: Who benefits? Who does not?* (pp. 217–224). Mahwah, NJ: Lawrence Erlbaum.

Furstenberg, F. F. (1990). Coming of age in a changing family system. In S. Feldman & G. Elliot (Eds.), *At the threshold: The developing adolescent* (pp. 147–170). Cambridge, MA: Harvard University Press.

Furstenberg, F. F., & Cherlin, A. (1991). *Divided families: What happens to children when parents part.* Cambridge, MA: Harvard University Press.

Furstenberg, F., & Teitler, J. O. (1994). Reconsidering the effects of marital disruption: What happens to the children of divorce in early adulthood. *Journal of Family Issues, 15,* 173–190.

Furukawa, S. (1994). The diverse living arrangements of children: Summer 1991. In U.S. Bureau of the Census, *Current Population Reports* (Series P70–38). Washington, DC: Government Printing Office.

Garasky, S. (1995). The effects of family structure on educational attainment: Do the effects vary by the age of the child? *American Journal of Economics and Sociology, 54*(1), 89–106.

Glick, P. C., & Lin, S.-L. (1986). Recent changes in divorce and remarriage. *Journal of Marriage and the Family, 48,* 737–747.

Gottman, J. M. (1994). *What predicts divorce? The relationship between marital processes and marital outcomes.* Hillsdale, NJ: Lawrence Erlbaum.

Hetherington, E. M. (1987). Family relations six years after divorce. In K. Pasley & M. Ihinger-Tallman (Eds.), *Remarriage and stepparenting: Current research and theory* (pp. 185–205). New York: Guilford.

Hetherington, E. M. (1988). Parents, children, and siblings six years after divorce. In R. Hinde & J. Stevenson-Hinde (Eds.), *Relationships within families* (pp. 311–331). Cambridge, UK: Clarendon.

Hetherington, E. M. (1993). An overview of the Virginia Longitudinal Study of Divorce and Remarriage with a focus on early adolescence. *Journal of Family Psychology, 7,* 39–56.

Hetherington, E. M., & Clingempeel, W. G., with Anderson, E., Deal, J., Hagan, M. S., Hollier, A., & Lindner, M. (1992). Coping with marital transitions: A family systems perspective. *Monographs of the Society for Research in Child Development, 57*(2-3, Serial No. 227), 1–14.

Hetherington, E., Cox, M., & Cox, R. (1982). Effects of parents and children. In M. Lamb (Ed.), *Nontraditional families: Parenting and child development* (pp. 233–288). Hillsdale, NJ: Lawrence Erlbaum.

Hoffman, S. D., & Duncan, G. D. (1988). What are the consequences of divorce? *Demography, 23,* 641–645.

Holden, K., & Smock, P. J. (1991). The economic costs of marital dissolution: Why do women bear a disproportionate cost? *Annual Review of Sociology, 17,* 51–78.

Kellan, S. (1995). Child custody and support. *Congressional Quarterly Researcher, 5*(2), 25–48.

Kitson, G. C., with Holmes, W. M. (1992). *Portrait of divorce: Adjustment to marital breakdown.* New York: Guilford.

Kitson, G., & Morgan, L. (1990). The multiple consequences of divorce: A decade review. *Journal of Marriage and the Family, 52,* 913–924.

Kurz, D. (1995). *For better or for worse: Mothers confront divorce.* New York: Routledge.

Maccoby, E. E., & Mnookin, R. H. (1992). *Dividing the child: Social and legal dilemmas of custody.* Cambridge, MA: Harvard University Press.

Martin, T. C., & Bumpass, L. L. (1989). Recent trends in marital disruption. *Demography,* *26*(1), 37–51.

Mills, D. M. (1988). Stepfamilies in context. In W. Beer (Ed.), *Relative strangers* (pp. 1–29). Totowa, NJ: Rowman & Littlefield.

Mnookin, R., Maccoby, E. E., Albiston, C. R., & Depner, C. E. (1990). Private ordering revisited: What custodial arrangements are parents negotiating? In S. Sugarman & H. H. Kay (Eds.), *Divorce reform at the crossroads* (pp. 37–74). New Haven, CT: Yale University Press.

Morrison, D. R., & Cherlin, F. J. (1995). The divorce process and young children's well-being: A perspective analysis. *Journal of Marriage and the Family, 57,* 800–812.

Papernow, P. L. (1988). Stepparent role development: From outsider to intimate. In W. Beer (Ed.), *Relative strangers* (pp. 54–82). Totowa, NJ: Rowman & Littlefield.

Powell, B., & Downey, D. B. (1995, August). *Well-being of adolescents in single-parent households: The case of the same-sex hypothesis.* Paper presented at the annual meeting of the American Sociological Association, Washington, DC.

Scoon-Rogers, L., & Lester, G. H. (1995). Child support for custodial mothers and fathers: 1991. In U.S. Bureau of the Census, *Current population reports* (Series P60–187). Washington, DC: Government Printing Office.

Seltzer, J. A. (1991). Relationships between fathers and children who live apart: The father's role after separation. *Journal of Marriage and the Family, 53,* 79–101.

Seltzer, J. A., & Bianchi, S. M. (1988). Children's contact with absent parents. *Journal of Marriage and the Family, 50,* 663–677.

Seltzer, J. A. & Garfinkel, I. (1990). Inequality in divorce settlements: An investigation of property settlements and child support awards. *Social Science Research, 19,* 82–111.

Shaw, D. S., Emery, R. E., & Tuer, M. D. (1993). Parental functioning and children's adjustment in families of divorce: A prospective study. *Journal of Abnormal Child Psychology, 21,* 119–134.

Teachman, J. (1991). Contributions to children by divorced fathers. *Social Problems, 38,* 358–371.

Thomson, E., McLanahan, S. S., & Curtin, R. B. (1992). Family structure, gender, and parental socialization. *Journal of Marriage and the Family, 54,* 368–378.

U.S. Bureau of the Census. (1995a). Child support for custodial mothers and fathers: 1991. In *Current population reports* (Series P60–187). Washington, DC: Government Printing Office.

U.S. Bureau of the Census. (1995b). *Statistical abstract of the United States, 1994.* Washington, DC: Government Printing Office.

Wallerstein, J., & Blakeslee, S. (1989). *Second chances: Men, women, and children a decade after divorce.* New York: Ticknor & Fields.

DISCUSSION QUESTIONS

1. What are the factors that research has identified as affecting children's well-being after divorce? Are there others that you would add?

2. Remarriage produces disruptions and transitions in family life that affect all family members. What are some of the processes that emerge in the creation of stepfamilies and how are different family members affected by these changes?

42

Caring for Our Young: Child Care in Europe and the United States

DAN CLAWSON AND NAOMI GERSTEL

This article contrasts child-care practices in the United States and Europe. The authors find that most parents in the United States struggle to find quality, affordable child care. The authors argue that you can identify what a society values by examining its child-care systems.

When a delegation of American child care experts visited France, they were amazed by the full-day, free *écoles maternelles* that enroll almost 100 percent of French three-, four- and five-year-olds:

> Libraries better stocked than those in many U.S. elementary schools. Three-year-olds serving one another radicchio salad, then using cloth napkins, knives, forks and real glasses of milk to wash down their bread and chicken. Young children asked whether dragons exist [as] a lesson in developing vocabulary and creative thinking.

In the United States, by contrast, working parents struggle to arrange and pay for private care. Publicly-funded child care programs are restricted to the poor. Although most U.S. parents believe (or want to believe) that their children receive quality care, standardized ratings find most of the care mediocre and much of it seriously inadequate.

Looking at child care in comparative perspective offers us an opportunity—almost requires us—to think about our goals and hopes for children, parents, education and levels of social inequality. Any child care program or funding system has social and political assumptions with far-reaching consequences. National systems vary in their emphasis on education; for three- to five-year-olds, some stress child care as preparation for school, while others take a more playful view of childhood. Systems vary in the extent to which they stress that children's early development depends on interaction with peers or some version of intensive mothering. They also vary in the extent to which they support policies promoting center-based care as opposed to time for parents to stay at home with their very young children. Each of these emphases entails different national assumptions, if only implicit, about children and parents, education, teachers, peers and societies as a whole.

From: Clawson, Dan, and Naomi Gerstel. "Caring for Our Young: Child Care in Europe and the United States." *Contexts* (Fall/Winter 2002): 28–35. Reprinted with permission.

What do we want, why and what are the implications? Rethinking these questions is timely because with changing welfare, employment, and family patterns, more U.S. parents have come to believe they want and need a place for their children in child care centers. Even parents who are not in the labor force want their children to spend time in preschool. In the United States almost half of children less than one year old now spend a good portion of their day in some form of non-parental care. Experts increasingly emphasize the potential benefits of child care. A recent National Academy of Sciences report summarizes the views of experts: "Higher quality care is associated with outcomes that all parents want to see in their children." The word in Congress these days, especially in discussions of welfare reform, is that child care is good—it saves money later on by helping kids through school (which keeps them out of jail), and it helps keep mothers on the job and families together. A generation ago, by contrast, Nixon vetoed a child care bill as a "radical piece of social legislation" designed to deliver children to "communal approaches to child rearing over and against the family-centered approach." While today's vision is clearly different, most attempts to improve U.S. child care are incremental efforts to get a little more money here or there, with little consideration for what kind of system is being created.

The U.S. and French systems offer sharp contrasts. Although many hold up the French system as a model for children three or older, it is only one alternative. Other European countries provide thought-provoking alternatives, but the U.S.-French contrast is a good place to begin.

FRANCE AND THE UNITED STATES: PRIVATE VERSUS PUBLIC CARE

Until their children start school, most U.S. parents struggle to find child care, endure long waiting lists, and frequently change locations. They must weave a complex, often unreliable patchwork in which their children move among relatives, informal settings and formal center care, sometimes all in one day. Among three- to four-year-old children with employed mothers, more than one out of eight are in three or more child care arrangements, and almost half are in two or more arrangements. A very small number of the wealthy hire nannies, often immigrants; more parents place their youngest children with relatives, especially grandmothers, or work alternate shifts so fathers can share child care with mothers (these alternating shifters now include almost one-third of families with infants and toddlers). Many pay kin to provide child care—sometimes not because they prefer it, but because they cannot afford other care, and it is a way to provide jobs and income to struggling family members. For children three and older, however, the fastest-growing setting in the United States is child care centers—almost half of three-year-olds (46 percent) and almost two-thirds of four-year-olds (64 percent) now spend much of their time there.

In France, participation in the *école maternelle* system is voluntary, but a place is guaranteed to every child three to six years old. Almost 100 percent of parents enroll their three-year-olds. Even non-employed parents enroll their children, because they believe it is best for the children. Schools are open from 8:30 A.M. to 4:30 P.M. with an extended lunch break, but care is available at modest cost before and after school and during the lunch break.

Integrated with the school system, French child care is intended primarily as early education. All children, rich and poor, immigrant or not, are part of the same national system, with the same curriculum, staffed by teachers paid good wages by the same national ministry. No major political party or group opposes the system.

When extra assistance is offered, rather than targeting poor children (or families), additional resources are provided to geographic areas. Schools in some zones, mostly in urban areas, receive extra funding to reduce class size, give teachers extra training and a bonus, provide extra materials and employ special teachers. By targeting an entire area, poor children are not singled out (as they are in U.S. free lunch programs).

Staff in the French *écoles maternelles* have master's degrees and are paid teachers' wages; in 1998, U.S. preschool teachers earned an average of $8.32 an hour, and child care workers earned $6.61, not only considerably less than (underpaid) teachers but also less than parking lot attendants. As a consequence, employee turnover averages 30 percent a year, with predictably harmful effects on children.

What are the costs of these two very different systems? In almost every community across the United States, a year of child care costs more than a year at a public university—in some cases twice as much. Subsidy systems favor the poor, but subsidies (unlike tax breaks) depend on the level of appropriations. Congress does not appropriate enough money and, therefore, most of the children who qualify for subsidies do not receive them. In 1999, under federal rules 15 million children were eligible to receive benefits, but only 1.8 million actually received them. Middle- and working-class families can receive neither kind of subsidy. An Urban Institute study suggests that some parents place their children in care they consider unsatisfactory because other arrangements are just too expensive. The quality of care thus differs drastically depending on the parents' income, geographic location, diligence in searching out alternatives and luck.

The French system is not cheap. According to French government figures, the cost for a child in Paris was about $5,500 per year in 1999. That is only slightly more than the average U.S. parent paid for the care of a four-year-old in a center ($5,242 in 2000). But in France child care is a social responsibility, and thus free to parents, while in the United States parents pay the cost. Put another way, France spends about 1 percent of its Gross Domestic Product (GDP) on government-funded early education and care programs. If the United States devoted the same share of its GDP to preschools, the government would spend about $100 billion a year. Current U.S. government spending is less than $20 billion a year ($15 billion federal, $4 billion state).

OTHER EUROPEAN ALTERNATIVES

When the American child care community thinks about European models, the French model is often what they have in mind. With its emphasis on education, the French system has an obvious appeal to U.S. politicians, educators and child care advocates. Politicians' central concern in the United States appears to be raising children's test scores; in popular and academic literature, this standard is often cited as the major indicator of program success. But such an educational model is by no means the only alternative. Indeed, the U.S. focus on the French system may itself be a telling indicator of U.S. experts' values as well as their assessments of political realities. Many advocates insist that a substantial expansion of the U.S. system will be possible only if the system is presented as improving children's education. These advocates are no longer willing to use the term "child care," insisting on "early education" instead. The French model fits these priorities: it begins quasi-school about three years earlier than in the United States. Although the French obviously assist employed parents and children's center activities are said to be fun, the system is primarily touted and understood as educational—intended to treat children as pupils, to prepare them to do better in school.

The 11 European nations included in a recent Organization for Economic Cooperation and Development study (while quite different from one another) all have significantly better child care and paid leave than the United States. Each also differs significantly from France. Offering alternatives, these models challenge us to think even more broadly about childhood, parenting and the kind of society we value.

NON-SCHOOL MODEL: DENMARK

From birth to age six most Danish children go to child care, but most find that care in non-school settings. Overseen by the Ministry of Social Affairs (rather than the Ministry of Education), the Danish system stresses "relatively unstructured curricula" that give children time to "hang out." Lead staff are pedagogues, not teachers. Although pedagogues have college degrees and are paid teachers' wages, their role is "equally important but different" from that of the school-based teacher. "Listening to children" is one of the government's five principles, and centers emphasize "looking at everything from the child's perspective."

The Danish model differs from the French system in two additional ways that clarify its non-school character. First, in the Danish system, pedagogues care for very young children (from birth to age three as well as older children ages three to six). The French preschool (*école maternelle*) model applies only to children three and older. Before that, children of working parents can attend *crèches*. *Crèche* staff, however, have only high school educations and are paid substantially less than the (master's degree-trained) *écoles maternelles* teachers. Second, while the *écoles maternelles* are available to all children, the Danish system (like the French *crèches*) is only available to children with working parents because it is intended to aid working parents, not to educate children.

The Danish system is decentralized, with each individual center required to have a management board with a parent majority. But the system receives most of its money from public funding, and parents contribute only about one-fifth of total costs.

Given its non-school emphasis, age integration, and the importance it assigns to local autonomy, the Danish system might be appealing to U.S. parents, especially some people of color. To be sure, many U.S. parents—across race and class—are ambivalent about child care for their youngest children. Especially given the growing emphasis on testing, they believe that preschool might give them an edge, but they also want their children to have fun and play—to have, in short, what most Americans still consider a childhood. Some research suggests that Latina mothers are especially likely to feel that center-based care, with its emphasis on academic learning, does not provide the warmth and moral guidance they seek. They are, therefore, less likely to select center-based care than either white or African-American parents, relying instead on kin or family child care providers whom they know and trust. U.S. experts' emphasis on the French model may speak not only to political realities but also to the particular class, and, even more clearly, race preferences framing those realities.

MOTHERS OR PEERS

The United States, if only implicitly, operates on a mother-substitute model of child care. Because of a widespread assumption in the United States that all women naturally have maternal feelings and capacities, child care staff, who are almost all women (about 98 percent), are not required to have special training (and do not need to be well paid). Even for regulated providers, 41 out of 50 states require no pre-service training beyond orientation. Consequently, in the United States the child-staff ratio is one of the most prominent measures used to assess quality and is central to most state licensing systems. The assumption, based on the mother-substitute model, is that emotional support can be given and learning can take place only with such low ratios.

Considering the high quality and ample funding of many European systems, it comes as a surprise that most have much higher child-staff ratios than the United States. In the French *écoles maternelles,* for example, there is one teacher and one half-time aide for every 25 children. In Italy, in a center with one adult for every eight children (ages one to three years) the early childhood workers see no need for additional adults and think the existing ratios are appropriate. Leading researchers Sheila Kamerman and Alfred Kahn report that in Denmark, "what is particularly impressive is that children are pretty much on their own in playing with their peers. An adult is present all the time but does not lead or play with the children." In a similar vein, a cross-national study of academic literature found substantial focus on adult-child ratios in the United States, but very little literature on the topic in German-, French- or Spanish-language publications. Why not? These systems have a different view of children and learning. Outside the United States systems often center around the peer group. In Denmark the role of staff is to work "alongside children, rather than [to be] experts or leaders who teach children." Similarly, the first director of the early childhood services in Reggio, Italy, argues that children learn through conflict and that placing children in groups facilitates learning through "attractive," "advantageous," and "constructive" conflict "because among children there are not strong relationships of authority and dependence." In a non-European example, Joseph Tobin, David Wu, and Dana Davidson argue that in Japan the aim is ratios that "keep teachers from being too mother-like in their interactions with students . . . Large class sizes and large ratios have become increasingly important strategies for promoting the Japanese values of groupism and selflessness." Such practices contrast with the individualistic focus in U.S. child care.

FAMILY LEAVES AND WORK TIME

When we ask how to care for children, especially those younger than three, one answer is for parents to stay home. Policy that promotes such leaves is what one would expect from a society such as the United States, which emphasizes a mothering model of child care. It is, however, European countries that provide extensive paid family leave, usually universal, with not only job protection but also substantial income replacement. In Sweden, for example, parents receive a full year and a half of paid parental leave (with 12 months at 80 percent of prior earnings) for each child. Because so many parents (mostly mothers) use family leave, fewer than 200 children under one year old in the entire country are in public care. Generous programs are common throughout Europe (although the length, flexibility and level of payment they provide vary).

The United States provides far less in the way of family leaves. Since its passage in 1993, the Family and Medical Leave Act (FMLA) has guaranteed a 12-week job-protected leave to workers of covered employers. Most employers (95 percent) and many workers (45 percent), however, are not covered. And all federally mandated leaves are unpaid.

The unpaid leaves provided by the FMLA, like the private system of child care, accentuate the inequality between those who can afford them and those who can't. Although the FMLA was touted as a "gender neutral" piece of legislation, men (especially white men) are unlikely to take leaves; it is overwhelmingly women (especially those who are married) who take them. As a result, such women pay a wage penalty when they interrupt their careers. To address such inequities, Sweden and Norway have introduced a "use it or lose it" policy. For each child, parents may divide up to a year of paid leave (say nine months for the mother, three for the father), *except* that the mother may not use more than eleven months total. One month is reserved for the father; if he does not use the leave, the family loses the month.

Finally, although not usually discussed as child care policy in the United States, policy makers in many European countries now emphasize that the number of hours parents work clearly shapes the ways young children are cared for and by whom. Workers in the United States, on average, put in 300 hours more per year than workers in France (and 400 more than those in Sweden).

CONCLUSION

The child care system in the United States is a fragmentary patchwork, both at the level of the individual child and at the level of the overall system. Recent research suggests that the quality of care for young children is poor or fair in well over half of child care settings. This low quality of care, in concert with a model of intensive mothering, means that many anxious mothers privately hunt for high-quality substitutes while trying to ensure they are not being really replaced. System administrators need to patch together a variety of funding streams, each with its own regulations and paperwork. Because the current system was fashioned primarily for the affluent at one end and those being pushed off welfare at the other, it poorly serves most of the working class and much of the middle class.

Most efforts at reform are equally piecemeal, seeking a little extra money here or there in ways that reinforce the existing fragmentation. Although increasing numbers of advocates are pushing for a better system of child care in the United States, they rarely step back to think about the characteristics of the system as a whole. If they did, what lessons could be learned from Europe?

The features that are common to our peer nations in Europe would presumably be a part of a new U.S. system. The programs would be publicly funded and universal, available to all, either at no cost or at a modest cost with subsidies for low-income participants. The staff would be paid about the same as public school teachers. The core programs would cover at least as many hours as the school day, and "wrap-around" care would be available before and after this time. Participation in the programs would be voluntary, but the programs would be of such a high quality that a majority of children would enroll. Because the quality of the programs would be high, parents would feel much less ambivalence about their children's participation, and the system would enjoy strong public support. In addition to child care centers, parents would be universally offered a significant period of paid parental leave. Of course, this system is expensive. But as the National Academy of Science Report makes clear, not caring for our children is in the long term, and probably even in the short term, even more expensive.

Centers in all nations emphasize education, peer group dynamics, and emotional support to some extent. But the balance varies. The varieties of European experience pose a set of issues to be considered if and when reform of the U.S. system is on the agenda:

- To what degree should organized care approximate school and at what age, and to what extent is the purpose of such systems primarily educational?

- To what extent should we focus on adult-child interactions that sustain or substitute for mother care as opposed to fostering child-child interactions and the development of peer groups?

- To what extent should policies promote parental time with children versus high-quality organized care, and what are the implications for gender equity of either choice?

These are fundamental questions because they address issues of social equality and force us to rethink deep-seated images of children and parents.

DISCUSSION QUESTIONS

1. What are the social values in the United States that result in the child-care practices that Clawson and Gerstel identify? How would they have to change to produce practices more like those found in European nations?

2. Child-care workers in the United States are among the poorest paid workers. Why?

http://infotrac.thomsonlearning.com

InfoTrac College Edition

BONUS READING

Nayak, Madhabika B., Christina A. Byrne, Mutsumi K. Martin, and Anna George Abraham. "Attitudes toward Violence against Women: A Cross-Nation Study." *Sex Roles* 49 (October 2003): 333–42.

This article examines attitudes toward violence against women in four countries: India, Japan, Kuwait, and the United States. Variations across nations are explained by differences in the political, social, economic, and historical context influencing gender roles.

SEARCH TERMS

child-care policy
child custody
no-fault divorce
stepfamilies
work and motherhood

43

The Protestant Ethic
and the Spirit of Capitalism

MAX WEBER

Max Weber's classic analysis of the Protestant ethic and the spirit of capitalism shows how cultural belief systems, such as a religious ethic, can support the development of specific economic institutions. His multidimensional analysis shows how capitalism became morally defined as something more than pursuing monetary interests and, instead, has been culturally defined as a moral calling because of its consistency with Protestant values.

The impulse to acquisition, pursuit of gain, of money, of the greatest possible amount of money, has in itself nothing to do with capitalism. This impulse exists and has existed among waiters, physicians, coachmen, artists, prostitutes, dishonest officials, soldiers, nobles, crusaders, gamblers, and beggars. One may say that it has been common to all sorts and conditions of men at all times and in all countries of the earth, wherever the objective possibility of it is or has been given. It should be taught in the kindergarten of cultural history that this naïve idea of capitalism must be given up once and for all. Unlimited greed for gain is not in the least identical with capitalism, and is still less its spirit. Capitalism may even be identical with the restraint, or at least a rational tempering, of this irrational impulse. But capitalism is identical with the pursuit of profit, and forever renewed profit, by means of continuous, rational, capitalistic enterprise. . . .

If any inner relationship between certain expressions of the old Protestant spirit and modern capitalistic culture is to be found, we must attempt to find it, for better or worse, not in its alleged more or less materialistic or at least anti-ascetic joy of living, but in its purely religious characteristics. . . .

In the title of this study is used the somewhat pretentious phrase, the *spirit* of capitalism. What is to be understood by it? The attempt to give anything like a definition of it brings out certain difficulties which are in the very nature of this type of investigation.

If any object can be found to which this term can be applied with any understandable meaning, it can only be an historical individual, i.e. a complex of elements associated in historical reality which we unite into a conceptual whole from the standpoint of their cultural significance. . . .

From: Weber, Max. *The Protestant Ethic and the Spirit of Capitalism,* translated by Talcott Parsons, 17–27, 44–83, 157–83. New York: Scribner, 1958. Reprinted with permission.

"Remember, that *time* is money. He that can earn ten shillings a day by his labour, and goes abroad, or sits idle, one half of that day, though he spends but sixpence during his diversion or idleness, ought not to reckon *that* the only expense; he has really spent, or rather thrown away, five shillings besides.

"Remember, that *credit* is money. If a man lets his money lie in my hands after it is due, he gives me the interest, or so much as I can make of it during that time. This amounts to a considerable sum where a man has good and large credit, and makes good use of it. . . .

"The most trifling actions that affect a man's credit are to be regarded. The sound of your hammer at five in the morning, or eight at night, heard by a creditor, makes him easy six months longer; but if he sees you at a billiard-table, or hears your voice at a tavern, when you should be at work, he sends for his money the next day; demands it, before he can receive it, in a lump." . . .

Truly what is here preached is not simply a means of making one's way in the world, but a peculiar ethic. The infraction of its rules is treated not as foolishness but as forgetfulness of duty. That is the essence of the matter. It is not mere business astuteness, that sort of thing is common enough, it is an ethos. *This* is the quality which interests us.

When Jacob Fugger, in speaking to a business associate who had retired and who wanted to persuade him to do the same, since he had made enough money and should let others have a chance, rejected that as pusillanimity and answered that "he (Fugger) thought otherwise, he wanted to make money as long as he could," the spirit of his statement is evidently quite different from that of Franklin.[1] What in the former case was an expression of commercial daring and a personal inclination morally neutral, in the latter takes on the character of an ethically coloured maxim for the conduct of life. The concept spirit of capitalism is here used in this specific sense, it is the spirit of modern capitalism. For that we are here dealing only with Western European and American capitalism is obvious from the way in which the problem was stated. Capitalism existed in China, India, Babylon, in the classic world, and in the Middle Ages. But in all these cases, as we shall see, this particular ethos was lacking. . . .

And in truth this peculiar idea, so familiar to us today, but in reality so little a matter of course, of one's duty in a calling, is what is most characteristic of the social ethic of capitalistic culture, and is in a sense the fundamental basis of it. It is an obligation which the individual is supposed to feel and does feel towards the content of his professional activity, no matter in what it consists, in particular no matter whether it appears on the surface as a utilization of his personal powers, or only of his material possessions (as capital). . . .

Rationalism is an historical concept which covers a whole world of different things. It will be our task to find out whose intellectual child the particular concrete form of rational thought was, from which the idea of a calling and the devotion to labour in the calling has grown, which is, as we have seen, so irrational from the standpoint of purely eudæmonistic self-interest, but which has been and still is one of the most characteristic elements of our capitalistic culture. We are here particularly interested in the origin of precisely the irrational element which lies in this, as in every conception of a calling. . . .

. . . Like the meaning of the word, the idea is new, a product of the Reformation. This may be assumed as generally known. It is true that certain suggestions of the positive valuation of routine activity in the world, which is contained in this conception of the calling, had already existed in the Middle Ages, and even in late Hellenistic antiquity. We shall speak of that later. But at least one thing was unquestionably new: the valuation of the fulfillment of duty in worldly affairs as the highest form which the moral activity of the individual could assume. This it was which inevitably gave every-day worldly activity a religious significance, and which first created the conception of a calling in this sense. . . . Late Scholasticism, is, from a capitalistic viewpoint, definitely backward. Especially, of course, the doctrine of the sterility of money which Anthony of Florence had already refuted.

. . . For, above all, the consequences of the conception of the calling in the religious sense for worldly conduct were susceptible to quite different interpretations. The effect of the Reformation as such was only that, as compared with the Catholic attitude, the moral emphasis on and the religious sanction of, organized worldly labour in a calling was mightily increased. . . .

The real moral objection is to relaxation in the security of possession, the enjoyment of wealth with the consequence of idleness and the temptations of the flesh, above all of distraction from the pursuit of a righteous life. In fact, it is only because possession involves this danger of relaxation that it is objectionable at all. For the saints' everlasting rest is in the next world; on earth man must, to be certain of his state of grace, "do the works of him who sent him, as long as it is yet day." Not leisure and enjoyment, but only activity serves to increase the glory of God, according to the definite manifestations of His will.

Waste of time is thus the first and in principle the deadliest of sins. The span of human life is infinitely short and precious to make sure of one's own election. Loss of time through sociability, idle talk, luxury, even more sleep than is necessary for health, six to at most eight hours, is worthy of absolute moral condemnation. It does not yet hold, with Franklin, that time is money, but the proposition is true in a certain spiritual sense. It is infinitely valuable because every hour lost is lost to labour for the glory of God. Thus inactive contemplation is also valueless, or even directly reprehensible if it is at the expense of one's daily work. . . .

It is true that the usefulness of a calling, and thus its favour in the sight of God, is measured primarily in moral terms, and thus in terms of the importance of the goods produced in it for the community. But a further, and, above all, in practice the most important, criterion is found in private profitableness. For if that God, whose hand the Puritan sees in all the occurrences of life, shows one of His elect a chance of profit, he must do it with a purpose. Hence the faithful Christian must follow the call by taking advantage of the opportunity. "If God shows you a way in which you may lawfully get more than in another way (without wrong to your soul or to any other), if you refuse this, and choose the less gainful way, you cross one of the ends of your calling, and you refuse to be God's steward, and to accept His gifts and use them for Him when He requireth it: you may labour to be rich for God, though not for the flesh and sin."

Wealth is thus bad ethically only in so far as it is a temptation to idleness and sinful enjoyment of life, and its acquisition is bad only when it is with the purpose of later living merrily and without care. But as a performance of duty in a calling it is not only morally permissible, but actually enjoined. . . .

Let us now try to clarify the points in which the Puritan idea of the calling and the premium it placed upon ascetic conduct was bound directly to influence the development of a capitalistic way of life. As we have seen, this asceticism turned with all its force against one thing: the spontaneous enjoyment of life and all it had to offer. . . .

On the side of the production of private wealth, asceticism condemned both dishonesty and impulsive avarice. What was condemned as covetousness, Mammonism, etc., was the pursuit of riches for their own sake. For wealth in itself was a temptation. But here asceticism was the power "which ever seeks the good but ever creates evil"; what was evil in its sense was possession and its temptations. For, in conformity with the Old Testament and in analogy to the ethical valuation of good works, asceticism looked upon the pursuit of wealth as an end in itself as highly reprehensible; but the attainment of it as a fruit of labour in a calling was a sign of God's blessing. And even more important: the religious valuation of restless, continuous, systematic work in a worldly calling, as the highest means to asceticism, and at the same time the surest and most evident proof of rebirth and genuine faith, must have been the most powerful conceivable lever for the expansion of that attitude toward life which we have here called the spirit of capitalism.

When the limitation of consumption is combined with this release of acquisitive activity, the inevitable practical result is obvious: accumulation of capital through ascetic compulsion to save. The restraints which were imposed upon the consumption of wealth naturally served to increase it by making possible the productive investment of capital. . . .

One of the fundamental elements of the spirit of modern capitalism, and not only of that but of all modern culture: rational conduct on the basis of the idea of the calling, was born—that is what this discussion has sought to demonstrate—from the spirit of Christian asceticism. . . .

The Puritan wanted to work in a calling; we are forced to do so. For when asceticism was carried out of monastic cells into everyday life, and began to dominate worldly morality, it did its part in building the tremendous cosmos of the modern economic order. This order is now bound to the technical and economic conditions of machine production which today determine the lives of all the individuals who are born into this mechanism, not only those directly concerned with economic acquisition, with irresistible force. . . .

Since asceticism undertook to remodel the world and to work out its ideals in the world, material goods have gained an increasing and finally an inexorable power over the lives of men as at no previous period in history. Today the spirit of religious asceticism—whether finally, who knows? has escaped from the cage. But victorious capitalism, since it rests on mechanical foundations, needs its support no longer. The rosy blush of its laughing heir, the Enlightenment, seems also to be irretrievably fading, and the idea of duty in one's calling prowls about in our lives like the ghost of dead religious beliefs. Where the fulfillment of the calling cannot directly be related to the highest spiritual and cultural values, or when, on the other hand, it need not be felt simply as economic compulsion, the individual generally abandons the attempt to justify it at all. In the field of its highest development, in the United States, the pursuit of wealth, stripped of its religious and ethical meaning, tends to become associated with purely mundane passions, which often actually give it the character of sport.

No one knows who will live in this cage in the future, or whether at the end of this tremendous development entirely new prophets will arise, or there will be a great rebirth of old ideas and ideals, or, if neither, mechanized petrification, embellished with a sort of convulsive self-importance. For of the last stage of this cultural development, it might well be truly said: "Specialists without spirit, sensualists without heart; this nullity imagines that it has attained a level of civilization never before achieved." . . .

The modern man is in general, even with the best will, unable to give religious ideas a significance for culture and national character which they deserve. But it is, of course, not my aim to substitute for a one-sided materialistic an equally one-sided spiritualistic causal interpretation of culture and of history. Each is equally possible, but each, if it does not serve as the preparation, but as the conclusion of an investigation, accomplishes equally little in the interest of historical truth.

NOTE

1. The quotations are attributed to Benjamin Franklin.

DISCUSSION QUESTIONS

1. Weber is known for developing a multidimensional view of human society. What role does he see the Protestant ethic as playing in the development of capitalism?

2. Weber's analysis sees western capitalists as not pursuing money just for the sake of money, but because of the moral calling invoked by the Protestant ethic. Given the place of consumerism in contemporary society, how do you think Weber might modify his argument were he writing now? In other words, are there still remnants of the Protestant ethic in our beliefs about stratification? If so, how do they fit with contemporary capitalist values?

44

Abiding Faith

MARK CHAVES AND DIANNE HAGAMAN

These authors discuss the fact that the United States is an increasingly secular society; that means it is characterized by beliefs and actions that are not explicitly religious. Yet, religious belief remains high. The authors analyze this seeming contradiction, finding it to be not so contradictory after all.

God is dead—or God is taking over. Depending on the headlines of the day, soothsayers pronounce the end of religion or the ascendancy of religious extremists. What is really going on?

Taking stock of religion is almost as old as religion itself. Tracking religious trends is difficult, however, when religion means so many different things. Should we look at belief in the supernatural? Frequency of formal religious worship? The role of faith in major life decisions? The power of individual religious movements? These different dimensions of religion can change in different ways. Whether religion is declining or not depends on the definition of religion and what signifies a decline.

Perhaps the most basic manifestation of religious observance is piety: individual belief and participation in formal religious worship. Recent research on trends in American piety supports neither simple secularization nor staunch religious resilience in the face of modern life. Instead, Americans seem to believe as much but practice less.

RELIGIOUS BELIEF

Conventional Judeo-Christian religious belief remains very high in the United States, and little evidence suggests it has declined in recent decades. Gallup polls and other surveys show that more than 90 percent of Americans believe in a higher power, and more than 60 percent are certain that God exists. Approximately 80 percent believe in miracles and in life after death, 70 percent believe in heaven, and 60 percent believe in hell. Far fewer Americans—from two in three in 1963 to one in three today—believe the Bible is the literal Word of God. The number who say the Bible is either the inerrant or the inspired Word of God is still impressively high, however—four of every five.

From: Chaves, Mark, and Dianne Hagaman, "Abiding Faith." *Contexts* 1 (Summer 2002): 19–26. Reprinted with permission.

Religious faith in the United States is more broad than deep, and it has been for as long as it has been tracked. Of Americans who say the Bible is either the actual or the inspired Word of God, only half can name the first book in the Bible and only one-third can say who preached the Sermon on the Mount. More than 90 percent believe in a higher power, but only one-third say they rely more on that power than on themselves in overcoming adversity. People who claim to be born-again or evangelical Christians are no less likely than others to believe in ideas foreign to traditional Christianity, such as reincarnation (20 percent of all Americans), channeling (17 percent), or astrology (26 percent), and they are no less likely to have visited a fortune teller (16 percent).

Despite the superficiality of belief among many, the percentage of Americans expressing religious faith is still remarkably high. How should we understand this persistent religious belief? High levels of religious belief in the United States seem to show that, contrary to widespread expectations of many scholars, industrialization, urbanization, bureaucratization, advances in science and other developments associated with modern life do not automatically undermine religious belief. In part this is because modernization does not immunize people against the human experiences that inspire religious sentiment. As anthropologist Mary Douglas points out, scientific advances do not make us less likely to feel awe and wonder when we ponder the universe and its workings. . . . Likewise, bureaucracy does not demystify our world—on the contrary, it may make us feel more helpless and confused in the face of powers beyond our control. When confronted with large and complex bureaucracies, modern people may not feel any more in control of the world around them than a South Pacific Islander confronted with the prospect of deep-sea fishing for shark. Modern people still turn to religion in part because certain experiences—anthropologist Clifford Geertz emphasizes bafflement, pain and moral dilemmas—remain part of the human condition.

That condition cannot, however, completely explain the persistence of religious belief. It is clearly possible to respond in nonreligious ways to these universal human experiences, and many people do, suggesting that religiosity is a feature of some responses to these experiences, not an automatic consequence of the experiences themselves. From this perspective, attempting to explain religion's persistence by the persistence of bafflement, pain and moral paradox sidesteps a key question: Why do so many people continue to respond to these experiences by turning to religion?

Another, more sociological explanation of the persistence of religious belief emphasizes the fact that religion—like language and ethnicity—is one of the main ways of delineating group boundaries and collective identities. As long as who we are and how we differ from others remains a salient organizing principle for social movements and institutions, religion can be expected to thrive. Indeed, this identity-marking aspect of religion may also explain why religious belief often seems more broad than deep. If affirming that the Bible is the inerrant Word of God serves in part to identify oneself as part of the community of Bible-believing Christians, it is not so important to know in much detail what the Bible actually says.

The modern world is not inherently inhospitable to religious belief, and many kinds of belief have not declined at all over the past several decades. Certain aspects of modernity, however, do seem to reduce levels of religious observance. In a recent study of 65 countries, Ronald Inglehart and Wayne Baker find that people in industrialized and wealthy nations are typically less religious than others. That said, among advanced industrial democracies the United States still stands out for its relatively high level of religious belief. When asked to rate the importance of God in their lives on a scale of 1 to 10, 50 percent of Americans say "10," far higher than the 28 percent in Canada, 26 percent in Spain, 21 percent in Australia, 16 percent in Great Britain and Germany and 10 percent in France. Among advanced industrial democracies, only Ireland, at 40 percent, approaches the U.S. level of religious conviction.

RELIGIOUS PARTICIPATION

Cross-national comparisons also show that Americans participate in organized religion more often than do people in other affluent nations. In the United States, 55 percent of those who are asked say they attend religious services at least once a month, compared with 40 percent in Canada, 38 percent in Spain, 25 percent in Australia, Great Britain and West Germany, and 17 percent in France.

The trends over time, however, are murkier. Roger Finke and Rodney Stark have argued that religious participation has increased over the course of American history. This claim is based mainly on increasing rates of church membership. In 1789 only 10 percent of Americans belonged to churches, with church membership rising to 22 percent in 1890 and reaching 50 to 60 percent in the 1950s. Today, about two-thirds of Americans say they are members of a church or a synagogue. These rising figures should not, however, be taken at face value, because churches have become less exclusive clubs than they were earlier in our history. Fewer people attend religious services today than claim formal membership in religious congregations, but the opposite was true in earlier times. The long-term trend in religious participation is difficult to discern.

Although we have much more evidence about recent trends in religious participation, it still is difficult to say definitively whether religious-service attendance—the main way Americans participate collectively in religion—has declined or remained stable in recent decades. The available evidence is conflicting. Surveys using the traditional approach of asking people directly about their attendance mainly show stability over time, confirming the consensus that attendance has not declined much.

New evidence, however, points toward decline. Drawing on time-use records, which ask individuals to report everything they do on a given day, Stanley Presser and Linda Stinson find that weekly religious-service attendance has declined over the past 30 years from about 40 percent in 1965 to about 25 percent in 1994. . . .

Additional evidence of declining activity comes from political scientist Robert Putnam's book on civic engagement in the United States, *Bowling Alone.*

Combining survey data from five different sources, Putnam finds some decline in religious participation. Perhaps more important, because of the context they provide, are Putnam's findings about a range of civic and voluntary association activities that are closely related to religious participation. Virtually every type of civic engagement declined in the last third of the 20th century: voting, attending political, public and club meetings, serving as officer or committee member in local clubs and organizations, belonging to national organizations, belonging to unions, playing sports and working on community projects. If religious participation has indeed remained constant, it would be virtually the only type of civic engagement that has not declined in recent decades. Nor did the events of September 11, 2001, alter attendance patterns. If there was a spike in religious service attendance immediately following September 11, it was short-lived.

Overall, the following picture emerges from recent research. Since the 1960s, Americans have engaged less frequently in religious activities, but they have continued to believe just as much in the supernatural and to be just as interested in spirituality. This pattern characterizes many other countries around the world as well. . . .

Important differences among subgroups remain nonetheless. Blacks are more religiously active than whites, and women are more active than men. There is little reason to think, however, that the recent declines in participation vary among subgroups. . . .

Overall, the current knowledge of individual piety in the United States does not conform to expectations that modernity is fundamentally hostile to religion. Many conventional religious beliefs remain popular, showing no sign of decline. That said, research on individual piety neither points to stability on every dimension nor implies that social changes associated with modernity leave religious belief and practice unimpaired. The evidence supports neither a simple version of secularization nor a wholesale rejection of secularization. Moreover, focusing on levels of religious piety diverts attention from what may be more important: the social significance of religion. . . .

The social significance of religious belief and participation depends on the institutional settings in which they occur. This is why the religious movements of our day with the greatest potential for increasing religion's influence are not those that simply seek new converts or spur belief and practice, no matter how successful they may be. The movements with the greatest such potential are those that seek to expand religion's authority or influence in other domains. In some parts of the contemporary world, this has meant religious leaders seeking and sometimes achieving the power to veto legislation, dictate university curricula, exclude girls from schooling and women from working in certain jobs and determine the kinds of art or literature offered to the public. In the United States, the most significant contemporary movement to expand religious influence probably is the effort to shape school curricula concerning evolution and creationism. Wherever they occur, when such movements succeed they change the meaning and significance of religious piety. Efforts like these reflect and shape the abiding role of religion in a society in ways that go beyond the percentages of people who believe in God, pray, or attend religious services.

DISCUSSION QUESTIONS

1. How can it be that religious attendance is declining at the same time that religious belief in the United States remains very high?
2. What social changes have influenced a decline in religious attendance over the years?

45

Are American Jews Vanishing Again?

CALVIN GOLDSCHEIDER

High rates of interfaith marriage between Jews and non-Jews have led some to worry that this will result in the disappearance of Judaism in the United States. Here the author explores this question, concluding that the likelihood is that Jewish faith will persist.

In 1964 *Look* magazine published an article on "The Vanishing American Jew," predicting the demise of North American Jewish communities by the end of the 20th century. *Look* magazine has vanished. The American Jewish community has not, nor is it likely to.

Preliminary results from the National Jewish Population Study of 2000–01 estimate a declining American Jewish population. Considerable skepticism has been expressed over these population estimates since it is unclear how Jews were defined, who was missed, and how the intermarried were treated. Ignoring the journalistic sensationalism of the phrase "decline and demographic erosion," the public relations release itself emphasizes that the U.S. Jewish population has been "fairly stable over a decade." And there is widespread consensus that size is the least important of the findings.

Between 5 and 6 million Americans, approximately 3 percent of the population, are Jewish and this number has remained relatively stable over the last several decades. But about half of all Jews now marry someone who is not Jewish,

From: Goldscheider, Calvin. "Are American Jews Vanishing Again?" *Contexts* 2, no. 1 (Winter 2003): 18–24. Reprinted with permission.

making it appear that a major reduction in the Jewish population is inevitable. Indeed, high rates of intermarriage have become an obsession with Jewish communal leaders, some social scientists and many Jewish parents. American Jews have been viewed as a "model" of economic success and acceptance by the larger society, but the worry is that they may also be a model of numerical decline and disappearance through intermarriage.

This fear of decline is exaggerated. A closer look at the numbers and, more importantly, at the quality of Jewish life shows that there is no inexorable mathematics of decline. By broadening Jewish life in America, Jewish institutions and families have ensured its continuity. It is an experience from which other ethnic groups facing assimilation—such as Hispanics and Asian Americans—might gain. . . .

The alarm is based on a mechanical and mistaken understanding of social life. Indeed, the American Jewish community exemplifies how white ethnic groups can retain their distinctiveness instead of disappearing into the melting pot. And Jewish intermarriage shows how ethnic assimilation may gain members rather than lose them. High intermarriage rates certainly mean that many Jews are touched by intermarriage. There is hardly a Jewish household in America that has not experienced the intermarriage of a family member, a neighbor or a friend. The irony is that at the same time that more Jews are intermarrying, their marriages are increasingly accepted by relatives and Jewish institutions.

The reduction of discrimination over the last 40 years has given Jews new choices of residence, jobs and marriage. Jews now have much more contact with non-Jews than in the 1950s, noticeably as neighbors and spouses. By 1997, according to a national poll carried out by Steven Cohen, just 10 percent of American Jews reported that all or almost all of their friends were Jewish; less than half reported that most of their friends were Jewish and even fewer reported that most of their neighbors were Jewish. These numbers were a vast change from a generation earlier. They have serious implications for intermarriage.

The key indicators of an ethnic community's strength, however, are not who marries whom, but the activities that their grandchildren engage in. A group's continuity depends on the ethnic and religious commitments of the family. Focusing on families and the ethnic commitments of the young redirects questions about assimilation away from biology and marriage and toward economic activities, cultural obligations and how parents pass on traditions to their children. In this regard, the American Jewish community is surviving, maybe even thriving.

THE QUESTION OF BEING JEWISH

In America, unlike "the old country." Jewish group membership is by and large voluntary, based on social criteria, not biological or religious-legal definitions. It is informal, not formal, group membership, but that makes it no less powerful.

Most American Jews continue to be Jewish by birth. But most also identify as Jews by ethnicity or community, not by narrow religious definitions such as orthodox practice or matrilineal descent. This is the new reality of Jews in the United States. Like others, Jews become increasingly religious and identify more closely with the religious community as they form families and educate their children.

Therefore, the issue is not intermarriage but Jewishness in the family. How do people raise their children and connect to their extended families and the broader community that is Jewish? Questions about intermarriage and Jewish continuity need to be redirected toward these family networks.

In the past, Jewish men intermarried more than Jewish women did, but that is no longer true. Because traditionally women supervised the home, their out-marriage raised concerns. Yet Jewish women who marry out may want to reinforce the Jewishness of the home. And as the children of mixed marriages are increasingly children of Jewish mothers rather than fathers and as Jewish institutions increasingly accept intermarriage, intermarriage is not likely to have the same meaning as it did in the older context of rejection and escape. Synagogues increasingly allow non-Jewish spouses and parents to participate in various ritual family activities, announce the intermarriages of members' children and celebrate the birth of their grandchildren. Many mixed families are welcomed in Jewish religious schools and in Jewish community centers. Only about one out of four American Jews say that they would oppose the marriage of their child to a non-Jewish person who does not plan to convert to Judaism; only one out of three Jews report that a "good" Jew must marry a Jew.

Over the last decade, as rates of intermarriage increased, so have conversion, community acceptance and the integration of intermarried couples. More children raised by intermarried parents remain Jewish in a variety of ways than ever before. A 1993 survey found that 45 percent of mixed couples raise their children as Jewish and another 25 percent raise them as Jewish and something else. Most children in intermarried Jewish households celebrate their bar or bat mitzvah (a coming-of-age ritual), share in their family's Passover meals and observe Hanukah. Significant proportions occasionally attend religious services (at least as often as those from families where both partners were raised as Jews.)

THE MATH OF INTERMARRIAGE

It is a simple exercise to show that high rates of intermarriage can be consistent with group continuity. . . . I have estimated the numbers based on available statistics to illustrate some of the popular misconceptions in understanding intermarriage rates. In this illustration, the community begins with 15 Jews-by-birth in the first generation and gains two spouses by conversion. (In reality, 20 to 25 percent of non-Jewish spouses convert in a religious ceremony; it is reasonable to suppose that another 15 percent identify themselves as Jewish without formally converting.) Even if we assume that only one of the six children from the remaining mixed marriages grows up Jewish—and research suggests the proportion is much higher, at least two of five—the result is a second generation of 15 Jews.

The lessons to be learned from this exercise are two-fold: High intermarriage rates may result in stable numbers when some spouses convert or informally identify themselves as Jewish. More importantly, numerical stability with intermarriage occurs when children are raised as Jews. Intermarriage is not the question; the Jewishness of homes is.

RAISING JEWISH CHILDREN

Formal conversions to Judaism or identification with the Jewish community are paths to raising Jewish children. Many people who were not born Jewish and have not undergone formal conversions identify themselves as Jews, and are identified as Jews by their families, friends and the Jewish community. They participate in family, communal and organizational activities that are primarily Jewish. Also, non-conversion at the time of marriage does not foreclose conversion to Judaism at a later point in time, Jewish identification and practices can expand over the life course. A 1998 New York study found that intermarried couples were about three times more likely to find Judaism more important over time than less important. Jewish identity, as well as association with Jewish institutions, increases as families make choices about the education of their children and their own life style. Growing up in a Jewish household and taking part in Jewish communal activities encourages children to be Jewish and thus encourages continuity.

What does Jewishness of the home mean? It is not limited to religious practices or ritual observances, even where both spouses are born Jewish. Rather, being enmeshed in family, friendship and community networks is the key. Institutions such as Jewish community centers, schools, day care programs and camps organize such networks. Many young Jews form ties with one another at college and professional schools. Such networks help provide the content of Jewish identity and are the sources of changing cultural values.

Even with respect to religion, the majority of intermarried Jewish couples, including those without religious conversion, identify with a synagogue, occasionally attend religious services, and perform seasonal rituals, such as holding Passover seders and lighting Hanukah candles, at only slightly lower levels than do Jewish-born couples.

A study of eight different Jewish communities in the major metropolitan areas of the United States found that 40 percent of the mixed married couples—compared to 50 percent of the Jewish-born couples—attend synagogue services at least a few times a year (primarily on holidays). More than half attend a Passover seder and over 60 percent contribute to Jewish charities. While less active than families where both spouses were born Jewish, these levels of engagement, even without conversion, indicate important formal and informal commitment to Jewishness in America.

Thus, intermarriage and disengagement from the Jewish community are no longer synonymous. Because those who intermarry are often no less attached to the Jewish community and no less Jewish in their behavior and commitments than those who marry Jews, increasing rates of intermarriage by themselves are poor indicators of the weakening quality of Jewish life. . . .

THE FUTURE OF AMERICAN JEWS AND OTHERS

Most Jews have long-term roots in America and have over many years developed life styles and institutions that enrich their ethnic and religious expressions. Their Judaism and their Jewishness are expressed in diverse and changing ways that challenge simple assumptions about the total assimilation of ethnic white minorities. Although Jews have assimilated and become secular in some ways, their communities have become more cohesive and viable in other ways; Jewish communities remain distinctive within American society. Jews have developed new expressions of Judaism in a secular context and of ethnic Jewishness in a diverse, pluralist society. These expressions include organizations pursuing justice and charity, showing concern for the poor and the disenfranchised, as well as those promoting Jewish culture, art, dance and music. Commitment to Israel and to the memory of the Holocaust powerfully expresses Jewishness. Even swimming together in Jewish community centers symbolizes new values and paths to Jewish involvement. Indeed, how could a community be disintegrating whose multiple and powerful institutions continuously remind its members that it is eroding?

The astounding fact is that most American Jews living in a voluntary and open society choose to be Jewish rather than something else. Most Jewish families want their children and grandchildren to be Jewish, at least in some ways. Instead of asking whether the great grandchildren of Eastern European Jewish immigrants to America are assimilating or whether they are surviving as a community (they are doing both), we should try to understand what sustains ethnic and religious continuity in the absence of overt discrimination and economic disadvantage and in the face of pressures that erase distinctiveness. Communal institutions and social and family networks are the core elements sustaining ethnic continuity. It is a sign of ethnic vitality when these institutions construct new forms of Jewish cultural uniqueness that redefine collective identity. Communal acceptance may be responsible for transforming the negative consequences of intermarriage for group continuity into positive ones.

Seeing intermarriage as a potential source of strength has implications for other minorities in America as they become incorporated into America's pluralism. Rates of intermarriage among ethnic and religious groups have increased and have been viewed by some as diluting cultural identities. Certainly, changes in ethnic and religious communities can be expected in the future. But a careful examination of the Jewish experience suggests that high rates of intermarriage can reinforce ethnic distinctiveness and ethnic culture when family and institutions incorporate the intermarried into their community. Whether changes in the community are seen as part of its vanishing or its transformation depends on how the community constructs its institutions and values. Issues of ethnic assimilation and the loss of ethnic identity may begin—but do not end—with calculating rates of intermarriage.

DISCUSSION QUESTIONS

1. Among people in your peer group, what are the attitudes toward interfaith marriage? What social factors influence these attitudes?

2. How is religious faith among Jews (and others, for that matter) tied to their sense of group identity? What social changes facilitate or impede continued identification as an ethnic minority?

http://infotrac.thomsonlearning.com

InfoTrac College Edition

BONUS READING

Woodward, Kenneth L. "A Peaceful Faith, a Fanatic Few." *Newsweek* (September 24, 2001): 66.

All religions have religious fanatics and, since 9/11, many have wondered how those who are devout members of the Islamic faith could commit such acts of violence as we have seen in contemporary terrorism. Here the author examines the Islamic faith in an attempt to understand how it fostered such violence. What other religions have generated violence and fanaticism? Why do religions tend to produce martyrs?

SEARCH TERMS

American dream
interfaith marriage
Protestant ethic
religious minority
secularization

46

School Girls

PEGGY ORENSTEIN

Peggy Orenstein gives a concise summary of the many ways that schools shortchange and cheat girls in the educational system from kindergarten through high school.

The bell rings, as it always does, at 8:30 sharp. Twenty-eight sixth graders file into their classroom at Everett Middle School in San Francisco, straggling a bit since this is the first warm day in months—warm enough for shorts and cutoffs, warm enough for Stüssy T-shirts.

The students take their seats. Heidi, who wears bright green Converse sneakers and a matching cap, pulls off her backpack and shouts, "Did everyone bring their permission slips? You have to bring them so we can have the pizza party."

"Pizza," moans Carrie, who has brown bangs and a permanently bored expression. "That has milk. I'm allergic to milk."

Heidi looks stunned. "You can't eat pizza?"

The drama is interrupted as Judy Logan, a comfortably built woman with gray-flecked hair and oversized glasses, steps to the front of the classroom. She tapes two four-foot lengths of butcher paper to the chalkboard. Across the top of one she writes: "MALES," across the other: "FEMALES."

Ms. Logan is about to begin the lesson from which her entire middle school curriculum flows, the exercise that explains why she makes her students bother to learn about women, why the bookshelves in her room are brimful with women's biographies, why her walls are covered with posters that tout women's achievements and draped with quilts that depict women through history and women in the students' own lives.

It's time for the gender journey.

"Ladies and gentlemen," Ms. Logan says, turning toward the children and clasping her hands. "I'd like you to put your heads down and close your eyes. We're going to take a journey back in time."

Ms. Logan's already soothing voice turns soft and dreamy. "Go back," she tells her students. "Forget about everything around you and go back to fifth grade. Imagine yourself in your classroom, at your desk, sitting in your chair. Notice who your teacher is, what you have on, who's sitting around you, who your friends are.

"Continue your journey backward in time to third grade. Picture your third-grade teacher, your place in class. Imagine yourself in your room at home. What

From: Orenstein, Peggy. "School Girls." In *School Girls: Young Women, Self-Esteem and the Confidence Gap*, xi–xix. Copyright © 1994 Peggy Orenstein and American Association of University Women used by permission of Doubleday, a division of Random House, Inc.

do you like to do when you have free time? What kind of toys do you play with? What books are you reading?"

The children go further back, to the first magical day of kindergarten, then further still, remembering preschool, remembering their discovery of language, remembering their first toddling steps. Then Ms. Logan asks her students to recall the moment of their birth, to imagine the excitement of their parents. And then, when the great moment arrives . . .

They are each born the opposite sex.

The class gasps.

"Gross," offers Jonathan with great enthusiasm.

"Yuck," adds Carrie. "That's worse than being allergic to milk."

"You are born the opposite sex," Ms. Logan repeats firmly, and then asks her students to imagine moving forward through their lives again, exactly as they were and are, except for that one crucial detail.

Again they imagine themselves walking on tiny, uncertain feet. Again they imagine speaking, entering kindergarten. Again they envision their clothes as third graders, their toys, their books, their friends.

"I can't do this," says Jonathan, who has braces and short blond hair "I just picture myself like I am now except in a pink dress."

"This is stupid," agrees Carrie. "It's too hard."

"Just try," says Ms. Logan. "Try to imagine yourself in fourth grade, in fifth grade." By the time thirty minutes have elapsed, the students are back in their classroom, safe and sound and relieved to find their own personal anatomies intact.

"Without talking," Ms. Logan says, "I'd like you to make a list, your own personal list, not to turn in, of everything that would be different if you were the opposite sex."

The students write eagerly, with only occasional giggles. When there is more horseplay than wordplay, Ms. Logan asks them to share items from their lists with the class. The offerings go up on the butcher paper.

"I wouldn't play baseball because I'd worry about breaking a nail," says Mark, who wears a San Jose Sharks jersey.

"My father would feel more responsibility for me, he'd be more in my life," says Dayna, a soft-spoken African American girl.

Luke virtually spits his idea. "My room would be *pink* and I'd think everything would be *cute.*"

"I'd have my own room," says a girl.

"I wouldn't care how I look or if my clothes matched," offers another.

"I'd have to spend lots of time in the bathroom on my hair and stuff," says a boy whose own hair is conspicuously mussed. The other boys groan in agreement.

"I could stay out later," ventures a girl.

"I'd have to help my mom cook," says a boy.

"I'd get to play a lot more sports," says Annie, a freckled, red-haired girl who looks uncomfortable with the entire proposition. Many of the students are, in fact, unsettled by this exercise. Nearly a third opt to pass when their turns come, keeping their lists to themselves.

"I'd have to stand around at recess instead of getting to play basketball," says George, sneering. "And I'd worry about getting pregnant."

Raoul offers the final, if most obvious comment, which cracks up the crowd. "I'd have to sit down to go to the bathroom."

At this point the bell rings, although it is not the end of the lesson. The students will return after a short break to assess the accuracy of their images of one another. But while they're gone, I scrutinize the two butcher paper lists. Almost all of the boys' observations about gender swapping involve disparaging "have to"s, whereas the girls seem wistful with longing. By sixth grade, it is clear that both girls and boys have learned to equate maleness with opportunity and femininity with constraint.

It was a pattern I'd see again and again as I undertook my own gender journey, spending a year observing eighth-grade girls in two other Northern California middle schools. The girls I spoke with were from vastly different family structures and economic classes, and they had achieved varying degrees of academic success. Yet all of them, even those enjoying every conceivable advantage, saw their gender as a liability.

Sitting with groups of five or six girls, I'd ask a variation on Ms. Logan's theme: what did they think was lucky about being a girl? The question was invariably followed by a pause, a silence. Then answers such as "Nothing, really. All kinds of bad things happen to girls, like getting your period. Or getting pregnant."

Marta, a fourteen-year-old Latina girl, was blunt. "There's nothing lucky about being a girl," she told me one afternoon in her school's cafeteria. "I wish I was a boy."

SHORTCHANGING GIRLS:
WHAT THE AAUW SURVEY REVEALS

Like many people, I first saw the results of the American Association of University Women's report *Shortchanging Girls, Shortchanging America* in my daily newspaper. The headline unfurled across the front page of the San Francisco *Examiner*: "Girls' Low Self-Esteem Slows Their Progress," and *The New York Times* proclaimed: "Girls' Self-Esteem Is Lost on the Way to Adolescence." And, like many people, as I read further, I felt my stomach sink.

This was the most extensive national survey on gender and self-esteem ever conducted, the articles said: three thousand boys and girls between the ages of nine and fifteen were polled on their attitudes toward self, school, family, and friends. As part of the project the students were asked to respond to multiple-choice questions, provide comments, and in some cases, were interviewed in focus groups. The results confirmed something that many women already knew too well. For a girl, the passage into adolescence is not just marked by menarche or a few new curves. It is marked by a loss of confidence in herself and her abilities, especially in math and science. It is marked by a scathingly critical attitude toward her body and a blossoming sense of personal inadequacy.

In spite of the changes in women's roles in society, in spite of the changes in their own mothers' lives, many of today's girls fall into traditional patterns of low self-image, self-doubt, and self-censorship of their creative and intellectual potential. Although all children experience confusion and a faltering sense of self at adolescence, girls' self-regard drops further than boys' and never catches up. They emerge from their teenage years with reduced expectations and have less confidence in themselves and their abilities than do boys. Teenage girls are more vulnerable to feelings of depression and hopelessness and are four times more likely to attempt suicide.

The AAUW discovered that the most dramatic gender gap in self-esteem is centered in the area of competence. Boys are more likely than girls to say they are "pretty good at a lot of things" and are twice as likely to name their talents as the thing they like most about themselves. Girls, meanwhile, cite an aspect of their physical appearance. Unsurprisingly, then, teenage girls are much more likely than boys to say they are "not smart enough" or "not good enough" to achieve their dreams.

The education system is supposed to provide our young people with opportunity, to encourage their intellectual growth and prepare them as citizens. Yet students in the AAUW survey reported gender bias in the classroom—and illustrated its effects—with the canniness of investigative reporters. Both boys and girls believed that teachers encouraged more assertive behavior in boys, and that, overall, boys receive the majority of their teachers' attention. The result is that boys will speak out in class more readily, and are more willing to "argue with my teachers when I think I'm right."

Meanwhile, girls show a more precipitous drop in their interest in math and science as they advance through school. Even girls who like the subjects are, by age fifteen, only half as likely as boys to feel competent in them. These findings are key: researchers have long understood that a loss of confidence in math usually *precedes* a drop in achievement, rather than vice versa. A confidence gap, rather than an ability gap, may help explain why the numbers of female physical and computer scientists actually went down during the 1980s. The AAUW also discovered a circular relationship between math confidence and overall self-confidence, as well as a link between liking math and aspiring to professional careers—a correlation that is stronger for girls than boys. Apparently girls who can resist gender-role stereotypes in the classroom resist them elsewhere more effectively as well.

Among its most intriguing findings, the AAUW survey revealed that, although all girls report consistently lower self-esteem than boys, the severity and the nature of that reduced self-worth vary among ethnic groups. Far more African American girls retain their overall self-esteem during adolescence than white or Latina girls, maintaining a stronger sense of both personal and familial importance. They are about twice as likely to be "happy with the way I am" than girls of other groups and report feeling "pretty good at a lot of things" at nearly the rate of white boys. The one exception for African American girls is their feelings about school: black girls are more pessimistic about both their teachers and their schoolwork than other girls. Meanwhile, Latina girls' self-esteem crisis is in many ways the most profound. Between the ages of nine and fifteen, the number of Latina girls who are "happy with the way I am" plunges by 38 percentage points, compared with a 33 percent drop for white girls and a 7 percent drop for black girls. Family disappears as a source of positive self-worth for Latina teens, and academic confidence, belief in one's talents, and a sense of personal importance all plummet. During the year in which *Shortchanging Girls, Shortchanging America* was conducted, urban Latinas left school at a greater rate than any other group, male or female.

DISCUSSION QUESTIONS

1. As in Orenstein's exercise, imagine that you were born the opposite gender, and then begin to list the many differences in your life in school that would likely result.
2. How would you change education in the United States to help reduce the gender gap in certain areas of academic achievement?

47

Race in American Public Schools
Rapidly Resegregating School Districts

ERICA FRANKENBERG AND CHUNGMEI LEE

Over fifty years ago, the Supreme Court decision, Brown v. Board of Education, *brought the promise of desegregating the nation's racially separate schools. Yet, as these researchers find, not only has that promise not been realized, but in recent years schools have actually been resegregating. Racial segregation has serious implications for numerous other social problems.*

In 1954, the U.S. Supreme Court handed down the historic *Brown v. Board of Education* decision outlawing state-mandated separate schools for black and white students. Since that decision, hundreds of American school districts, if not more, have attempted to implement desegregation plans. In the early years

From: Frankenberg, Erica, and Chungmei Lee. *"Race in American Public Schools: Rapidly Resegregating School Districts."* The Civil Rights Project, Harvard University, August 2002. Reprinted with permission.

of desegregation most of these plans focused on the South and resulted in the most integrated schools being located in the South by the early 1970s. From the late 1960s on, some districts in all parts of the country began implementing such plans although the courts made it much more difficult to win desegregation orders outside the South and the 1974 Supreme Court decision against city-suburban desegregation made real desegregation impossible in a growing number of overwhelming minority central cities. We are now almost 50 years from the initial Supreme Court ruling banning segregation and more than a decade into a period in which the U.S. Supreme Court has authorized termination of desegregation orders. These plans are being dissolved by court orders even in some communities that want to maintain, them; in addition, some federal courts are forbidding even voluntary desegregation plans. Given this context, it is crucial to continue to mark the progress of these policies and examine how their presence or absence affects the schooling experience for all students.

Nationally, segregation for blacks has declined substantially since the pre-*Brown* era and reached its lowest point in the late 1980s. For Latinos, the story has been one of steadily rising segregation since the 1960s and no significant desegregation efforts outside of a handful of large districts. These changes in segregation patterns are happening in the context of an increasingly diverse public school enrollment. In particular, the 2000 Census shows an extraordinary growth of Latino population in the past decade. This change in overall population is reflected in the school population as well. High birth rates, low levels of private school enrollment and increased immigration of Latinos have resulted in a rise of Latino public school enrollment, which is now more than 7 million. Nationwide, the Latino share of public school enrollment has almost tripled since 1968, compared to an increase of just 30% in black enrollment and a decrease of 17% in white enrollment during the same time period. A smaller percent of students attend private schools than a half-century ago and white private school enrollment is lowest in the South and West where whites are in school with higher proportions of minority students. Yet, little attention has been paid to the results of these two trends—rising segregation and increasing diversity—on the racial composition of our public schools.

RESEARCH QUESTIONS

. . . This study examines segregation trends in large school districts across the country and addresses the following key questions.

- Are metropolitan countywide districts, which had shown considerable integration through the mid-1980s, still integrated?

- To what extent are children in central city school districts segregated from children of other races?

- Are there effects of the dramatic increase in minority enrollment in large suburban systems? . . .

Patterns of segregation by race are strongly linked to segregation by poverty, and poverty concentrations are strongly linked to unequal opportunities and outcomes. Since public schools are the institution intended to create a common preparation for citizens in an increasingly multiracial society, this inequality can have serious consequences. Given that the largest school districts in this country (enrollment greater than 25,000) service one-third of all school-aged children, it is important to understand at a district level the ways in which school segregation, race, and poverty are intersecting and how they impact these students' lives. In our analysis we focus on two important components, race and segregation.

DATA AND METHODS

We analyze enrollment data collected by the U.S. Department of Education in the NCES Common Core of Data from the school year 2000–01, examining the 239 school districts with total enrollment greater than 25,000.

Using exposure indices, we calculate the racial isolation of both black and Latino students from white students; that is, we calculate the percent of white students in schools of typical black and Latino students. We also investigate the racial isolation of white students to determine whether their schooling experience is becoming more integrated as the minority share of the public school enrollment continues to increase. To do so, we calculate the percentage of black students and the percentage of Latino students in school of the average white student. We use this measure because it reports the actual racial composition of the school, and desegregated schools have been shown to have educational and diversity benefits for their students. This measure is not a measure of discrimination or of the feasibility of desegregation in a given district—just of the actual level of interracial exposure that existed in 2000–2001.

Additionally, this study looks specifically at districts that have, at various times, been under court-mandated desegregation plans. We examine districts in each of several categories pertaining to designs of desegregation plans: busing within city, magnet plans, city-suburban desegregation, no plan, court rejected city-suburban, and partial or complete unitary status declared by mid-1980s. We compare exposure of black students to white students, since most desegregation plans were primarily concerned with the segregation of blacks from whites. We compute the 2000 exposure indices for these districts to identify any trends among districts, based on the type of desegregation the district did (or did not) have, as well as to compare the 1988 and 2000 exposure indices.

FINDINGS

The racial trend in the school districts studied is substantial and clear: *virtually all* school districts analyzed are showing lower levels of inter-racial exposure since 1986, suggesting a trend towards resegregation, and in some districts, these declines are sharp. As courts across the country end long-running desegregation plans and, in some states, have forbidden the use of any racially-conscious student assignment plans, the last 10–15 years have seen a steady unraveling of almost 25 years worth of increased integration. From the early 1970s to the late 1980s, districts in the South had the highest levels of black-white desegregation in the nation; from 1986–2000, however, some of the most rapidly resegregating districts for black students' exposure to whites are in the South. Some of these districts maintained a very high level of integration for a quarter century or more until the desegregation policies were reversed.

Other findings include:

- Many of the districts experiencing the largest changes in black-white exposure are also having similar changes in Latino exposure to whites.

- Districts that show the least resegregation in black-white exposure are mostly in the South, likely due to lingering effects of desegregation plans in districts where the plans have been dissolved and the continuing impacts of plans still in place.

- The lowest levels of black-white exposure are in districts with either no desegregation plan or where the courts rejected a city-suburban plan. The highest exposure rates are in districts with city-suburban plans, even though all of these districts have since been declared unitary and show a trend toward resegregating.

- Despite an increasingly racially diverse public school enrollment, white students in over one-third of the districts analyzed became more segregated from black and/or Latino students.

As attention to civil rights issues is waning, it is even more important to document the segregation in our public schools in order to inform educational policy discussions on racial segregation and its related effects on public school children, particularly when these students attending racially isolated and unequal schools will be punished for not achieving at high levels.

We find that since 1986, in almost every district examined, black and Latino students have become more racially segregated from whites in their schools. The literature suggests that minority schools are highly correlated with high-poverty schools and these schools are also associated with low parental involvement, lack of resources, less experienced and credentialed teachers, and higher teacher turnovers—all of which combine to exacerbate educational inequality for minority students. Desegregation puts minority students in schools with better opportunities and higher achieving peer groups. . . .

CONCLUSION

While the public school enrollment reflects the country's growing diversity, our analysis of the nation's large school districts indicates a disturbing pattern of growing isolation. We find decreasing black and Latino exposure to white students is occurring in almost every large district as well as declining white exposure to blacks and Latinos in almost one-third of large districts. Black and Latino students display high levels of segregation from white students in many districts. This is due in part to small white percentages in these districts. However, even when white students are only a small percentage of total enrollment they tend to be concentrated in a few schools, which results in lower exposure of black and Latino students to white students even further.

The isolation of blacks and Latinos has serious ramifications: this isolation is highly correlated with poverty, which is often strongly related to striking inequalities in test scores, graduation rates, courses offered and college-going rates. Virtually no attention is being paid to this troubling pattern in the current discussion of educational reform even though it is very strongly related to many outcomes the reformers wish to change.

Recent Civil Rights Project studies of a number of cities have found important educational and civic benefits for students who attend diverse schools. However, the Supreme Court desegregation decisions of the 1990s relaxed the judicial standards school districts had to meet to be released from court oversight, and many school districts are no longer under desegregation plans. Further, school systems that wish to pursue voluntary desegregation measures by reducing racial isolation and/or to promote diversity in their schools must prove that this is both a "compelling governmental interest" and that the plan is narrowly tailored; lower court decisions have split as to whether these are compelling interests.

Many Americans believe that there is nothing that can be done about these problems and that desegregation efforts have failed. This report suggests that a great deal was done, particularly in the South, and that, after a series of court decisions sharply limiting desegregation rights, it is being undone, even in large districts where the desegregation was substantial and long lasting. Interracial exposure can simply not occur in districts that do not have different racial groups present. Perhaps it is time for communities, educational leaders and our courts to consider whether or not there is a better alternative to the system of increasingly separate and unequal schools we are creating in our large districts.

DISCUSSION QUESTIONS

1. What evidence have you seen of racial segregation or desegregation in your school district? How does this follow from patterns of housing segregation?

2. If different racial-ethnic groups have little contact with each other in neighborhoods and schools, what are the implications for the persistence of prejudice and racism?

48

Racial Desegregation
Magnet Schools, Vouchers, Privatization, and Home Schooling

LORETTA F. MEEKS, WENDELL A. MEEKS, AND CLAUDIA A. WARREN

Education has historically been seen as a public responsibility—because having an edu-cated population is essential to democratic societies. Education has also been seen as impor-tant in reducing inequality and promoting economic opportunity. Yet, sociological evidence reveals much inequality in education, resulting in numerous other problems. These authors examine policies such as school vouchers, magnet schools, homeschooling, and privatization, assessing whether they are likely to address problems of inequality in schooling.

The control and welfare of urban school continues to occupy the attention of mayors, governors, state legislatures, and local citizens. Public funding, media scrutiny, and the school's reflection of society are primary contribu-tors to this ongoing interest. The changing demographics of the urban setting and its effect on public policy have launched education into broader political, economic, and legal arenas. The major change in the education of urban students began with the 1954 ruling by the United States Supreme Court in *Brown v. Board of Education*. At that time, the Supreme Court recognized, through a unani-mous 9–0 decision, the significance of the fiscal, sociological, and psychological role of the public school and its significance to our democratic existence.

The empirical findings from Myrdal (1944) were the foundations of the psychological argument convincing the Court that segregation did in fact have a negative effect on African American students' potential for success (Orlich, 1991) and that there can be no equitable system of separate but equal schooling. Although the detrimental effects of segregation appeared clear to the Court a decade preceding the civil rights movement, neither *Brown* (1954) nor *Brown* (1955) provided prescriptive strategies to incorporate desegregation or to elimi-nate segregation. Some see this as synonymous with legalizing freedom but not abolishing slavery. However great, this omission has left a generation to continue to grapple with achieving a goal that has far-reaching underpinnings exacerbated by unforeseen circumstances, such as a changing national demographic,

From: Meeks, Loretta F., Wendell A. Meeks, and Claudia A. Warren. "Racial Desegregation: Magnet Schools, Vouchers, Privatization, and Home Schooling." *Education and Urban Society* 33, no. 1 (November 2000): 88–101. © 2000 Corwin Press, Inc. Reprinted with permission.

pervasiveness of racial separatism, unequal patterns of poverty, the political divisiveness of this educational issue, and the shift in the country's economic base from national to global.

Today, the United Slates is experiencing a shift in the demographic configurations of its cities and schools. Minority populations are growing at a faster pace than the majority. Some urban school districts have greater numbers of minority students than majority students, and the number of affected districts is predicted to increase (Stringfield, 1997). Neighborhood segregation and the imbalance of wealth, which influence the racial divide in this country, are factors that have not changed significantly since the inception of *Brown*. Reports indicate that the per capita cost of public schools is increasing, whereas scores on achievement measures are decreasing. In addition, the United States is evidencing a shift in its global positioning. It is no longer the monolithic power from previous decades but, rather, one of several major economic entities in a global market. The authors of *Brown* could not have anticipated these issues.

Although *Brown* continues to uphold the moral principles of desegregation, the legal significance is being debated and is eroding with current policies supported or ignored by the Court. The current Court has intentionally or unintentionally made no rulings on desegregation cases in nearly 10 years (Russo, Harris, & Sandidge, 1994). This failure to provide the legal framework for school reform may have given decision makers too much latitude, which many interpret to nullify the effects of *Brown*.

Since *Brown,* race has been a constant factor for parents in the selection of schools for their children. Brown and Hunter (1995) reported that, although 95% of White parents surveyed had no objection to their children attending a school in which a few of the children are Black, the proportion of objecting Whites grew steadily as a school population became increasingly Black. This resentment led to "White flight," leaving the urban schools primarily minority.

Although public perception of integration is clearly more positive now than nearly 50 years ago, the realization of legalized desegregation as a means toward this end has not been as readily endorsed. Court-ordered desegregation was accepted as a remedy for the "deliberate speed" implementation mandate for almost three decades, but now the federal courts are relieving most urban schools from court-ordered supervision of all efforts to desegregate schools.

The contradictory messages regarding the effects of desegregation from both races—perception of White parents that desegregated schools benefit Black students more so than Whites and the growing discontent of Blacks that desegregation has not had the intended positive effect on academic success for Black students—have led to the incorporation of politically inspired initiatives into the public educational arena. Consequently, the basis for assessment of these initiatives has primarily been political rather than educational because parent and student satisfaction are replacing traditional measures of educational achievement and program effectiveness. . . .

With the apparent end of court-ordered desegregation, the avenues of escape for White parents from enrolling their children in largely minority and poor schools have been identified as choice options. The most prevalent of these include magnet schools; vouchers; privatization of public schools or private, for-profit schools; and home schooling. These choice options initiated in the 1970s present an alternative to forced busing. They are particularly significant for middle-class families who cannot afford to reside in affluent neighborhoods with well-financed, predominately White schools nor afford the tuition of private schools (Glenn, 1998). Proponents of both political parties see some form of choice option as being the most efficient route to needed educational reform.

If the central focus of *Brown* (1954) was to create a school environment conducive to learning for all students, the central question in analyzing the choice option movement is whether this premise is still at the forefront of policy making and school options. Have the alternatives to traditional education promoted in the past decade lived up to their promise? The discussion that follows scrutinizes the alternatives based on this question.

ALTERNATIVE CHOICES

Magnet Schools

Magnet schools are defined as a selective, academically demanding public elementary or secondary school with superior facilities and programs that are more readily received by White citizens than is forced busing (Dejnozka & Kapel, 1991). They were established under the administrations of Presidents Richard Nixon and Ronald Reagan as a mechanism of a choice option for parents in lieu of forced busing to desegregate public schools. The assumption was that middle-class White parents could be lured back to inner-city schools with the assurance of an innovative and focused curriculum with locally tailored extras and that the financial base influenced by this infusion of students would revitalize the school. Magnet schools typically offer a nontraditional curriculum, incorporating thematic learning and technology that are governed by the school district and local school board (Metz, 1986).

The number of students enrolled in magnet schools has tripled in the past decade (Steel & Levine, 1994). Between 1985 and 1993, the federal Magnet School Assistance Program spent $739 million in school districts promoting magnet schools (Steele & Eaton, 1996). Typically, program offerings include an emphasis on basic skills, language immersion, humanities, and instructional approaches such as open classrooms, individualized instruction, and enriched curricula as well as career or vocational education, the arts, and gifted-talented programs. In other words, magnet schools specialize in programs that cater to the population that will support and control them. One important aspect of magnet schools is internal control that involves parents and teachers. The expectation is that the school will be responsive to its constituency because this group is in control. These schools are governed by administrative policies dictated through internal leadership. . . .

Although the practical, economic impetus for magnet schools is clear, from an inclusion perspective, the intent of magnet schools seems suspect. Archbald (1996) notes, "Variables related to parent socioeconomic status and proximity to magnet schools were found to be significant predictors of magnet school enrollment" (p. 152). This is further evidenced by Orfield and Eaton's (1996) study of magnet schools in Kansas City, Missouri, that showed the following:

- Magnet schools did little for integration.
- Magnet programs tend to help desegregate schools in middle-class communities with school districts and sizable minority and White populations.
- Magnet schools have less effect in large cities where Whites have fled to the suburbs. . . .

Vouchers

Vouchers are individual scholarships to parents that can be used to defray the cost of a child's tuition at any school—public or private, religious or secular—so long as that voucher is awarded on the basis of neutral secular criteria (Lewin, 1999). This plan provides public monies to parents to pay or supplement the cost of schooling. Parents rather than the government determine the schooling options for children financed by public funds. From the very beginning, constitutional challenges of vouchers have been at the forefront of discussions regarding their place in the educational arena. The challenge centers on the interpretation of the Establishment Clause of the First Amendment that prohibits public funds from being spent on religious activities or teaching. . . .

The assumption underlying vouchers is that parents, not the government, should have control over the selection of schools for their children. Gallup poll results indicate that the percentage of parents in favor of having this option is steadily increasing, especially among Black parents (Rose & Gallup, 1998). In the educational arena, the belief is common that parents are their child's best teacher, so the proposition that parents are the group best equipped to select an educational experience for their children also seems reasonable. According to proponents, this would be the natural result of providing parents with an opportunity to actively advocate for schools that are effective versus those that are ineffective. Theoretically, effective schools would be deemed as those schools that provide the greatest opportunity for achievement and have a history of proven achievement gains. Achievement results, not school demographics, would be the ultimate determiner of school selection. Therefore, the premise for vouchers is that all parents can choose schools based on academic success. Neither race nor socioeconomic makeup of the school would be a primary factor in selection increasing the opportunities for equitable access. . . .

. . . Based on final evaluation findings from the Milwaukee Parental Choice Program, the first publicly funded voucher program, no conclusive evidence supports the academic superiority of one system versus the other (Witte, Stern, & Thorn, 1995). . . .

The results do not lend support to the premise that race is not a primary factor in school selection, and Goldhaber (1997) suggests that race and socioeconomic status are primary factors in parents' selection of schools. Results such as these validate opponents' fears regarding vouchers' potential for greater economic and racial stratification of the schools. . . .

Privatization and Charter Schools

A third and most recent alternative to public schools that evolved during the latter part of the 20th century is the move toward privatization, engaging private enterprise in the management and operation of schools. The contracting strategy of privatization is the most popular, in which the public sector remains the financier of the school but delegates production or provision of services to the private sector (Murphy, 1996). The resulting schools from this approach are called charter schools. Although similar in intent, charter schools are different from magnet schools in that they are privately owned and managed by that entity rather than the local school board. They are chartered to produce achievement gains, with state and federal funding based on the delivery of results, and operate as tuition-free public schools. They were fueled by the nation's perception of the high cost and inefficiency of government and a renewed interest in private-market values (Murphy, 1996). Assumptions that emulating organizational models of the private sector and educational models of other industrialized countries that boast of high academic achievements can enhance education belie the move toward privatization. . . .

Although parent satisfaction with charter schools is high, the question remains of whether the original intent of charter schools to produce greater academic gains than public schools remains unanswered. . . .

Home Schooling

. . . Home-schooled or home educated, the teaching of one's own children at home is steadily becoming an accepted and respected alternative to public school education. Duffy (1998) reports recent estimates of home-schooled students to be about slightly more than a million.

Home schooling is not new, but the recognition of the phenomenon as a legitimate option to public school is receiving renewed attention. Also new is its position in the desegregation controversy. The right to home school is not questioned but, rather, the effect on public schools is the focus of discussions.

Researchers cite parents' concerns for safety, security, morality, and educational quality as primary reasons for home schooling (Dahm, 1996). Parents want decision-making authority to determine their child's teacher, classmates, and curriculum with access to extracurricular activities of the public schools.

Opponents of home schooling (e.g., Gorder, 1996; Mayberry, Knowles, Ray, & Marlow, 1995; Ramirez, 1998) cite issues regarding accreditation, parents' lack of formal training for teaching, comparable facilities and resources for schooling, lack of opportunity for socialization, and the deflection of students from public school as detractors. A primary area of concern is the lack of standardization of home schooling laws across states.

Duffy (1998) reports that the typical home schooling family is White and Protestant with two parents, three children, and above-average income and education. The mother is the primary instructor and religion is likely to be the most important—although seldom the only—reason for home schooling. Home schooling is typically not an option for urban parents primarily because of the same problems that plague urban schools (e.g., poverty, lower levels of parental education, parental involvement in the educational process). A single parent—primarily the mother—who is also the principal wage earner heads most urban families. Rather than a choice issue, however, home-schooled students do present another opportunity to divert funds from urban schools. Currently, with the percentages of home-schooled students being proportionally low, the threat appears minimal. But as the trend grows in popularity, the competition will be more evident. . . .

CONCLUSION

. . . The origin of the argument for choice options for public schools was a response to desegregation and based on the assumption that schools were ineffective and unresponsive to the varying needs of the population. The problems identified with the system were internal, requiring systemic changes at many levels. There is no evidence that choice option programs recognize or make any attempt to address this key factor. Rather, these options are focused on creating a separate system that will avoid the ills of its predecessor rather than address them. Consequently, an analysis of the problems that they attempt to avoid is critical to urban education.

Educational issues that confront urban public schools are low test scores, poor graduation rates, poor attendance, inequality, discipline, overcrowding, lack of parent involvement, violence, and poor teaching. Few studies have shown that, when the same populations of the public schools are provided the current choice options, these issues arc significantly influenced. These alternatives are providing an avenue for those that are dissatisfied with the current system to abandon it in lieu of a better choice. The literature has shown that there are inequities in income and education for those that are electing to take this route to school reform; therefore, one can conclude that these options are not flawless. . . .

. . . Parents who separate from the traditional system supposedly leave these same problems for another generation. Funding alternatives naturally deplete resources from the primary system. You cannot make one system equitable without making another more inequitable. This concept was struck down and found ineffective, unfair, and unconstitutional more than 50 years ago.

REFERENCES

Archbald, D. (1996). SES and demographic predictors of magnet school enrollment. *Journal of Research and Development in Education, 29*(3), 152–162.

Brown, F., & Hunter, R. C. (1995). Introduction privatization of public school services. *Education and Urban Society, 27*(2), 107–113.

Brown v. Board of Education, 347 U.S. 483 (1954).

Brown v. Board of Education, 335 U.S. 294 (1955).

Dahm, L. (1996). Education at home with help from school. *Educational Leadership, 54,* 68–71.

Dejnozka, E. L., & Kapel, D. E. (1991). *American Educator's Encyclopedia.* Westport, CT: Greenwood.

Duffy, J. (1998). Home schooling: A controversial alternative. *Principal, 77,* 23–26.

Glenn, C. L. (1988). Public school choice: Searching for direction. *Principal, 77*(5), 10–12.

Goldhaber, D. D. (1997). School choice as education reform. *Phi Delta Kappan, 79,* 143–147.

Gorder, C. (1996). *Home schools: An alternative,* Mesa, AZ: Blue Bird Publishing.

Lewin, N. (1999). Are vouchers constitutional? *Policy Review, 93,* 5–8.

Mayberry, M., Knowles, J., Ray, B. B., & Marlow, S. (1995). *Home schooling: Parents as educators.* Thousand Oaks, CA: Corwin Press.

Metz, M. H. (1986). *Different by design: The context and character of three magnet schools.* New York: Routledge Kegan Paul.

Murphy, J. (1996). Why privatization signals a sea of change in schooling. *Education Leadership, 54,* 60–62.

Myrdal, G. (1944). *An American Dilemma: The Negro problem and modern democracy.* New York: Harper.

Orfield, G., & Eaton, S. E. (1996). *Dismantling desegregation: The quiet reversal of* Brown v. Board of Education. New York: The New Press.

Orlich, D. C. (1991). *Brown v. Board of Education:* Time for a reassessment. *Phi Delta Kappan, 72,* 631–632.

Ramirez, A. (1998). Vouchers and voodoo economics. *Educational Leadership, 56,* 36–39.

Rose, L. C., & Gallup, A. M. (1998). The 30th Annual Phi Delta Kappa/Gallup Poll of the public's attitudes toward the public schools. *Phi Delta Kappan, 80,* 41–56.

Russo, C. J., Harris, J. J., III, & Sandidge, R. F. (1994). *Brown v. Board of Education* at 40: A legal history of equal educational opportunities in American public education. *Journal of Negro Education, 63,* 297–309.

Stringfield, S. (1997). Research on effective instruction for at-risk students: Implications for the St. Louis public schools. *Journal of Negro Education, 66,* 258–288.

Steel, L., & Levine, R. H. (1994). *Educational innovation in multiracial contexts: The growth of magnet schools in American education.* Washington, DC: U.S. Department of Education.

Steele, L., & Eaton, M. (1996). *Reducing, eliminating, and preventing minority isolation in American schools: The impact of the Magnet Schools Assistance Program.* Washington, DC: American Institutes for Research.

Witte, J. F., Sterr, T. D., & Thorn, C. A. (1995). *Fifth year report: Milwaukee parental choice program,* Madison, WI: Department of Political Science and the Robert M. LaFollette Institute of Public Affairs, University of Wisconsin-Madison.

DISCUSSION QUESTIONS

1. Why do these authors think that various "school choice" plans will undermine the quality of urban education?

2. What do research studies find in terms of the social and economic characteristics of those who have been able to exercise "choice" in their children's schooling?

http://infotrac.thomsonlearning.com

InfoTrac College Edition

BONUS READING

Donlevy, Jim. "Educational Reforms and High-Stakes Testing: Are Public Schools Still for the Public?" *International Journal of Instructional Media* 30 (Summer 2003): 225–31.

    This author examines the assumptions that lie behind the current move toward more standardized testing of students. What assumptions underlie the increasing use of such "high-stakes" tests? What are the consequences of such testing for students in different social classes? races? gender?

SEARCH TERMS

gender bias in education
privatization of education
race and education
school vouchers
single-sex schools
social class and education

49

The Service Society and the Changing Experience of Work

CAMERON LYNNE MACDONALD AND CARMEN SIRIANNI

The U.S. economy has changed from being based primarily on manufacturing to being based on service industries. The transition to more "service work" has changed the character of workplace control. The service economy is embedded in systems of race, gender, and class stratification that are revealed in patterns of employment and perceptions of who is most fit for particular jobs.

We live and work in a service society. Employment in the service sector currently accounts for 79 percent of nonagricultural jobs in the United States (U.S. Department of Labor 1994: 83). More important, 90 percent of the new jobs projected to be created by the year 2000 will be in service occupations, while the number of goods-producing jobs is projected to decline (Kutscher 1987: 5). Since the mid-nineteenth century the U.S. economy has been gradually transformed from an agriculture-based economy to a manufacturing-based economy to a service-based economy. Near the turn of the century, employment distribution among the three major economic sectors was equally divided at roughly one-third each. Since then, agriculture's labor market share has declined rapidly, now accounting for only about 3 percent of U.S. jobs, while the service sector provides over 70 percent and the goods-producing sector about 25 percent.

The decrease in proportion of manufacturing jobs occurred not because U.S. corporations manufacture fewer goods, but primarily because they use fewer workers to make the goods they produce (Albrecht and Zemke 1990). They use fewer workers due to increasing levels of automation and the exportation of manufacturing functions to low-wage job markets overseas. In addition, the feminization of the work force has created a self-fulfilling cycle in which the entrance of more women into the work force has led to increased demand for those consumer services once provided gratis by housewives (cleaning, cooking, child care, etc.), which in turn has produced more service jobs that are predominantly filled by women.

From: Macdonald, Cameron Lynne, and Carmen Sirianni, eds. "The Service Society and the Changing Experience of Work." In *Working in the Service Society*, 1–24. Philadelphia: Temple University Press, 1996. Reprinted with permission.

Still, these trends fail to account fully for the dominance of service work in the U.S. economy, since companies outside of the service sector also contain service occupations. For example, 13.2 percent of the employees in the manufacturing sector work in service occupations such as clerical work, customer service, telemarketing, and transportation (Kutscher 1987). Further, manufacturing and technical occupations are comprised increasingly of service components as U.S. firms adopt Total Quality Management (TQM) and other customer-focused strategies to generate a competitive edge in the global economy. When production efficiency and quality are maximized, the critical variable in the struggle for economic dominance is the quality of interactions with customers. As one business school professor remarks, "Sooner or later, new technology becomes available to everyone. Customer-oriented employees are a lot harder to copy or buy" (Schlesinger and Heskett 1991: 81). So whether one believes that U.S. manufacturing is going to Mexico, to automation, or to the dogs, it is clear that the United States is increasingly becoming a service society and that service work is here to stay.

What do we mean when we speak of "service work"? By definition, a service is intangible; it is produced and consumed simultaneously, and the customer generally participates in its production (Packham 1992). . . . Service work includes jobs in which face-to-face or voice-to-voice interaction is a fundamental element of the work. "Interactive service work" (Leidner 1993) generally requires some form of what Arlie Hochschild (1983) has termed "emotional labor," meaning the conscious manipulation of the workers' self-presentation either to display feeling states and/or to create feeling states in others. In addition, the guidelines, or "feeling rules," for this emotional labor are created by management and conveyed to the worker as a critical aspect of the job. . . .

Much managerial and professional work also entails emotional labor. For example, doctors are expected to display an appropriate "bedside manner," lawyers are expert actors in and out of the courtroom, and managers, at the most fundamental level, try to instill feeling states and thus promote action in others. However, there remains a critical distinction between white-collar work and work in the emotional proletariat: in management and in the professions, guidelines for emotional labor are generated collegially and, to a great extent, are self-supervised. In front-line service jobs, workers are given very explicit instructions concerning what to say and how to act, and both consumers and managers watch to ensure that these instructions are carried out. However, one could argue that even those in higher ranking positions increasingly experience the kinds of monitoring of their interactive labor encountered by those lower on the occupational ladder, be it by customers, supervisors, or employees. . . .

Given the rising dominance of service occupations in the labor force, what are the special difficulties and opportunities that workers encounter in a service society? A key problem seems to be how to inhabit the job. In the past there was a clear distinction between *careers,* which required a level of personalization, emotion management, authenticity in interaction, and general integration of personal and workplace identities, and *jobs,* which required the active engagement of the body and parts of the mind while the spirit and soul of the worker might be elsewhere. Workers in service occupations are asked to inhabit jobs in ways that were formerly limited to managers and professionals alone. They are required to bring some level of personal identity and self-expression into their work, even if it is only at the level of basic interactions, and even if the job itself is only temporary. The assembly-line worker could openly hate his job, despise his supervisor, and even dislike his co-workers, and while this might be an unpleasant state of affairs, if he completed his assigned tasks efficiently, his attitude was his own problem. For the service worker, inhabiting the job means, at the very least, pretending to like it, and, at most, actually bringing his whole self into the job, liking it, and genuinely caring about the people with whom he interacts.

This demand has several implications: who will be asked to fill what jobs, how they are expected to perform, and how they will respond to those demands. Because personal interaction is a primary component of all service occupations, managers continually strive to find ways to oversee and control those interactions, and worker responses to these attempts vary along a continuum from enthusiastic compliance to outright refusal. Hiring, control of the work process, and the stresses of bringing one's emotions to work are all shaped by the characteristics of the worker and the nature of the work. The self-presentation and other personal characteristics of the worker make up the work process and the work product, and are increasingly the domain of management-worker struggles (see Leidner 1993). In addition, because much of the labor itself is invisible, contests over control of the labor process are often more implicit than explicit. . . .

There are three trends emanating from the rising dominance of service work. First, the need to supervise the production of an intangible, good service, has given rise to particularly invasive forms of workplace control and has led managers to attempt to oversee areas of workers' personal and psychic lives that have heretofore been considered off-limits. Second, the fact that workers' personal characteristics are so firmly linked to their "suitability" for certain service occupations continues to lead to increasing levels of stratification within the service *labor* force. Finally, . . . how [do] workers respond to these and other aspects of working in the service society, and how might they build autonomy and dignity into their work, ensuring that service work does not equal servitude? . . .

GENDER, RACE, AND STRATIFICATION
IN THE SERVICE SECTOR

Service industries tend to produce two kinds of jobs: large numbers of low-skill, low-pay jobs and a smaller number of high-skill, high-income jobs, with very few jobs that could be classified in the middle. As Joel Nelson (1994) notes, "Service workers are more likely than manufacturing workers to have lower incomes, fewer opportunities for full-time employment, and greater inequality in earnings" (p. 240). A typical example of this kind of highly stratified work force can be found in fast food industries. These firms tend to operate with a small core of managers and administrators and a large, predominantly part-time work force who possess few skills and therefore are considered expendable (Woody 1989).

As a result, service jobs fall into two broad categories: those likely to be production-line jobs and those likely to be empowered jobs. This distinction not only refers to the level of responsibility and autonomy expected of workers, but also to wages, benefits, job security, and potential for advancement. While empowered service jobs are associated with full-time work, decent wages and benefits, and internal job ladders, production-line jobs offer none of these. Some researchers have described the distinction between empowered and production-line service jobs as one between "core" and "periphery" jobs in the service economy (Hirschhorn 1988; Walsh 1990; Wood 1989). . . .

The core/periphery distinction may be a misleading characterization of functions in service industries, however. In many firms contingent workers perform functions essential to the operation of the firm and can comprise up to two-thirds of a firm's labor force while "core" workers perform nonessential functions (Walsh 1990). For example, a majority of key functions in industries such as hospitality, food service, and retail sales are performed by workers who, based on their level of benefits, pay, and job security, would be considered periphery workers. In low-skill service positions, job tenure has no relation to output or productivity. Therefore, employers can rely on contingent workers to provide high-quality service at low costs. As T. J. Walsh (1990: 527) points out, it is therefore likely that the poor compensation afforded these workers is due not to their productivity level but to the perceptions of their needs, level of commitment, and availability.

Given the proliferation of service jobs in the United States, key questions for labor analysts are what kinds of jobs are service industries producing, and who is likely to fill them? At the high end, service industries demand educated workers who can rapidly adapt to changing economic conditions. This means that employers may demand a college degree or better for occupations that formerly required only a high school diploma,

> even though many of the job-holders' activities have not changed, or appear relatively simple, because they want workers to be more responsive to the general situation in which they are working and the broader purposes of their work. (Hirschhorn 1988: 35)

In addition, core workers are frequently expected to take on more responsibilities, work longer hours, and intensify their output.

At the low end of the spectrum of service occupations are periphery workers who are frequently classified as part-time, temporary, contract, or contingent. These flexible-use workers act as a safety valve for service firms, allowing managers to redeploy labor costs in response to market conditions. In addition, they allow managers to minimize overhead because they rarely qualify for benefits and generally receive low wages. In 1988, 86 percent of all part-time workers worked in service sector industries, and this trend has continued since (Tilly 1992: p. 30). Labor analysts argue that the bulk of the expansion in part-time and contingent work is due to the expansion of the service sector, which has always used shift and part-time workers as cost-control mechanisms. . . .

Service sector expansion has also sparked the rapid growth of the temporary help industry. Over the past decade, the number of temporary workers has tripled (Kilborn 1995). As the chairman of Manpower, Inc., the largest single employer in the United States, remarked, "The U.S. is going from just-in-time manufacturing to just-in-time employment. The employer tells us, 'I want them delivered exactly when I want them, as many as I need, and when I don't need them, I don't want them here'" (quoted in Castro 1993: 44). Like part-time workers, temporary workers carry no "overhead" costs in terms of taxes and benefits, and they are on call as needed. Since they experience little workplace continuity, they are less likely than full-time, continuous employees to organize or advocate changes in working conditions.

Women, youth, and minorities comprise the bulk of the part-time and contingent work force in the service sector. For example, Karen Brodkin Sacks (1990) has noted that the health care industry is "so stratified by race and gender that the uniforms worn to distinguish the jobs and statuses of health care workers are largely redundant" (p. 188). Patients respond to the signals implicitly transmitted via gender and race and act accordingly, offering deference to some workers and expecting it from others. Likewise in domestic service, race and gender determine who gets which jobs. White American and European women are most likely to be hired for domestic jobs defined primarily as child care, while women of color predominate in those defined as house cleaning (Rollins 1985), regardless of what the actual allocation of work might be. The same kinds of stratification can be found in secretarial work, food service, hotels, and sales occupations (Hochschild 1983; Leidner 1991). In all of these occupational groups, demographic characteristics of the worker determine the job title and thus other factors such as status, pay, benefits, and degree of autonomy on the job.

As a result, the shift to services has had a differential impact on various sectors of the labor market, increasing stratification between the well employed and the underemployed. The service sector work force is highly feminized, especially at the bottom. Within personal and business services, for example, Bette Woody (1989) found that "men are concentrated in high-ticket, high commission sales jobs, and women in retail and food service" (p. 57). And although the decline in manufacturing forced male workers to move into service industries, Jon Lorence (1992: 150) notes that within the service sector, occupations are highly stratified by gender. From 1950 to 1990, 60 percent of all new service sector employment and 74 percent of all new low-skill jobs were filled by women. . . .

Service sector employment is equally stratified by race. For example, Woody (1989) finds that the shift to services has affected black women in two important ways. First, it has meant higher rates of unemployment and underemployment for black men, which increased pressure on black women to be the primary breadwinners for their families and contributed to the overall reduction in black family income. Second, although the increase in service sector jobs has meant greater opportunity for black women and has allowed them to move out of domestic service into the formal economy, as Evelyn Nakano Glenn (1996) also notes, they have remained at the lowest rung of the service employment ladder. Unlike some white women who "moved up" to male-intensive occupations with the shift to services, black women "moved over" to sectors traditionally employing white women (Woody 1989: 54). These traditionally male occupations are not only low-security, low-pay jobs, but they lack internal career ladders. As Ruth Needleman and Anne Nelson (1988) note, "There is no progression from nurse to physician, from secretary to manager" (p. 297).

Service work differs most radically from manufacturing, construction, or agricultural work in the relationship between worker characteristics and the job. Even though discrimination in hiring, differential treatment, differential pay, and other forms of stratification exist in all labor markets, service occupations are the only ones in which the producer in some sense equals the product. In no other area of wage labor are the personal characteristics of the workers so strongly associated with the nature of work. Because at least part of the job in all service occupations is to "manufacture social relations" (Filby 1992; 37), traits such as gender, race, age, and sexuality serve a signaling function, indicating to the customer/employer important cues about the tone of the interaction. Women are expected to be more nurturing and empathetic than men and to tolerate more offensive behavior from customers (Hochschild 1983; Leidner 1991; Pierce 1996, Sutton and Rafaeli 1989). Similarly, both women and men of color are expected to be deferential and to take on more demeaning tasks (Rollins 1985; Woody 1989). In addition, a given task may be viewed as more or less demeaning depending on who is doing it.

These occupations are so stratified that worker characteristics such as race and gender determine not only who is considered desirable or even eligible to fill certain jobs, but also who will want to fill certain jobs and how the job itself is performed. Worker characteristics shape what is expected of a worker by management and customers, how that worker adapts to the job, and what aspects of the job he or she will resist or embrace. The strategies workers use to adapt to the demands of service jobs are likely to differ according to gender and other characteristics. Women are more likely to embrace the emotional demands (e.g., nurturing, care giving) of certain types of service jobs because these demands generally fit their notion of gender-appropriate behavior. As Pierce notes, heterosexual men tend to resist these demands because they find "feminine" emotional labor demeaning; in response they either reframe the nature of the job to emphasize traditional masculine qualities or distance themselves by providing service by rote, making it clear that they are acting under duress.

All of these interconnections between worker, work, and product result in tendencies toward very specific types of labor market stratification. A long and heated debate has raged concerning the ultimate impact of deindustrialization on the structure of the labor market. On one side are those who argue that a shift to a service-based economy will produce skill upgrading and a leveling of job hierarchies as information and communications technologies reshape the labor market (see, for example, Bell 1973). Others take a more pessimistic view, arguing that the shift to services will give rise to two trends: "towards polarization and towards the proliferation of low-wage jobs" (Bluestone and Harrison 1988: 126). In a sense, both positions are correct but for different segments of the labor market.

Overall, the transition to a service-based economy will likely mean a more stratified work force in which more part-time and contingent jobs are filled predominantly by women, minorities, and workers without college degrees. In jobs lacking internal career ladders, these workers have little chance for upward mobility. Contingent workers are also less likely to organize successfully due to their tenuous attachment to specific employers and to the labor force in general. At the opposite end of the service sector occupational spectrum are highly educated managerial, professional, and paraprofessional workers, who will have equally weak attachments to employers, but who, due to their highly marketable skills, will move with relative ease from one well-paying job to another. Given this divided and economically segregated work force, what are the opportunities for workers to advocate, collectively or individually, for greater security, better working conditions, and a voice in shaping their work? . . .

REFERENCES

Albrecht, Karl, and Ron Zemke. 1990. *Service America!: Doing Business in the New Economy.* New York: Warner Books.

Bell, Daniel. 1973. *The Coming of Post-Industrial Society: A Venture in Social Forecasting.* New York: Basic Books.

Bluestone, Barry, and Bennett Harrison. 1988. *The Great U-Turn: Corporate Restructuring and the Polarizing of America.* New York: Basic Books.

Castro, Janice. 1993. "Disposable Workers." *Time Magazine,* March 29, pp. 43–57.

Filby, M. P. 1992, " 'The Figures, the Personality, and the Bums': Service Work and Sexuality." *Work, Employment, and Society* 6 (March): 23–42.

Glenn, Evelyn Nakano. 1996. "From Servitude to Service Work: Historical Continuities in the Racial Division of Paid Reproductive Labor." In Cameron Lynne Macdonald and Carmen Sirianni, eds., *Working in the Service Society,* pp. 115–156. Philadelphia: Temple University Press.

Hirschhorn, Larry. 1988. "The Post-Industrial Economy: Labor, Skills, and the New Mode of Production." *The Service Industries Journal* 8: 19–38.

Hochschild, Arlie Russell. 1983. *The Managed Heart: Commercialization of Human Feeling.* Berkeley: University of California Press.

Kilborn, Peter T. 1995. "In New Work World, Employers Call All the Shots: Job Insecurity, a Special Report." *New York Times,* July 3, p. A1.

Kutscher, Ronald E. 1987. "Projections 2000: Overview and Implications of the Projections to 2000." *Monthly Labor Review* (September): 3–9.

Leidner, Robin. 1993. *Fast Food, Fast Talk: Service Work and the Routinization of Everyday Life.* Berkeley: University of California Press.

———. 1991. "Serving Hamburgers and Selling Insurance: Gender, Work, and Identity in Interactive Service Jobs." *Gender & Society* 5: 154–77.

Lorence, Jon. 1992. "Service Sector Growth and Metropolitan Occupational Sex Segregation." *Work and Occupations* 19: 128–56.

Needleman, Ruth, and Anne Nelson. 1988. "Policy Implications: The Worth of Women's Work." In Anne Starham, Eleanor M. Miller, and Hans O. Mauksch, eds., *The Worth of Women's Work: A Qualitative Synthesis* (pp. 293–308). Albany: SUNY Press.

Nelson, Joel I. 1994. "Work and Benefits: The Multiple Problems of Service Sector Employment." *Social Problems* 41: 240–55.

Packham, John. 1992. "The Organization of Work on the Service-Sector Shop Floor." Unpublished paper.

Pierce, Jennifer L. 1996. *Gender Trials: Emotional Lives in Contemporary Law Firms.* Berkeley: University of California Press.

Rollins, Judith. 1985. *Between Women: Domestics and Their Employers.* Philadelphia: Temple University Press.

Sacks, Karen Brodkin. 1990. "Does It Pay to Care?" In Emily K. Abel and Margaret K. Nelson, eds., *Circles of Care: Work and Identity in Women's Lives* (pp. 188–206). Albany: SUNY Press.

Schlesinger, Leonard, and James Heskett. 1991. "The Service-Driven Service Company." *Harvard Business Review* (September–October): 71–81.

Sutton, Robert I., and Anat Rafaeli. 1989. "The Expression of Emotion in Organizational Life." *Research in Organizational Behavior* 11: 1–42.

Tilly, Chris. 1992. "Short Hours, Short Shrift: The Causes and Consequences of Part-Time Employment." In Virginia L. Du Rivage, ed., *New Policies for the Part-Time and Contingent Work-Force* (pp. 15–43).Armonk,NY: M. E. Sharpe, Economic Policy Institute Series.

U.S. Department of Labor, Bureau of Labor Statistics. 1994. *Monthly Labor Review* (July): 74–83.

Walsh,T. J. 1990. "Flexible Labor Utilization in the Private Service Sector." *Work, Employment, and Society* 4: 517–30.

Wood, Stephen, ed. 1989. *The Transformation of Work? Skill, Flexibility, and the Labor Process.* London: Unwin Hyman.

Woody, Bette. 1989. "Black Women in the Emerging Services Economy." *Sex Roles* 21: 45–67.

DISCUSSION QUESTIONS

1. How has the transition to a service-based economy changed employment patterns in the contemporary economy? What implications does this have for the relationship between management and labor?

2. How do race and gender stratification influence the perceptions of some workers as suitable for particular forms of service work? What evidence have you witnessed of this phenomenon in your experiences in a service-based society?

50

Toward a 24-Hour Economy

HARRIET B. PRESSER

Analysis of employment data indicates that few in the United States are working a "typical" forty-hour work week. This change is being driven by the changing economy, the changing demography of the workplace, and changing technology. Harriet Presser discusses the effects of the increase in working hours on families.

Americans are moving toward a 24-hour, 7-day-a-week economy. Two-fifths of all employed Americans work mostly during the evenings or nights, on rotating shifts, or on weekends. Much more attention has been given to the number of hours Americans work[1,2] than to the issue of which hours—or days—Americans work. Yet the widespread prevalence of nonstandard work schedules is a significant social phenomenon, with important implications for the health and well-being of individuals and their families and for the implementation of social policies. Here I discuss recent national data on the widespread prevalence of nonstandard work schedules, explain why this has come about, and highlight some of the important social implications.

PREVALENCE

As of 1997, only 29.1% of employed U.S. citizens worked a "standard work week," defined as 35 to 40 hours a week, Monday through Friday, on a fixed daytime schedule. For employed men, the proportion is 26.5%; for employed women, 32.8%. Only 54.4%—a bare majority—regularly work a fixed daytime schedule, all five weekdays, for any number of hours.

These figures are derived from the May 1997 Current Population Survey (CPS), a representative sample of about 48,000 U.S. households. I selected for further study a subset of about 50,000 employed Americans ages 18 and over in these households with nonagricultural occupations and who reported on their specific work hours and/or work days.

Of the people in this group, one in five work other than on a fixed daytime schedule, and one in three work on weekends (and, for most, on weekdays as well). Men and women are similar in their prevalence of evening employment, but a somewhat higher proportion of men than women work fixed nights, rotating and

From: Reprinted with permission from Presser, Harriet B. "Toward a 24-Hour Economy." *Science* 284 (June 1999): 1778–79. Copyright © 1999 AAAS.

variable hours, and weekends. The most marked differences are between those working full-time and part-time. More part-timers work other than a fixed day (29.6%) than do full-timers (17.0%); evening employment is especially high among part-timers. The difference between full- and part-timers is less marked for weekend employment (30.7% and 34.7%, respectively).

For the modal U.S. family—the two-earner couple—the prevalence of nonstandard work schedules is especially high, because either the husband or wife may be working nondays or weekends. (Rarely do both work the same nonstandard schedules.) Among two-earner couples, 27.8% include at least one spouse who works other than a fixed daytime schedule, and 54.6% include at least one spouse working weekends. When children under age 14 are in the household the respective percentages are 31.1 and 46.8%. Indeed of all two-earner couples with children, those with both spouses working fixed daytime schedules and weekdays are a minority; 57.3% do not fit this description. Thus, the temporal context in which millions of American couples are raising their children today is diverse and is likely to become even more so in the future.

ORIGINS AND CAUSES

At least three interrelated factors are increasing the demand for Americans to work late hours and weekends: a changing economy, changing demography, and changing technology. With regard to the changing economy, an important aspect is the growth of the service sector with its high prevalence of nonstandard work schedules relative to the goods-producing sector. In the 1960s, employees in manufacturing greatly exceeded those in service industries, whereas by 1995 the percentage was about twice as high in services as in manufacturing.[3] In particular, there is an interaction between the growth of women's employment and the growth of the service sector because there is a disproportionately high percentage of female occupations in this sector. In turn, the increasing participation of women in the labor force contributes to the growth of the service economy. For example, the decline in full-time homemaking has generated an increase in family members eating out and purchasing other services. Moreover, women's increasing daytime labor force participation has generated a demand for services during nondaytime hours and weekends.[4]

Demographic changes also have contributed. The postponement of marriage, along with the rise in real family income resulting from two earners, has increased the demand for recreation and entertainment during late hours and weekends. The aging of the population has increased the demand for medical services over a 24-hour day, 7 days a week.

Finally, technological change, along with reduced costs, has moved us to a global 24-hour economy. The ability to be "on call" at all hours of the day and night to others around the world at low cost generates a need to do so. For example, the rise of multinational corporations, along with the use of computers, faxes, and other forms of rapid communication, increases the demand for branch offices to operate at the same time that corporate headquarters are open.

Similarly, international financial markets are expanding their hours of operation. Express mailing companies such as United Parcel Service require round-the-clock workers all days of the week.

We do not have precise national estimates of the amount of growth over recent decades in the prevalence of nonstandard work schedules as a consequence of these changes. Questions on work hours have been asked differently by the Bureau of Labor Statistics in each of the CPSs since 1980; questions on work days were not even asked until 1991.

Most of the top 10 occupations projected by the Bureau of Labor Statistics to have the largest job growth between 1996 and 2006 are service occupations.[5] Using the May 1997 CPS data. I calculated the percentages in the top growth occupations for which nonstandard schedules are prevalent and considered their gender and racial composition.

The data suggest that not only will future job growth generate an increase in employment during nonstandard hours and weekends, but also that this increase will be experienced disproportionately by females and blacks. Many of the top growth occupations that tend to have nonstandard work schedules also have high percentages of female workers: cashiers, registered nurses, retail salespersons, nurses' aides, orderlies, and attendants combined with home health aides. The top growth occupations that disproportionately include blacks and tend to have nonstandard work schedules are cashiers, truck drivers, nurses' aides, orderlies, and attendants combined with home health aides.

Although nonstandard work schedules are pervasive throughout the occupational structure, such schedules are disproportionately concentrated in jobs low in the occupational hierarchy.[6] This fact, combined with the expectation that women and blacks will disproportionately increase their participation in nonstandard work schedules, suggests that this phenomenon will increasingly affect the working poor.

Effects on Families

The physical consequences of working nonstandard hours, particularly night and rotating hours, have been well documented.[7] Such work schedules alter one's circadian rhythms, often leading to sleep disturbances, gastrointestinal disorders, and chronic malaise. The social consequences of such employment have received less attention, although working nonstandard schedules may be significantly altering the structure and stability of family life. Some of the consequences can be viewed as positive, others negative, and both may vary by family member. Moreover, short-term benefits may be offset by long-term costs and vice versa.

Consider, for example, the care of children among dual-earner couples. As noted above, one-third of such couples with preschool-aged children are split-shift couples with one spouse working days and the other evenings, nights, or rotating schedules. A national study of American couples with preschool-aged children showed that in virtually all cases in which mothers and fathers are employed different hours and neither are on rotating schedules, fathers are the primary caregivers of children when their wives are employed.[8] Insofar as we view the greater involvement of fathers in child care as desirable, and considering the economic benefits to the family of reduced child care expenses resulting from this arrangement, such split-shift parenting may be a positive outcome.

However, these gains may be more than offset by the longer term costs to the marriage. New research shows that among couples with children, when men work nights (and are married less than 5 years) the likelihood of separation or divorce 5 years later is some six times that when men work days. When women work nights (and are married more than 5 years) the odds of divorce or separation are three times as high. Moreover, the data suggest that the increased tendency for divorce is not because spouses in troubled marriages are more likely to opt for night work; the causality seems in the opposite direction.[9]

Single as well as married mothers often engage in a split-shift caregiving arrangement with grandmothers. More than one-third of grandmothers who provide care for preschool-aged children are otherwise employed.[10] Here, too, there may be both positive and negative aspects of such arrangements, but this has not been studied. The observation that single mothers are more likely than married mothers to work long as well as nonstandard hours and are more likely to be among the working poor[11,12] suggests that the problems of managing time and money are especially stressful for such mothers.

Policymakers and scholars must take a more realistic view of the temporal nature of family life among Americans. With regard to welfare reform, for example, close to half (43.3%) of employed mothers with a high school education or less, ages 18 to 34, work other than a fixed daytime schedule, weekdays only.[13] If mothers on welfare are to move into jobs similar to these mothers, a key policy issue is how to improve the fit between the availability of child care and these working mothers' schedules. Expanding day care alone will not be satisfactory.

The movement toward a 24-hour economy is well underway, and will continue into the next century. Although driven by factors external to individual families, it will affect the lives of family members in profound ways. The home-time structure of families is becoming temporally very complex. We need to change our conception of family life to include such complexities. This should help to improve social policies that seek to ease the economic and social tensions that often result from the dual demands of work and family, particularly among the working poor.

REFERENCES AND NOTES

1. J. Schor, *The Overworked American* (Basic Books, New York, 1991).

2. J. P. Robinson and G. Godbey. *Time for Life: The Surprising Ways Americans Use Their Time* (Pennsylvania State Univ. Press, University Park, PA, 1997).

3. J. R. Meisenheimer II, *Mon. Labor Rev.* 121, 22 (February 1998).

4. H. B. Presser, *Demography* 26, 523 (1989).

5. G. T. Silvestri, *Mon. Labor Rev.* 120, 58 (November 1997).

6. H. B. Presser, *Demography* 32, 577 (1995).

7. *Biological Rhythms: Implications for the Worker* (OTA-BA-463, Office of Technological Assessment, Washington, DC, 1991).

8. H. B. Presser, *J. Marr. Fam.* 50, 133 (1988).

9. ———. *ibid.,* in press.

10. ———. *ibid.,* 51, 581 (1989).

11. A. G. Cox, thesis, University of Maryland (1994).

12. ———. and H. B. Presser, in *Work and Family: Research Informing Policy.* T. Parcel and D. B. Cornfield, Eds. (Sage. Thousand Oaks, CA, in press.).

13. H. B. Presser and A. G. Cor, *Mon. Labor Rev.* 120, 25 (April 1997).

DISCUSSION QUESTIONS

1. What factors does Presser identify as leading to increases in the working hours of employed people? What impact are these changes likely to have on the labor force experiences of two-earner households?

2. Sociologists have often argued that work and family structures are mutually interdependent. How is this evidenced by the effects of increasing work hours on families' experiences? What evidence of this have you seen in your own life?

51

Nickel-and-Dimed: On (Not) Getting By in America

BARBARA EHRENREICH

Barbara Ehrenreich, a journalist and faculty member, "posed" for several weeks as an unskilled worker to find out how people survive doing low-wage work. Her account, later published in the best-selling book, Nickel and Dimed, *is a compelling portrait of the conditions of work for millions of people in the United States.*

At the beginning of June 1998 I leave behind everything that normally soothes the ego and sustains the body—home, career, companion, reputation, ATM card—for a plunge into the low-wage workforce. There, I become another, occupationally much diminished "Barbara Ehrenreich"—depicted

From: Ehrenreich, Barbara. "Nickel-and-Dimed: On (Not) Getting By in America."
Harper's Magazine 298 (January 1999): 37 ff. Reprinted with permission.

on job-application forms as a divorced homemaker whose sole work experience consists of housekeeping in a few private homes. I am terrified, at the beginning, of being unmasked for what I am: a middle-class journalist setting out to explore the world that welfare mothers are entering, at the rate of approximately 50,000 a month, as welfare reform kicks in. Happily, though, my fears turn out to be entirely unwarranted: during a month of poverty and toil, my name goes unnoticed and for the most part unuttered. In this parallel universe where my father never got out of the mines and I never got through college, I am "baby," "honey," "blondie," and, most commonly, "girl."

My first task is to find a place to live. I figure that if I can earn $7 an hour—which, from the want ads, seems doable—I can afford to spend $500 on rent, or maybe, with severe economies, $600. In the Key West area, where I live, this pretty much confines me to flophouses and trailer homes—like the one, a pleasing fifteen-minute drive from town, that has no air-conditioning, no screens, no fans, no television, and, by way of diversion, only the challenge of evading the landlord's Doberman pinscher. The big problem with this place, though, is the rent, which at $675 a month is well beyond my reach. All right, Key West is expensive. But so is New York City, or the Bay Area, or Jackson Hole, or Telluride, or Boston, or any other place where tourists and the wealthy compete for living space with the people who clean their toilets and fry their hash browns. Still, it is a shock to realize that "trailer trash" has become, for me, a demographic category to aspire to.

So I decide to make the common trade-off between affordability and convenience, and go for a $500-a-month efficiency thirty miles up a two-lane highway from the employment opportunities of Key West, meaning forty-five minutes if there's no road construction and I don't get caught behind some sundazed Canadian tourists. I hate the drive, along a roadside studded with white crosses commemorating the more effective head-on collisions, but it's a sweet little place—a cabin, more or less, set in the swampy back yard of the converted mobile home where my landlord, an affable TV repairman, lives with his bartender girlfriend. Anthropologically speaking, a bustling trailer park would be preferable, but here I have a gleaming white floor and a firm mattress, and the few resident bugs are easily vanquished.

Besides, I am not doing this for the anthropology. My aim is nothing so mistily subjective as to "experience poverty" or find out how it "really feels" to be a long-term low-wage worker. I've had enough unchosen encounters with poverty and the world of low-wage work to know it's not a place you want to visit for touristic purposes; it just smells too much like fear. And with all my real-life assets—bank account, IRA, health insurance, multiroom home—waiting indulgently in the background, I am, of course, thoroughly insulated from the terrors that afflict the genuinely poor.

No, this is a purely objective, scientific sort of mission. The humanitarian rationale for welfare reform—as opposed to the more punitive and stingy impulses that may actually have motivated it—is that work will lift poor women out of poverty while simultaneously inflating their self-esteem and hence their future value in the labor market. Thus, whatever the hassles involved in finding child care, transportation, etc., the transition from welfare to work will end happily, in greater prosperity for all. Now there are many problems with this comforting prediction, such as the fact that the economy will inevitably undergo a downturn, eliminating many jobs. Even without a downturn, the influx of a million former welfare recipients into the low-wage labor market could depress wages by as much as 11.9 percent, according to the Economic Policy Institute (EPI) in Washington, D.C.

But is it really possible to make a living on the kinds of jobs currently available to unskilled people? Mathematically, the answer is no, as can be shown by taking $6 to $7 an hour, perhaps subtracting a dollar or two an hour for child care, multiplying by 160 hours a month, and comparing the result to the prevailing rents. According to the National Coalition for the Homeless, for example, in 1998 it took, on average nationwide, an hourly wage of $8.89 to afford a one-bedroom apartment, and the Preamble Center for Public Policy estimates that the odds against a typical welfare recipient's landing a job at such a "living wage" are about 97 to 1. If these numbers are right, low-wage work is not a solution to poverty and possibly not even to homelessness.

It may seem excessive to put this proposition to an experimental test. As certain family members keep unhelpfully reminding me, the viability of low-wage work could be tested, after a fashion, without ever leaving my study. I could just pay myself $7 an hour for eight hours a day, charge myself for room and board, and total up the numbers after a month. Why leave the people and work that I love? But I am an experimental scientist by training. In that business, you don't just sit at a desk and theorize; you plunge into the everyday chaos of nature, where surprises lurk in the most mundane measurements. Maybe, when I got into it, I would discover some hidden economies in the world of the low-wage worker. After all, if 30 percent of the workforce toils for less than $8 an hour, according to the EPI, they may have found some tricks as yet unknown to me. Maybe—who knows?—would even be able to detect in myself the bracing psychological effects of getting out of the house, as promised by the welfare wonks at places like the Heritage Foundation. Or, on the other hand, maybe there would be unexpected costs—physical, mental, or financial—to throw off all my calculations. Ideally, I should do this with two small children in tow, that being the welfare average, but mine are grown and no one is willing to lend me theirs for a month-long vacation in penury. So this is not the perfect experiment, just a test of the best possible case: an unencumbered woman, smart and even strong, attempting to live more or less off the land.

On the morning of my first full day of job searching, I take a red pen to the want ads, which are auspiciously numerous. Everyone in Key West's booming "hospitality industry" seems to be looking for someone like me—trainable, flexible, and with suitably humble expectations as to pay. I know I possess certain traits that might be advantageous—I'm white and, I like to think, well-spoken and poised—but I decide on two rules: One, I cannot use any skills derived from my education or usual work—not that there are a lot of want ads for satirical essayists anyway. Two, I have to take the best-paid job that is offered me and of course do my best to hold it; no Marxist rants or sneaking off to read novels in the ladies' room. In addition, I rule out various occupations for one reason or another: Hotel front-desk clerk, for example, which to my surprise is regarded as unskilled and pays around $7 an hour, gets eliminated because it involves standing in one spot for eight hours a day. Waitressing is similarly something I'd like to avoid, because I remember it leaving me bone tired when I was eighteen, and I'm decades of varicosities and back pain beyond that now. Telemarketing, one of the first refuges of the suddenly indigent, can be dismissed on grounds of personality. This leaves certain supermarket jobs, such as deli clerk, or housekeeping in Key West's thousands of hotel and guest rooms. Housekeeping is especially appealing, for reasons both atavistic and practical: it's what my mother did before I came along, and it can't be too different from what I've been doing part-time, in my own home, all my life.

So I put on what I take to be a respectful-looking outfit of ironed Bermuda shorts and scooped-neck T-shirt and set out for a tour of the local hotels and supermarkets. Best Western, Econo Lodge, and HoJo's all let me fill out application forms, and these are, to my relief, interested in little more than whether I am a legal resident of the United States and have committed any felonies. My next stop is Winn-Dixie, the supermarket, which turns out to have a particularly onerous application process, featuring a fifteen-minute "interview" by computer since, apparently, no human on the premises is deemed capable of representing the corporate point of view. I am conducted to a large room decorated with posters illustrating how to look "professional" (it helps to be white and, if female, permed) and warning of the slick promises that union organizers might try to tempt me with. The interview is multiple choice: Do I have anything, such as child-care problems, that might make it hard for me to get to work on time? Do I think safety on the job is the responsibility of management? Then, popping up cunningly out of the blue: How many dollars' worth of stolen goods have I purchased in the last year? Would I turn in a fellow employee if I caught him stealing? Finally, "Are you an honest person?"

Apparently, I ace the interview, because I am told that all I have to do is show up some doctor's office tomorrow for a urine test. This seems to be a fairly general rule: if you want to stack Cheerio boxes or vacuum hotel rooms in chemically fascist America, you have to be willing to squat down and pee in front of some health worker (who has no doubt had to do the same thing herself). The wages Winn-Dixie is offering—$6 and a couple of dimes to start with—are not enough, I decide, to compensate for this indignity.

I lunch at Wendy's, where $4.99 gets you unlimited refills at the Mexican part of the Superbar, a comforting surfeit of refried beans and "cheese sauce." A teenage employee, seeing me studying the want ads, kindly offers me an application form, which I fill out, though here, too, the pay is just $6 and change an hour. Then it's off for a round of the locally owned inns and guest-houses. At "The Palms," let's call it, a bouncy manager actually takes me around to see the rooms and meet the existing housekeepers, who, I note with satisfaction, look pretty much like me—faded ex-hippie types in shorts with long hair pulled back in braids. Mostly, though, no one speaks to me or even looks at me except to proffer an application form. At my last stop, a palatial B&B, I wait twenty minutes to meet "Max," only to be told that there are no jobs now but there should be one soon, since "nobody lasts more than a couple weeks." (Because none of the people I talked to knew I was a reporter, I have changed their names to protect their privacy and, in some cases perhaps, their jobs.)

Three days go by like this, and, to my chagrin, no one out of the approximately twenty places I've applied calls me for an interview. I had been vain enough to worry about coming across as too educated for the jobs I sought, but no one even seems interested in finding out how overqualified I am. Only later will I realize that the want ads are not a reliable measure of the actual jobs available at any particular time. They are, as I should have guessed from Max's comment, the employers' insurance policy against the relentless turnover of the low-wage work-force. Most of the big hotels run ads almost continually, just to build a supply of applicants to replace the current workers as they drift away or are fired, so finding a job is just a matter of being at the right place at the right time and flexible enough to take whatever is being offered that day. This finally happens to me at a one of the big discount hotel chains, where I go, as usual, for housekeeping and am sent, instead, to try out as a waitress at the attached "family restaurant," a dismal spot with a counter and about thirty tables that looks out on a parking garage and features such tempting fare as "Polish [sic] sausage and BBQ sauce" on 95-degree days. Phillip, the dapper young West Indian who introduces himself as the manager, interviews me with about as much enthusiasm as if he were a clerk processing me for Medicare, the principal questions being what shifts can I work and when can I start. I mutter something about being woefully out of practice as a waitress, but he's already on to the uniform: I'm to show up tomorrow wearing black slacks and black shoes; he'll provide the rust-colored polo shirt with HEARTHSIDE embroidered on it, though I might want to wear my own shirt to get to work, ha ha. At the word "tomorrow," something between fear and indignation rises in my chest. I want to say, "Thank you for your time, sir, but this is just an experiment, you know, not my actual life."

So begins my career at the Hearthside, I shall call it, one small profit center within a global discount hotel chain, where for two weeks I work from 2:00 till 10:00 P.M. for $2.43 an hour plus tips. In some futile bid for gentility, the management has barred employees from using the front door, so my first day I enter through the kitchen, where a red-faced man with shoulder-length blond hair is throwing frozen steaks against the wall and yelling, "Fuck this shit!" "That's just Jack," explains Gail, the wiry middle-aged waitress who is assigned to train me. "He's on the rag again"—a condition occasioned, in this instance, by the fact that the cook on the morning shift had forgotten to thaw out the steaks. For the next eight hours, I run after the agile Gail, absorbing bits of instruction along with fragments of personal tragedy. All food must be trayed, and the reason she's so tired today is that she woke up in a cold sweat thinking of her boyfriend, who killed himself recently in an upstate prison. No refills on lemonade. And the reason he was in prison is that a few DUIs caught up with him, that's all, could have happened to anyone. Carry the creamers to the table in a monkey bowl, never in your hand. And after he was gone she spent several months living in her truck, peeing in a plastic pee bottle and reading by candlelight at night, but you can't live in a truck in the summer, since you need to have the windows down, which means anything can get in, from mosquitoes on up.

At least Gail puts to rest any fears I had of appearing overqualified. From the first day on, I find that of all the things I have left behind, such as home and identity, what I miss the most is competence. Not that I have ever felt utterly competent in the writing business, in which one day's success augers nothing at all for the next. But in my writing life I at least have some notion of procedure: do the research, make the outline, rough out a draft, etc. As a server, though, I am beset by requests like bees: more iced tea here, ketchup over there, a to-go box for table fourteen, and where are the high chairs, anyway? Of the twenty-seven tables, up to six are usually mine at any time, though on slow afternoons or if Gail is off, I sometimes have the whole place to myself. There is the touch-screen computer-ordering system to master, which is, I suppose, meant to minimize server-cook contact, but in practice requires constant verbal fine-tuning: "That's gravy on the mashed, okay? None on the meatloaf," and so forth—while the cook scowls as if I were inventing these refinements just to torment him. Plus, something I had forgotten in the years since I was eighteen: about a third of a server's job is "side work" that's invisible to customers—sweeping, scrubbing, slicing, refilling, and restocking. If it isn't all done, every little bit of it, you're going to face the 6:00 P.M. dinner rush defenseless and probably go down in flames. I screw up dozens of times at the beginning, sustained in my shame entirely by Gail's support—"It's okay, baby, everyone does that sometime"—because, to my total surprise and despite the scientific detachment I am doing my best to maintain, I care.

The whole thing would be a lot easier if I could just skate through it as Lily Tomlin in one of her waitress skits, but I was raised by the absurd Booker T. Washingtonian precept that says: If you're going to do something, do it well. In fact, "well" isn't good enough by half. Do it better than anyone has ever done it before. Or so said my father, who must have known what he was talking about because he managed to pull himself, and us with him, up from the mile-deep copper mines of Butte to the leafy suburbs of the Northeast, ascending from boilermakers to martinis before booze beat out ambition. As in most endeavors I have encountered in my life, doing it "better than anyone" is not a reasonable goal. Still, when I wake up at 4:00 A.M. in my own cold sweat, I am not thinking about the writing deadlines I'm neglecting; I'm thinking about the table whose order I screwed up so that one of the boys didn't get his kiddie meal until the rest of the family had moved on to their Key Lime pies. That's the other power-ful motivation I hadn't expected—the customers, or "patients," as I can't help thinking of them on account of the mysterious vulnerability that seems to have left them temporarily unable to feed themselves. After a few days at the Hearth-side, I feel the service ethic kick in like a shot of oxytocin, the nurturance hor-mone. The plurality of my customers are hard-working locals—truck drivers, construction workers, even housekeepers from the attached hotel—and I want them to have the closest to a "fine dining" experience that the grubby circum-stances will allow. No "you guys" for me; everyone over twelve is "sir" or "ma'am." I ply them with iced tea and coffee refills; I return, mid-meal, to inquire how everything is; I doll up their salads with chopped raw mushrooms, summer squash slices, or whatever bits of produce I can find that have survived their sojourn in the cold-storage room mold-free. . . .

Ten days into it, this is beginning to look like a livable lifestyle. I like Gail, who is "looking at fifty" but moves so fast she can alight in one place and then another without apparently being anywhere between them. I clown around with Lionel, the teenage Haitian busboy, and catch a few fragments of conversation with Joan, the svelte fortyish hostess and militant feminist who is the only one of us who dares to tell Jack to shut the fuck up. I even warm up to Jack when, on a slow night and to make up for a particularly unwarranted attack on my abilities, or so I imagine, he tells me about his glory days as a young man at "coronary school"—or do you say "culinary"?—in Brooklyn, where he dated a knock-out Puerto Rican chick and learned everything there is to know about food. I finish up at 10:00 or 10:30, depending on how much side work I've been able to get done during the shift, and cruise home to the tapes I snatched up at random when I left my real home—Marianne Faithfull, Tracy Chapman, Enigma, King Sunny Ade, the Violent Femmes—just drained enough for the music to set my cranium resonating but hardly dead. Midnight snack is Wheat Thins and Monterey Jack, accompanied by cheap white wine on ice and whatever AMC has to offer. To bed by 1:30 or 2:00, up at 9:00 or 10:00, read for an hour while my uniform whirls around in the landlord's washing machine, and then it's another eight hours spent following Mao's central instruction, as laid out in the Little Red Book, which was: Serve the people.

I could drift along like this, in some dreamy proletarian idyll, except for two things. One is management. If I have kept this subject on the margins thus far it is because I still flinch to think that I spent all those weeks under the surveillance of men (and later women) whose job it was to monitor my behavior for signs of sloth, theft, drug abuse, or worse. Not that managers and especially "assistant managers" in low-wage settings like this are exactly the class enemy. In the restaurant business, they are mostly former cooks or servers, still capable of pinch-hitting in the kitchen or on the floor, just as in hotels they are likely to be former clerks, and paid a salary of only about $400 a week. But everyone knows they have crossed over to the other side, which is, crudely put, corporate as opposed to human. Cooks want to prepare tasty meals; servers want to serve them graciously; but managers are there for only one reason—to make sure that money is made for some theoretical entity that exists far away in Chicago or New York, if a corporation can be said to have a physical existence at all. Reflecting on her career, Gail tells me ruefully that she had sworn, years ago, never to work for a corporation again. "They don't cut you no slack. You give and you give, and they take."

Managers can sit—for hours at a time if they want—but it's their job to see that no one else ever does, even when there's nothing to do, and this is why, for servers, slow times can be as exhausting as rushes. You start dragging out each little chore, because if the manager on duty catches you in an idle moment, he will give you something far nastier to do. So I wipe, I clean, I consolidate ketchup bottles and recheck the cheesecake supply, even tour the tables to make sure the customer evaluation forms are all standing perkily in their places—wondering all the time how many calories I burn in these strictly theatrical exercises. When, on a particularly dead afternoon, Stu finds me glancing at a USA Today a customer has left behind, he assigns me to vacuum the entire floor with the broken vacuum cleaner that has a handle only two feet long, and the only way to do that without incurring orthopedic damage is to proceed from spot to spot on your knees.

On my first Friday at the Hearthside there is a "mandatory meeting for all restaurant employees," which I attend, eager for insight into our overall marketing strategy and the niche (your basic Ohio cuisine with a tropical twist?) we aim to inhabit. But there is no "we" at this meeting. Phillip, our top manager except for an occasional "consultant" sent out by corporate headquarters, opens it with a sneer: "The break room—it's disgusting. Butts in the ashtrays, newspapers lying around, crumbs." This windowless little room, which also houses the time clock for the entire hotel, is where we stash our bags and civilian clothes and take our half-hour meal breaks. But a break room is not a right, he tells us. It can be taken away. We should also know that the lockers in the break room and whatever is in them can be searched at any time. Then comes gossip; there has been gossip; gossip (which seems to mean employees talking among themselves) must stop. Off-duty employees are henceforth barred from eating at the restaurant, because "other servers gather around them and gossip." When Phillip has exhausted his agenda of rebukes, Joan complains about the condition of the ladies' room and I throw in my two bits about the vacuum cleaner. But I don't see any backup coming from my fellow servers, each of whom has subsided into her own personal funk; Gail, my role model, stares sorrowfully at a point six inches from her nose. The meeting ends when Andy, one of the cooks, gets up, muttering about breaking up his day off for this almighty bullshit.

Just four days later we are suddenly summoned into the kitchen at 3:30 P.M., even though there are live tables on the floor. We all—about ten of us—stand around Phillip, who announces grimly that there has been a report of some "drug activity" on the night shift and that, as a result, we are now to be a "drug-free" workplace, meaning that all new hires will be tested, as will possibly current employees on a random basis. I am glad that this part of the kitchen is so dark, because I find myself blushing as hard as if I had been caught toking up in the ladies' room myself: I haven't been treated this way—lined up in the corridor, threatened with locker searches, peppered with carelessly aimed accusations—since junior high school. Back on the floor, Joan cracks, "Next they'll be telling us we can't have sex on the job." When I ask Stu what happened to inspire the crackdown, he just mutters about "management decisions" and takes the opportunity to upbraid Gail and me for being too generous with the rolls. From now on there's to be only one per customer, and it goes out with the dinner, not with the salad. He's also been riding the cooks, prompting Andy to come out of the kitchen and observe—with the serenity of a man whose customary implement is a butcher knife—that "Stu has a death wish today."

Later in the evening, the gossip crystallizes around the theory that Stu is himself the drug culprit, that he uses the restaurant phone to order up marijuana and sends one of the late servers out to fetch it for him. The server was caught, and she may have ratted Stu out or at least said enough to cast some suspicion on him, thus accounting for his pissy behavior. Who knows? Lionel, the busboy, entertains us for the rest of the shift by standing just behind Stu's back and sucking deliriously on an imaginary joint.

The other problem, in addition to the less-than-nurturing management style, is that this job shows no sign of being financially viable. You might imagine, from a comfortable distance, that people who live, year in and year out, on $6 to $10 an hour have discovered some survival stratagems unknown to the middle class. But no. It's not hard to get my co-workers to talk about their living situations, because housing, in almost every case, is the principal source of disruption in their lives, the first thing they fill you in on when they arrive for their shifts. After a week, I have compiled the following survey:

- Gail is sharing a room in a well-known down-town flophouse for which she and a roommate pay about $250 a week. Her roommate, a male friend, has begun hitting on her, driving her nuts, but the rent would be impossible alone.

- Claude, the Haitian cook, is desperate to get out of the two-room apartment he shares with his girlfriend and two other, unrelated, people. As far as I can determine, the other Haitian men (most of whom only speak Creole) live in similarly crowded situations.

- Annette, a twenty-year-old server who is six months pregnant and has been abandoned by her boyfriend, lives with her mother, a postal clerk.

- Marianne and her boyfriend are paying $170 a week for a one-person trailer.

- Jack, who is, at $10 an hour, the wealthiest of us, lives in the trailer he owns, paying only the $400-a-month lot fee.

- The other white cook, Andy, lives on his dry-docked boat, which, as far as I can tell from his loving descriptions, can't be more than twenty feet long. He offers to take me out on it, once it's repaired, but the offer comes with inquiries as to my marital status, so I do not follow up on it.

- Tina and her husband are paying $60 a night for a double room in a Days Inn. This is because they have no car and the Days Inn is within walking distance of the Hearthside. When Marianne, one of the breakfast servers, is tossed out of her trailer for subletting (which is against the trailer-park rules), she leaves her boyfriend and moves in with Tina and her husband.

- Joan, who had fooled me with her numerous and tasteful outfits (hostesses wear their own clothes), lives in a van she parks behind a shopping center at night and showers in Tina's motel room. The clothes are from thrift shops.

It strikes me, in my middle-class solipsism, that there is gross improvidence in some of these arrangements. When Gail and I are wrapping silverware in napkins—the only task for which we are permitted to sit—she tells me she is thinking of escaping from her roommate by moving into the Days Inn herself. I am astounded: How can she even think of paying between $40 and $60 a day? But if I was afraid of sounding like a social worker, I come out just sounding like a fool. She squints at me in disbelief, "And where am I supposed to get a month's rent and a month's deposit for an apartment?" I'd been feeling pretty smug about my $500 efficiency, but of course it was made possible only by the $1,300 I had allotted myself for start-up costs when I began my low-wage life: $1,000 for the first month's rent and deposit, $100 for initial groceries and cash in my pocket, $200 stuffed away for emergencies. In poverty, as in certain propositions in physics, starting conditions are everything.

There are no secret economies that nourish the poor; on the contrary, there are a host of special costs. If you can't put up the two months' rent you need to secure an apartment, you end up paying through the nose for a room by the week. If you have only a room, with a hot plate at best, you can't save by cooking up huge lentil stews that can be frozen for the week ahead. You eat fast food, or the hot dogs and styrofoam cups of soup that can be microwaved in a convenience store. If you have no money for health insurance—and the Hearthside's niggardly plan kicks in only after three months—you go without routine care or prescription drugs and end up paying the price. Gail, for example, was fine until she ran out of money for estrogen pills. She is supposed to be on the company plan by now, but they claim to have lost her application form and need to begin the paperwork all over again. So she spends $9 per migraine pill to control the headaches she wouldn't have, she insists, if her estrogen supplements were covered. Similarly, Marianne's boyfriend lost his job as a roofer because he missed so much time after getting a cut on his foot for which he couldn't afford the prescribed antibiotic.

My own situation, when I sit down to assess it after two weeks of work, would not be much better if this were my actual life. The seductive thing about waitressing is that you don't have to wait for payday to feel a few bills in your pocket, and my tips usually cover meals and gas, plus something left over to stuff into the kitchen drawer I use as a bank. But as the tourist business slows in the summer heat, I sometimes leave work with only $20 in tips (the gross is higher, but servers share about 15 percent of their tips with the busboys and bartenders). With wages included, this amounts to about the minimum wage of $5.15 an hour. Although the sum in the drawer is piling up, at the present rate of accumulation it will be more than a hundred dollars short of my rent when the end of the month comes around. Nor can I see any expenses to cut. True, I haven't gone the lentil-stew route yet, but that's because I don't have a large cooking pot, pot holders, or a ladle to stir with (which cost about $30 at Kmart, less at thrift stores), not to mention onions, carrots, and the indispensable bay leaf. I do make my lunch almost every day—usually some slow-burning, high-protein combo like frozen chicken patties with melted cheese on top and canned pinto beans on the side. Dinner is at the Hearthside, which offers its employees a choice of BLT, fish sandwich, or hamburger for only $2. The burger lasts longest, especially if it's heaped with gut-puckering jalapenos, but by midnight my stomach is growling again.

So unless I want to start using my car as a residence, I have to find a second, or alternative, job. I call all the hotels where I filled out housekeeping applications weeks ago—the Hyatt, Holiday Inn, Econo Lodge, Hojo's, Best Western, plus a half dozen or so locally run guesthouses. Nothing. Then I start making the rounds again, wasting whole mornings waiting for some assistant manager to show up, even dipping into places so creepy that the front-desk clerk greets you from behind bulletproof glass and sells pints of liquor over the counter. But either someone has exposed my real-life housekeeping habits—which are, shall we say, mellow—or I am at the wrong end of some infallible ethnic equation: most, but by no means all, of the working housekeepers I see on my job searches are African Americans, Spanish-speaking, or immigrants from the Central European post-Communist world, whereas servers are almost invariably white and monolingually English-speaking. When I finally get a positive response, I have been identified once again as server material. Jerry's, which is part of a well-known national family restaurant chain and physically attached here to another budget hotel chain, is ready to use me at once. The prospect is both exciting and terrifying, because, with about the same number of tables and counter seats, Jerry's attracts three or four times the volume of customers as the gloomy old Hearthside. . . .

I start out with the beautiful, heroic idea of handling the two jobs at once, and for two days I almost do it: the breakfast/lunch shift at Jerry's, which goes till 2:00, arriving at the Hearthside at 2:10, and attempting to hold out until 10:00. In the ten minutes between jobs, I pick up a spicy chicken sandwich at the Wendy's drive-through window, gobble it down in the car, and change from khaki slacks to black, from Hawaiian to rust polo. There is a problem, though. When during the 3:00 to 4:00 P.M. dead time I finally sit down to wrap silver, my flesh seems to bond to the seat. I try to refuel with a purloined cup of soup, as I've seen Gail and Joan do dozens of times, but a manager catches me and hisses "No eating!" though there's not a customer around to be offended by the sight of food making contact with a server's lips. So I tell Gail I'm going to quit, and she hugs me and says she might just follow me to Jerry's herself.

But the chances of this are minuscule. She has left the flophouse and her annoying roommate and is back to living in her beat-up old truck. But guess what? she reports to me excitedly later that evening: Phillip has given her permission to park overnight in the hotel parking lot, as long as she keeps out of sight, and the parking lot should be totally safe, since it's patrolled by a hotel security guard! With the Hearthside offering benefits like that, how could anyone think of leaving?

Gail would have triumphed at Jerry's, I'm sure, but for me it's a crash course in exhaustion management. Years ago, the kindly fry cook who trained me to waitress at a Los Angeles truck stop used to say: Never make an unnecessary trip; if you don't have to walk fast, walk slow; if you don't have to walk, stand. But at Jerry's the effort of distinguishing necessary from unnecessary and urgent from whenever would itself be too much of an energy drain. The only thing to do is to treat each shift as a one-time-only emergency: you've got fifty starving people out there, lying scattered on the battlefield, so get out there and feed them! Forget that you will have to do this again tomorrow, forget that you will have to be alert enough to dodge the drunks on the drive home tonight—just burn, burn, burn! Ideally, at some point you enter what servers call "a rhythm" and psychologists term a "flow state," in which signals pass from the sense organs directly to the muscles, bypassing the cerebral cortex, and a Zen-like emptiness sets in.

A male server from the Hearthside's morning shift tells me about the time he "pulled a triple"—three shifts in a row, all the way around the clock—and then got off and had a drink and met this girl, and maybe he shouldn't tell me this, but they had sex right then and there, and it was like, beautiful.

But there's another capacity of the neuromuscular system, which is pain. I start tossing back drugstore-brand ibuprofen pills as if they were vitamin C, four before each shift, because an old mouse-related repetitive-stress injury in my upper back has come back to full-spasm strength, thanks to the tray carrying. In my ordinary life, this level of disability might justify a day of ice packs and stretching. Here I comfort myself with the Aleve commercial in which the cute blue-collar guy asks: If you quit after working four hours, what would your boss say? And the not-so-cute blue-collar guy, who's lugging a metal beam on his back, answers: He'd fire me, that's what. But fortunately, the commercial tells us, we workers can exert the same kind of authority over our painkillers that our bosses exert over us. If Tylenol doesn't want to work for more than four hours, you just fire its ass and switch to Aleve.

True, I take occasional breaks from this life, going home now and then to catch up on e-mail and for conjugal visits (though I am careful to "pay" for anything I eat there), seeing The Truman Show with friends and letting them buy my ticket. And I still have those what-am-I-doing-here moments at work, when I get so homesick for the printed word that I obsessively reread the six-page menu. But as the days go by, my old life is beginning to look exceedingly strange. The e-mails and phone messages addressed to my former self come from a distant race of people with exotic concerns and far too much time on their hands. The neighborly market I used to cruise for produce now looks forbiddingly like a Manhattan yuppie emporium. And when I sit down one morning in my real home to pay bills from my past life, I am dazzled at the two- and three-figure sums owed to outfits like Club BodyTech and Amazon.com. . . .

I make the decision to move closer to Key West. First, because of the drive. Second and third, also because of the drive: gas is eating up $4 to $5 a day, and although Jerry's is as high-volume as you can get, the tips average only 10 percent, and not just for a newbie like me. Between the base pay of $2.15 an hour and the obligation to share tips with the busboys and dishwashers, we're averaging only about $7.50 an hour. Then there is the $30 I had to spend on the regulation tan slacks worn by Jerry's servers—a setback it could take weeks to absorb. (I had combed the town's two downscale department stores hoping for something cheaper but decided in the end that these marked-down Dockers, originally $49, were more likely to survive a daily washing.) Of my fellow servers, everyone who lacks a working husband or boyfriend seems to have a second job: Nita does something at a computer eight hours a day; another welds. Without the forty-five-minute commute, I can picture myself working two jobs and having the time to shower between them. . . .

I can do this two-job thing, is my theory, if I can drink enough caffeine and avoid getting distracted by George's ever more obvious suffering. . . .

Then it comes, the perfect storm. Four of my tables fill up at once. Four tables is nothing for me now, but only so long as they are obligingly staggered. As

I bev table 27, tables 25, 28, and 24 are watching enviously. As I bev 25, 24 glowers because their bevs haven't even been ordered. Twenty-eight is four yuppyish types, meaning everything on the side and agonizing instructions as to the chicken Caesars. Twenty-five is a middle-aged black couple, who complain, with some justice, that the iced tea isn't fresh and the tabletop is sticky. But table 24 is the meteorological event of the century: ten British tourists who seem to have made the decision to absorb the American experience entirely by mouth. Here everyone has at least two drinks—iced tea and milk shake, Michelob and water (with lemon slice, please)—and a huge promiscuous orgy of breakfast specials, mozz sticks, chicken strips, quesadillas, burgers with cheese and without, sides of hash browns with cheddar, with onions, with gravy, seasoned fries, plain fries, banana splits. Poor Jesus! Poor me! Because when I arrive with their first tray of food—after three prior trips just to refill bevs—Princess Di refuses to eat her chicken strips with her pancake-and-sausage special, since, as she now reveals, the strips were meant to be an appetizer. Maybe the others would have accepted their meals, but Di, who is deep into her third Michelob, insists that everything else go back while they work on their "starters." Meanwhile, the yuppies are waving me down for more decaf and the black couple looks ready to summon the NAACP.

Much of what happened next is lost in the fog of war. . . .

I leave. I don't walk out, I just leave. I don't finish my side work or pick up my credit-card tips, if any, at the cash register or, of course, ask Joy's permission to go. And the surprising thing is that you can walk out without permission, that the door opens, that the thick tropical night air parts to let me pass, that my car is still parked where I left it. There is no vindication in this exit, no fuck-you surge of relief, just an overwhelming, dank sense of failure pressing down on me and the entire parking lot. I had gone into this venture in the spirit of science, to test a mathematical proposition, but somewhere along the line, in the tunnel vision imposed by long shifts and relentless concentration, it became a test of myself, and clearly I have failed. Not only had I flamed out as a housekeeper/server, I had even forgotten to give George my tips, and, for reasons perhaps best known to hardworking, generous people like Gail and Ellen, this hurts. I don't cry, but I am in a position to realize, for the first time in many years, that the tear ducts are still there, and still capable of doing their job. . . .

In one month, I had earned approximately $1,040 and spent $517 on food, gas, toiletries, laundry, phone, and utilities. If I had remained in my $500 efficiency, I would have been able to pay the rent and have $22 left over (which is $78 less than the cash I had in my pocket at the start of the month). During this time I bought no clothing except for the required slacks and no prescription drugs or medical care (I did finally buy some vitamin B to compensate for the lack of vegetables in my diet). Perhaps I could have saved a little on food if I had gotten to a supermarket more often, instead of convenience stores, but it should be noted that I lost almost four pounds in four weeks, on a diet weighted heavily toward burgers and fries.

How former welfare recipients and single mothers will (and do) survive in the low-wage workforce, I cannot imagine. Maybe they will figure out how to condense their lives—including child-raising, laundry, romance, and meals—into the couple of hours between full-time jobs. Maybe they will take up residence in their vehicles, if they have one. All I know is that I couldn't hold two jobs and I couldn't make enough money to live on with one. And I had advantages unthinkable to many of the long-term poor—health, stamina, a working car, and no children to care for and support. Certainly nothing in my experience contradicts the conclusion of Kathryn Edin and Laura Lein, in their recent book *Making Ends Meet: How Single Mothers Survive Welfare and Low-Wage Work,* that low-wage work actually involves more hardship and deprivation than life at the mercy of the welfare state. In the coming months and years, economic conditions for the working poor are bound to worsen, even without the almost inevitable recession. As mentioned earlier, the influx of former welfare recipients into the low-skilled workforce will have a depressing effect on both wages and the number of jobs available. A general economic downturn will only enhance these effects, and the working poor will of course be facing it without the slight, but nonetheless often saving, protection of welfare as a backup.

The thinking behind welfare reform was that even the humblest jobs are morally uplifting and psychologically buoying. In reality they are likely to be fraught with insult and stress. But I did discover one redeeming feature of the most abject low-wage work—the camaraderie of people who are, in almost all cases, far too smart and funny and caring for the work they do and the wages they're paid. The hope, of course, is that someday these people will come to know what they're worth, and take appropriate action.

DISCUSSION QUESTIONS

1. What were the major lessons that Ehrenreich learned during her "experiment" as a low-wage worker and how do they inform your understanding of social class?

2. Having read Ehrenreich's article, how do you understand the relationship between work, stress, and health? What workplace policies could be implemented to improve worker health (mental and physical)?

http://infotrac.thomsonlearning.com

InfoTrac College Edition

BONUS READING

Lafer, Gordon. "Captive Labor: America's Prisoners as Corporate Workforce." *The American Prospect* 10 (September 1, 1999): 66ff.

Prison labor is increasingly being used by many U.S. companies for a variety of labor force needs. On the one hand, supporters would argue that this can provide prisoners with skills needed for re-entering the labor market on release. Critics contend that such labor is exploitative and takes away jobs from other potential workers. What issues around prison labor does Lafer identify here and what would it lead you to conclude about such work programs?

SEARCH TERMS

contingent work
gender division of labor
low-wage work
social speedup
working poor

52

The Power Elite

C. WRIGHT MILLS

C. Wright Mills's classic book, The Power Elite, *first published in 1956, remains an important analysis of the system of power in the United States. He argues that national power is located in three particular institutions: the economy, politics, and the military. An important point in his article is that the power of elites is derived from their institutional location, not their individual attributes.*

The powers of ordinary men are circumscribed by the everyday worlds in which they live, yet even in these rounds of job, family, and neighborhood they often seem driven by forces they can neither understand nor govern. "Great changes" are beyond their control, but affect their conduct and outlook none the less. The very framework of modern society confines them to projects not their own, but from every side, such changes now press upon the men and women of the mass society, who accordingly feel that they are without purpose in an epoch in which they are without power.

But not all men are in this sense ordinary. As the means of information and of power are centralized, some men come to occupy positions in American society from which they can look down upon, so to speak, and by their decisions mightily affect, the everyday worlds of ordinary men and women. They are not made by their jobs; they set up and break down jobs for thousands of others; they are not confined by simple family responsibilities; they can escape. They may live in many hotels and houses, but they are bound by no one community. They need not merely "meet the demands of the day and hour"; in some part, they create these demands, and cause others to meet them. Whether or not they profess their power, their technical and political experience of it far transcends that of the underlying population. What Jacob Burckhardt said of "great men," most Americans might well say of their elite: "They are all that we are not."

The power elite is composed of men whose positions enable them to transcend the ordinary environments of ordinary men and women; they are in positions to make decisions having major consequences. Whether they do or do not make such decisions is less important than the fact that they do occupy such pivotal positions: their failure to act, their failure to make decisions, is itself an act that is often of greater consequence than the decisions they do make. For they are in command of the major hierarchies and organizations of modern society.

They rule the big corporations. They run the machinery of the state and claim its prerogatives. They direct the military establishment. They occupy the strategic command posts of the social structure, in which are now centered the effective means of the power and the wealth and the celebrity which they enjoy.

The power elite are not solitary rulers. Advisers and consultants, spokesmen and opinion-makers are often the captains of their higher thought and decision. Immediately below the elite are the professional politicians of the middle levels of power, in the Congress and in the pressure groups, as well as among the new and old upper classes of town and city and region. Mingling with them, in curious ways which we shall explore, are those professional celebrities who live by being continually displayed but are never, so long as they remain celebrities, displayed enough. If such celebrities are not at the head of any dominating hierarchy, they do often have the power to distract the attention of the public or afford sensations to the masses, or, more directly, to gain the ear of those who do occupy positions of direct power. More or less unattached, as critics of morality and technicians of power, as spokesmen of God and creators of mass sensibility, such celebrities and consultants are part of the immediate scene in which the drama of the elite is enacted. But that drama itself is centered in the command posts of the major institutional hierarchies.

The truth about the nature and the power of the elite is not some secret which men of affairs know but will not tell. Such men hold quite various theories about their own roles in the sequence of event and decision. Often they are uncertain about their roles, and even more often they allow their fears and their hopes to affect their assessment of their own power. No matter how great their actual power, they tend to be less acutely aware of it than of the resistances of others to its use. Moreover, most American men of affairs have learned well the rhetoric of public relations, in some cases even to the point of using it when they are alone, and thus coming to believe it. The personal awareness of the actors is only one of the several sources one must examine in order to understand the higher circles. Yet many who believe that there is no elite, or at any rate none of any consequence, rest their argument upon what men of affairs believe about themselves, or at least assert in public.

There is, however, another view: those who feel, even if vaguely, that a compact and powerful elite of great importance does now prevail in America often base that feeling upon the historical trend of our time. They have felt, for example, the domination of the military event, and from this they infer that generals and admirals, as well as other men of decision influenced by them, must be enormously powerful. They hear that the Congress has again abdicated to a handful of men decisions clearly related to the issue of war or peace. They know that the bomb was dropped over Japan in the name of the United States of America, although they were at no time consulted about the matter. They feel that they live in a time of big decisions; they know that they are not making any. Accordingly, as they consider the present as history, they infer that at its center, making decisions or failing to make them, there must be an elite of power.

On the one hand, those who share this feeling about big historical events assume that there is an elite and that its power is great. On the other hand, those who listen carefully to the reports of men apparently involved in the great decisions often do not believe that there is an elite whose powers are of decisive consequence.

Both views must be taken into account, but neither is adequate. The way to understand the power of the American elite lies neither solely in recognizing the historic scale of events nor in accepting the personal awareness reported by men of apparent decision. Behind such men and behind the events of history, linking the two, are the major institutions of modern society. These hierarchies of state and corporation and army constitute the means of power; as such they are now of a consequence not before equaled in human history—and at their summits, there are now those command posts of modern society which offer us the sociological key to an understanding of the role of the higher circles in America.

Within American society, major national power now resides in the economic, the political, and the military domains. Other institutions seem off to the side of modern history, and, on occasion, duly subordinated to these. No family is as directly powerful in national affairs as any major corporation; no church is as directly powerful in the external biographies of young men in America today as the military establishment; no college is as powerful in the shaping of momentous events as the National Security Council. Religious, educational, and family institutions are not autonomous centers of national power; on the contrary, these decentralized areas are increasingly shaped by the big three, in which developments of decisive and immediate consequence now occur.

Families and churches and schools adapt to modern life; governments and armies and corporations shape it; and, as they do so, they turn these lesser institutions into means for their ends. Religious institutions provide chaplains to the armed forces where they are used as a means of increasing the effectiveness of its morale to kill. Schools select and train men for their jobs in corporations and their specialized tasks in the armed forces. The extended family has, of course, long been broken up by the industrial revolution, and now the son and the father are removed from the family, by compulsion if need be, whenever the army of the state sends out the call. And the symbols of all these lesser institutions are used to legitimate the power and the decisions of the big three.

The life-fate of the modern individual depends not only upon the family into which he was born or which he enters by marriage, but increasingly upon the corporation in which he spends the most alert hours of his best years; not only upon the school where he is educated as a child and adolescent, but also upon the state which touches him throughout his life; not only upon the church in which on occasion he hears the word of God, but also upon the army in which he is disciplined.

If the centralized state could not rely upon the inculcation of nationalist loyalties in public and private schools, its leaders would promptly seek to modify the decentralized educational system. If the bankruptcy rate among the top five hundred corporations were as high as the general divorce rate among the thirty-seven million married couples, there would be economic catastrophe on an international scale. If members of armies gave to them no more of their lives than do believers to the churches to which they belong, there would be a military crisis.

Within each of the big three, the typical institutional unit has become enlarged, has become administrative, and, in the power of its decisions, has become centralized. Behind these developments there is a fabulous technology, for as institutions, they have incorporated this technology and guide it, even as it shapes and paces their developments.

The economy—once a great scatter of small productive units in autonomous balance—has become dominated by two or three hundred giant corporations, administratively and politically interrelated, which together hold the keys to economic decisions.

The political order, once a decentralized set of several dozen states with a weak spinal cord, has become a centralized, executive establishment which has taken up into itself many powers previously scattered, and now enters into each and every cranny of the social structure.

The military order, once a slim establishment in a context of distrust fed by state militia, has become the largest and most expensive feature of government, and, although well versed in smiling public relations, now has all the grim and clumsy efficiency of a sprawling bureaucratic domain.

In each of these institutional areas, the means of power at the disposal of decision makers have increased enormously; their central executive powers have been enhanced; within each of them modern administrative routines have been elaborated and tightened up.

As each of these domains becomes enlarged and centralized, the consequences of its activities become greater, and its traffic with the others increases. The decisions of a handful of corporations bear upon military and political as well as upon economic developments around the world. The decisions of the military establishment rest upon and grievously affect political life as well as the very level of economic activity. The decisions made within the political domain determine economic activities and military programs. There is no longer, on the one hand, an economy, and, on the other hand, a political order containing a military establishment unimportant to politics and to money-making. There is a political economy linked, in a thousand ways, with military institutions and decisions. On each side of the world-split running through central Europe and around the Asiatic rimlands, there is an ever-increasing interlocking of economic, military, and political structures. If there is government intervention in the corporate economy, so is there corporate intervention in the governmental process. In the structural sense, this triangle of power is the source of the *interlocking directorate* that is most important for the historical structure of the present. . . .

At the pinnacle of each of the three enlarged and centralized domains, there have arisen those higher circles which make up the economic, the political, and the military elites. At the top of the economy, among the corporate rich, there are the chief executives; at the top of the political order, the members of the political directorate; at the top of the military establishment, the elite of soldier-statesmen clustered in and around the Joint Chiefs of Staff and the upper echelon. As each of these domains has coincided with the others, as decisions tend to become total in their consequence, the leading men in each of the three domains of power—the warlords, the corporation chieftains, the political directorate—tend to come together, to form the power elite of America. . . .

By the powerful we mean, of course, those who are able to realize their will, even if others resist it. No one, accordingly, can be truly powerful unless he has access to the command of major institutions, for it is over these institutional means of power that the truly powerful are, in the first instance, powerful. Higher politicians and key officials of government command such institutional power; so do admirals and generals, and so do the major owners and executives of the larger corporations. Not all power, it is true, is anchored in and exercised by means of such institutions, but only within and through them can power be more or less continuous and important.

Wealth also is acquired and held in and through institutions. The pyramid of wealth cannot be understood merely in terms of the very rich; for the great inheriting families, as we shall see, are now supplemented by the corporate institutions of modern society: every one of the very rich families has been and is closely connected—always legally and frequently managerially as well—with one of the multi-million dollar corporations.

The modern corporation is the prime source of wealth, but, in latter-day capitalism, the political apparatus also opens and closes many avenues to wealth. The amount as well as the source of income, the power over consumer's goods as well as over productive capital, are determined by position within the political economy. If our interest in the very rich goes beyond their lavish or their miserly consumption, we must examine their relations to modern forms of corporate property as well as to the state; for such relations now determine the chances of men to secure big property and to receive high income. . . .

If we took the one hundred most powerful men in America, the one hundred wealthiest, and the one hundred most celebrated away from the institutional positions they now occupy, away from their resources of men and women and money, away from the media of mass communication that are now focused upon them—then they would be powerless and poor and uncelebrated. For power is not of a man. Wealth does not center in the person of the wealthy. Celebrity is not inherent in any personality. To be celebrated, to be wealthy, to have power requires access to major institutions, for the institutional positions men occupy determine in large part their chances to have and to hold these valued experiences. . . .

DISCUSSION QUESTIONS

1. What evidence do you see of the presence of the power elite in today's economic, political, and military institutions? Suppose that Mills were writing his book today; what might he change about his essay?

2. Mills argues that the power elite use institutions such as religion, education, and the family as the means to their ends. Find an example of this from the daily news and explain how Mills would see this institution as being shaped by the power elite.

53

Diversity in the Power Elite

RICHARD ZWEIGENHAFT AND G. WILLIAM DOMHOFF

Richard Zweigenhaft and G. William Domhoff ask here whether the power elite has changed by incorporating more diverse groups into these higher circles. They find limited evidence that women, Jews, African Americans, Latinos, Asian Americans, and gays and lesbians have entered the power elite. Those who have tend to share the perspectives and values of those already in power.

Since the 1870s the refrain about the new diversity of the governing circles has been closely intertwined with a staple of American culture created by Horatio Alger Jr., whose name has become synonymous with upward mobility in America. Born in 1832 to a patrician family—Alger's father was a Harvard graduate, a Unitarian minister, and a Massachusetts state senator—Alger graduated from Harvard at the age of nineteen. There followed a series of unsuccessful efforts to establish himself in various careers. Finally, in 1864 Alger was hired as a Unitarian minister in Brewster, Massachusetts. Fifteen months later, he was dismissed from this position for homosexual acts with boys in the congregation.

Alger returned to New York, where he soon began to spend a great deal of time at the Newsboys' Lodging House, founded in 1853 for footloose youngsters between the ages of twelve and sixteen and home to many youths who had been mustered out of the Union Army after serving as drummer boys. At the Newsboys' Lodging House Alger found his literary niche and his subsequent claim to fame: writing books in which poor boys make good. His books sold by the hundreds of thousands in the last third of the nineteenth century, and by 1910 they were enjoying annual sales of more than one million in paperback.[1]

The deck is not stacked against the poor, according to Horatio Alger. When they simply show a bit of gumption, work hard, and thereby catch a break or two, they can become part of the American elite. The persistence of this theme, reinforced by the annual Horatio Alger Awards to such well-known personalities as Ronald Reagan, Bob Hope, and Billy Graham (who might not have been so eager to accept them if they had known of Alger's shadowed past), suggests that we may be dealing once again with a cultural myth. In its early versions, of course, the story concerned the great opportunities available for poor white boys willing to work their way to the top. More recently, the story has featured black Horatio Algers who started in the ghetto, Latino Horatio Algers who started in

From: Zweigenhaft, Richard, and G. William Domhoff. *Diversity in the Power Elite: Have Women and Minorities Reached the Top?*, 1–7, 192–94. New Haven, CT: Yale University Press, 1998. Reprinted with permission.

the barrio, Asian-American Horatio Algers whose parents were immigrants, and female Horatio Algers who seem to have no class backgrounds—all of whom now sit on the boards of the country's largest corporations.

But is any of this true? Can anecdotes and self-serving autobiographical accounts about diversity, meritocracy, and upward social mobility survive a more systematic analysis? Have very many women and previously excluded minorities made it to the top? Has class lost its importance in shaping life chances?

. . . We address these and related questions within the framework provided by the iconoclastic sociologist C. Wright Mills in his hard-hitting classic *The Power Elite,* published in 1956 when the media were in the midst of what Mills called the Great American Celebration. In spite of the Depression of the 1930s, Americans had pulled together to win World War II, and the country was both prosperous at home and influential abroad. Most of all, according to enthusiasts, the United States had become a relatively classless and pluralistic society, where power belonged to the people through their political parties and public opinion. Some groups certainly had more power than others, but no group or class had too much. The New Deal and World War II had forever transformed the corporate-based power structure of earlier decades.

Mills challenged this celebration of pluralism by studying the social backgrounds and career paths of the people who occupied the highest positions in what he saw as the three major institutional hierarchies in postwar America—the corporations, the executive branch of the federal government, and the military. He found that almost all the members of this leadership group, which he called the power elite, were white Christian males who came from "at most, the upper third of the income and occupational pyramids," despite the many Horatio Algeresque claims to the contrary.[2] A majority came from an even narrower stratum, the 11 percent of U.S. families headed by businesspeople or highly educated professionals like physicians and lawyers. Mills concluded that power in the United States in the 1950s was just about as concentrated as it had been since the rise of the large corporations, although he stressed that the New Deal and World War II had given political appointees and military chieftains more authority than they had exercised previously.

It is our purpose, therefore, to take a detailed look at the social, educational, and occupational backgrounds of the leaders of these three institutional hierarchies to see whether they have become more diverse in terms of gender, race, ethnicity, and sexual orientation, and also in terms of socioeconomic origins. Unlike Mills, we think the power elite is more than a set of institutional leaders. It is also the leadership group for the small upper class of owners and managers of large income-producing properties, the 1 percent of Americans who in 1992 possessed 37.2% of all net worth.[3] But that theoretical difference is not of great moment here. The important commonality is the great wealth and power embodied in these institutional hierarchies and the people who lead them. . . .

In addition to studying the extent to which women and minorities have risen in the system, we focus on whether they have followed different avenues to the top than their predecessors did, and on any special roles they may play. Are they in the innermost circles of the power elite, or are they more likely to serve as buffers and go-betweens? Do they go just so far and no farther? What obstacles does each group face?

We also examine whether or not the presence of women and minorities affects the power elite itself. Do those women and minorities who become part of the power elite influence it in a more liberal direction, or do they end up endorsing traditional conservative positions, such as opposition to trade unions, taxes, and government regulation of business? In addition, . . . we consider the possibility that the diversity forced on the power elite has had the ironic effect of strengthening it, at least in the short run, by providing it with people who can reach out to the previously excluded groups and by showing that the American system can deliver on its most important promise, an equal opportunity for every individual.

These are not simple issues, and the answers to some of the questions we ask vary greatly depending on which previously disadvantaged group we are talking about. Nonetheless, in the course of our research, a few general patterns emerged. . . .

1. The power elite now shows considerable diversity, at least as compared with its state in the 1950s, but its core group continues to be wealthy white Christian males, most of whom are still from the upper third of the social ladder. They have been filtered through a handful of elite schools in law, business, public policy, and international relations.

2. In spite of the increased diversity of the power elite, high social origins continue to be a distinct advantage in making it to the top. There are relatively few rags-to-riches stories in the groups we studied, and those we did find tended to come through the electoral process, usually within the Democratic Party. In general, it still takes at least three generations to rise from the bottom to the top in the United States.

3. The new diversity within the power elite is transcended by common values and a sense of hard-earned class privilege. The newcomers to the power elite have found ways to signal that they are willing to join the game as it has always been played, assuring the old guard that they will call for no more than relatively minor adjustments, if that. There are few liberals and fewer crusaders in the power elite, despite its new multiculturalism. Class backgrounds, current roles, and future aspirations are more powerful in shaping behavior in the power elite than gender, ethnicity, or race.

4. Not all the groups we studied have been equally successful in contributing to the new diversity in the power elite. Women, blacks, Latinos, Asian Americans, and openly homosexual men and women are all under-represented, but to varying degrees and with different rates of increasing representation. . . .

5. Although the corporate, political, and military elites accepted diversity only in response to pressure from minority activists and feminists, these elites have benefited from the presence of new members. Some serve either a buffer or a liaison function with such groups and institutions as consumers, angry neighborhoods, government agencies, and wealthy foreign entrepreneurs.

6. There is greater diversity in Congress than in the power elite, and the majority of the female and minority elected officials are Democrats. . . .

The power elite has been strengthened because diversity has been achieved primarily by the selection of women and minorities who share the prevailing perspectives and values of those already in power. The power elite is not "multi-cultural" in any full sense of the concept, but only in terms of ethnic or racial origins. This process has been helped along by those who have called for the inclusion of women and minorities without any consideration of criteria other than sex, race, or ethnicity. Because the demand was strictly for a woman on the Supreme Court, President Reagan could comply by choosing a conservative upper-class corporate lawyer, Sandra Day O'Connor. When pressure mounted to have more black justices, President Bush could respond by appointing Clarence Thomas, a conservative black Republican with a law degree from Yale University. It is yet another irony that appointments like these served to undercut the liberal social movements that caused them to happen.[4]

It is not surprising, therefore, that when we look at the business practices of the women and minorities who have risen to the top of the corporate world, we find that their perspectives and values do not differ markedly from those of their white male counterparts. When Linda Wachner, one of the few women to become CEO of a *Fortune*-level company, the Warnaco Group, concluded that one of Warnaco's many holdings, the Hathaway Shirt Company, was unprofitable, she decided to stop making Hathaway shirts and to sell or close down the factory. It did not matter to Wachner that Hathaway, which started making shirts in 1837, was one of the oldest companies in Maine, that almost all of the five hundred employees at the factory were working-class women, or even that the workers had given up a pay raise to hire consultants to teach them to work more effectively and, as a result, had doubled their productivity. The bottomline issue was that the company was considered unprofitable, and the average wage of the Hathaway workers, $7.50 an hour, was thought to be too high. (In 1995 Wachner was paid $10 million in salary and stock, and Warnaco had a net income of $46.5 million.) "We did need to do the right thing for the company and the stockholders," explained Wachner.[5]

Nor did ethnic background matter to Thomas Fuentes, a senior vice president at a consulting firm in Orange County, California, a director of Fleetwood Enterprises, and chairman of the Orange County Republican Party. Fuentes targeted fellow Latinos who happened to be Democrats when he sent uniformed security guards to twenty polling places in 1988 "carrying signs in Spanish and English warning people not to vote if they were not U.S. citizens." The security firm ended up paying $60,000 in damages when it lost a lawsuit stemming from this intimidation.[6]

We also recall that the Fanjuls, the Cuban-American sugar barons, have had no problem ignoring labor laws in dealing with their migrant labor force, and that the Sakioka family illegally gave short-handled hoes to its migrant farm workers. These people were acting as employers, not as members of ethnic groups. That is, members of the power elite of both genders and all ethnicities have practiced class politics, making it possible for the power structure to weather the challenge created by the social movements that began in the 1960s.

Those who challenged Christian white male homogeneity in the power structure during the 1960s not only sought to create civil rights and new job opportunities for men and women who had previously been mistreated, important though these goals were. They also hoped that new perspectives in the boardrooms and the halls of government would bring greater openness throughout the society. The idea was both to diversify the power elite and to shift some of its power to previously excluded groups and social classes. The social movements of the 1960s were strikingly successful in increasing the individual rights and freedoms available to all Americans, especially African Americans. As we have shown, they also created pressures that led to openings at the top for individuals from groups that had previously been excluded.

But as the concerns of social movements, political leaders, and the courts came to focus more and more on individual rights, the emphasis on social class and "distributive justice" was lost. The age-old American commitment to individualism, reinforced at every turn by members of the power elite, won out over the commitment to greater equality of income and wealth that had been one strand of New Deal liberalism and a major emphasis of left-wing activists in the 1960s.

We therefore have to conclude on the basis of our findings that the diversification of the power elite did not generate any changes in an underlying class system in which the top 1 percent have 45.6 percent of all financial wealth, the next 19 percent have 46.7 percent, and the bottom 80 percent have 7.8 percent.[7] The values of liberal individualism embedded in the Declaration of Independence, the Bill of Rights, and the civic culture were renewed by vigorous and courageous activists, but despite their efforts, the class structure remains a major obstacle to individual fulfillment for the overwhelming majority of Americans. This fact is more than an irony. It is a dilemma. It combines with the dilemma of race to create a nation that celebrates equal opportunity but is, in reality, a bastion of class privilege and conservatism.

NOTES

1. See Richard M. Huber, *The American Idea of Success* (New York: McGraw Hill, 1971), 44–46; Gary Scharnhorst, *Horatio Alger, Jr.* (Boston: Twayne, 1980), 24, 29, 141.

2. C. Wright Mills, *The Power Elite* (New York: Oxford University Press, 1956), 279. For Mills's specific findings, see 104–105, 128–129, 180–181, 393–394, and 400–401.

3. Edward N. Wolff, *Top Heavy* (New York: New Press, 1996), 67.

4. In addition, evidence from experimental work in social psychology suggests that tokenism has the effect of undercutting the impetus for collective action by the excluded group. See, for example, Stephen C. Wright, Donald M. Taylor, and Fathali M. Moghaddam, "Responding to Membership in a Disadvantaged Group: From Acceptance to Collective Protest," *Journal of Personality and Social Psychology* 58, no. 6 (1990), 994–1003. See also Bruce R. Hare, "On the Desegregation of the Visible Elite; or, Beware of the Emperor's New Helpers: He or She May Look Like You or Me," *Sociological Forum* 10, no. 4 (1995): 673–678.

5. Sara Rimer, "Fall of a Shirtmaking Legend Shakes Its Maine Hometown," *New York Times,* May 15, 1996. See also Floyd Norris, "Market Place," *New York Times,* June 7, 1996; Stephanie Strom, "Double Trouble at Linda Wachner's Twin Companies," *New York Times,* August 4, 1996. Strom's article reveals that Hathaway Shirts "got a reprieve" when an investor group stepped in to save it.

6. Claudia Luther and Steven Churm, "GOP Official Says He OK'd Observers at Polls," *Los Angeles Times,* November 12, 1988; Jeffrey Perlman, "Firm Will Pay $60,000 in Suit over Guards at Polls," *Los Angeles Times,* May 31, 1989.

7. Edward N. Wolff, *Top Heavy* (New York: New Press, 1996), 67.

DISCUSSION QUESTIONS

1. To what extent do Zweigenhaft and Domhoff see diverse groups as becoming a part of the power elite? What factors do they identify as important in gaining entrance to the power elite and how does this challenge the myth of upward mobility typically symbolized by the Horatio Alger story?

2. At the heart of Zweigenhaft and Domhoff's analysis is the question, "Does having more women and people of color in positions of power change institutions?" Using empirical evidence, how would you answer this question?

54

Forever Seen as New

Latino Participation in American Elections

LOUIS DESIPIO AND RODOLFO O. DE LA GARZA

The authors here identify a number of changes that are likely to result in an increase in the political influence of Latinos in years ahead. Like other groups, however, the extent of voting participation among Latinos is also shaped by factors such as age, social class, and region of residence. The authors show how changes in the Latino population are linked to patterns of political participation and influence.

. . . The rapid growth of the Latino population in the past thirty years has raised popular expectations for its political impact. Now numbering more than 35 million and soon to overtake African Americans as the nation's largest minority, Latinos are often the subject of naive predictions for their imminent domination of politics and society (de la Garza 1996a). The electorate, of course, is much smaller than the total population: just one in six Latinos votes.

From: DeSipio, Louis, and Rodolfo O. de la Garza. "Forever Seen as New: Latino Participation in American Elections." In *Latinos: Remaking America,* edited by Marcelo M. Suarez-Orozco and Mariela Paez, 398–409. Berkeley: University of California Press, 2002. Reprinted with permission.

Much of the gap between rates of electoral participation for Latino and Anglo citizens can be explained by simple demographics. Among all populations, the young, the less well educated, and the low-income are less likely to vote. All of these groups are disproportionately represented among Latinos. Remedies for the impact on turnout of youth, limited education, and low-income are less clear. Community-wide mobilization, such as that which the African American community experienced as a result of the civil rights movement, can overcome these impediments to participation. Without such mobilization, however, Latinos are likely to continue to experience lower-than-average rates of participation, despite the steady growth in the number of Latino voters from election to election.

One of these traits—youth—has a natural remedy, aging. At present, Latinos are approximately nine years younger than the average non-Hispanic white. Immigration and high birth rates will keep Latino voting rates low, but increasing numbers of Latinos will enter their forties and fifties, the ages that see peak voting in all populations. Today, approximately 6.3 million Latinos are forty-five or older. This number will increase to 11.0 million by 2010 and to 20.7 million by 2030 (U.S. Bureau of the Census 1996). Income and education levels are also rising, particularly among the U.S.-born (de la Garza 1996b). However, these slow gains could disappear if the U.S. economy were to deteriorate. For the time being, the steady aging of the Latino population and the growth of an educated middle class spur an increase of 10 to 15 percent every four years in the number of Latinos who turn out to vote.

Electoral growth, however, does not guarantee increased Latino influence. In 1996, for example, the Latino electorate had grown by 16 percent over 1992, and the impact of this growth in the Latino share of the electorate was magnified by a decline in the Anglo electorate. Yet Latino voters were influential in more states in 1992 when the election was closer and more states where many Latinos lived were crucial to the outcome. Thus, increasing the size of the electorate is an important goal, but it is not directly related to influence.

Over time, an increase in the size of the Latino electorate could bring people with different positions or interests into the electorate. This happened, for instance, as Cuban Americans began to vote, reducing the Democratic share of the Latino electorate. Looking to the future, electoral growth spurred by increasing incomes among Latinos would probably increase the Republican share of the electorate. On the other hand, as Latinos age, health care and Social Security could assume a central position in the issue agenda of Latino voters in a way that they do not today. These have long been Democratic issues, and their increased salience could strengthen Latino Democratic ties. . . .

NATURALIZED CITIZENS

Whereas demographic limits on electoral participation affect all groups, a second characteristic disproportionately affects Latinos. Noncitizens total 39 percent of Latino adults and are the largest potential new electorate among Latinos (DeSipio 1996). Any effort to mobilize them, however, must begin by encouraging naturalization, because noncitizens are barred from voting in virtually all elections.

. . . Approximately 2.4 million Latinos became naturalized citizens between 1995 and 2001 (they were joined by approximately 2.6 million non-Latino new citizens). These 2.4 million new Latino citizens are potentially influential for several reasons. First, if they were all to join the electorate (an unlikely scenario), they would add almost 50 percent to the existing Latino vote overnight. Second, the newly naturalized are concentrated in a few states, so their impact would be magnified. Finally, many became naturalized at least in part in response to government efforts to limit the rights of immigrants, so there was a political dimension to the decisions made by many such citizens. . . .

NON-NATURALIZED IMMIGRANTS

What of the remaining (non-naturalized) Latino immigrants? Do they offer the foundation for an expanding new electorate? The answer is yes, but their impact will be felt slowly. Approximately 1.1 million Latinos (including 7.7 million adults) were not U.S. citizens in 1999. Of these, more than 4 million were undocumented, so there were approximately four million permanent residents. Of these 1.6 million immigrated in the past five years and thus are presently ineligible for naturalization. The remaining 2.4 million naturalization-eligible Latino immigrants are a pool for further growth in the Latino electorate.

In all likelihood, however, this remaining pool of Latino immigrants will be slow to become naturalized. A unique set of pressures and incentives encouraged most of the eligible who were immigrants interested in citizenship to seek naturalization in the late 1990s. Those who did not do so will require added encouragement to pursue citizenship now. . . .

CONCENTRATION

Latinos are more geographically concentrated than Anglos and are becoming even more concentrated. Some argue that this concentration boosts Latino empowerment. We find that its minuses may well outweigh its plusses. Let's consider both.

Instead of having a national election for the presidency, the electoral college reflects fifty state elections, in which most states award all of their delegates to the candidate who receives the most votes. Any concentrated electorate that can secure a state for a candidate exerts a form of influence that would be lost if each vote were tallied nationally. Latinos benefit from concentrating most of their numbers in just nine states and from having cohesive voting patterns in each. Latino advocates are quick to observe that Latinos are concentrated in states that elect three-quarters of the electors needed to win the presidency.

Concentration also entails a cost in national elections. No campaign runs equal efforts in all fifty states. Rather, campaigns calculate how they can best spend their money. Little is spent in states that are probable losses *or* in states where victory is very likely. Money and time are focused on the competitive states so that a winning margin of 270 electoral votes can be earned. . . .

In the short run, then, concentration is an advantage only in states or other electoral districts that are competitive. In the longer run, areas of concentration can become the sites of sustained multiyear mobilization efforts targeted not just at a specific campaign or election cycle. Efforts such as these, which make concentration an advantage, require leadership and resources that have been absent from most Latino outreach in recent years.

ELITE RESOURCES AND INSTITUTIONS

Although it has been little studied, one genuinely new phenomenon in Latino electoral participation is the rise of a new cadre of Latino elites and new institutions to shape candidate outreach to Latinos. The new elites are made up primarily of young, highly educated Latinos who have begun to populate campaigns. In the process, they have drawn attention to Latino issues and have educated non-Latino candidates about Latino communities. Institutional development has been slower but will probably expand in the next decade. These institutions include Latino political action committees (PACs) and Latino organizations within the national and state political parties. . . .

The increase in the number of Latino votes and the opportunities for these votes to prove influential in electoral outcomes pave the way for expansion in the role of elite resources and institutions. This growth will be spurred by the talent pool of skilled Latino politicos who seek both influence in campaigns and a voice on Latino issues. Thus, although the phenomenon of elite institutions and resources is not new, its potential for influence is, and it merits continuing scholarly appraisal. . . .

CONCLUSION

In the continual search for what is new in the Latino electorate, there is a tendency to neglect the incremental but steady changes that shape Latino politics. Latino votes have become increasingly sought by candidates and parties. These votes do occasionally determine outcomes, although when they do, they are usually determined by factors exogenous to the Latino community. Both parties have designed Latino-specific outreach strategies and dedicated resources to winning Latino votes, and Latinos have become more centrally positioned in campaigns and elite networks. As a result, candidates are somewhat less likely to speak from ignorance when they address Latino issues. Although we have not discussed it here, the number of Latino elected and appointed officials has grown. When the Voting Rights Act was extended to Latinos in 1975, they were not a nationally influential electorate; today they are. What we find, however, is not a single, dramatic change but incremental change and growth in the electorate. We expect this pattern to continue, but we do not anticipate a sudden burst of new Latino electoral empowerment.

This increase in the importance of the Latino electorate is not the result of mobilization among all Latinos or all Latino citizens. Instead, Latinos have followed the pattern of Anglos: voting is more common among the educationally and economically advantaged. For a truly new Latino electorate to emerge, this pattern would have to change, and mobilization would have to extend to all segments of the Latino electorate. Survey data indicate that even if this were to occur, such a broad-based mobilization would not appreciably change the issue focus of Latino electorates, just the likelihood of their influencing electoral outcomes.

Will Latinos play a central role in the outcomes of upcoming national elections? The answer to this question is found primarily outside of the Latino community. The incremental growth in Latino electorates makes it, all other factors remaining constant, slightly more likely each presidential election cycle. But the final answer will be determined each election year. Latinos will be important if the race comes down to the states where their numbers are concentrated, if one or both of the candidates see Latino votes as central to their ability to win those states, and if one or both of the candidates dedicate resources to winning Latinos' votes (or to preventing the other candidate from winning their votes). In this scenario, their votes could make the difference. The candidates and parties now possess the expertise to structure a campaign to win their votes. In this scenario, their strong partisanship and state-level cohesion make them an inviting target relative to other electorates.

REFERENCES

de la Garza, Rodolfo O. 1996a. "El Cuento de los Números and Other Latino Political Myths." In *Su Voto es Su Voz: Latino Politics in California,* ed. Aníbal Yáñez-Chávez, 11–32. San Diego: Center for U.S. Mexican Studies, University of California, San Diego.

———. 1996b. "The Effects of Primordial Claims, Immigration, and the Voting Rights Act on Mexican American Sociopolitical Incorporation." In *The Politics of Minority Coalitions: Race, Ethnicity, and Shared Uncertainties,* ed. Wilbur C. Rich, 163–76. Westport, CT: Praeger Publishers.

DeSipio, Louis. 1996. *Counting on the Latino Vote: Latinos as a New Electorate.* Charlottesville, VA: University Press of Virginia.

U.S. Bureau of the Census. 1996. *Population Projections of the United States by Age, Sex, Race, and Hispanic Origin: 1995 to 2050.* Current Population Reports, Series P-25, No. 1130.

DISCUSSION QUESTIONS

1. What factors that are particular to the Latino population are likely to influence voting patterns in the future? How will the Latino vote be likely to shape election issues in the future?

2. How might the political participation of Latinos compare to other groups, such as African Americans, Whites, and Asian Americans? How does this show the influence of social factors on democratic participation?

http://infotrac.thomsonlearning.com

InfoTrac College Edition

BONUS READING

Hines, Revathi. "The Silent Voices: 2000 Presidential Election and the Minority Vote in Florida." *The Western Journal of Black Studies* 26 (Summer 2002): 71–74.

The Voting Rights Act of 1965 protects the rights of all citizens to vote; yet, there is ample evidence that discriminatory practices continue that result in the disenfranchisement of many, particularly members of racial minority groups. What does the author show about the presidential election of 2000 that disenfranchised African Americans and other groups and what reforms would be needed to ensure the full rights of citizenship for all?

SEARCH TERMS

Black corporate leaders
corporate power
interlocking directorates
Latino vote
military–industrial complex
power elite
women in corporations

55

Latinos' Access to Employment-Based Health Insurance

E. RICHARD BROWN AND HONGJIAN YU

Access to good health care is a precursor to good health. Yet, there are many disparities in who has access to care. As these authors show, Latinos are even less likely than non-Latinos to have health insurance. The authors examine why and suggest new social policies to alleviate this problem.

More than one in every three Latinos (37 percent) in the United States is without any public or private health insurance; this is more than two and one-half times the uninsured rate of 14 percent among non-Latino whites (Brown et al. 2000). Lack of health insurance creates significant barriers to obtaining needed health services, exacerbating disparities in access and health status between Latinos and non-Latino whites.

Latinos' Health Status and Access to Health Care

Many Latino immigrants come to the United States with a more favorable health status than would be expected given their economic circumstances . . . but residency in their new country is associated with increasingly adverse risk factors and poorer health status (Vega and Amaro 1994; Fuentes-Afflick and Lurie 1997; Abraaido-Lanza et al. 1999). . . . This decline is particularly notable among those who have lived longer in the United States and among the second- and the third-plus-generation, compared with first-generation, immigrants (Hernandez and Charney 1998). . . .

Perhaps as a result of lower incomes and less wealth, more hazardous social and physical environments in which they live and work, and acculturation into American culture linked to social class, Latinos overall are more likely than non-Latino whites of the same age group to report being in fair or poor health (Hajat, Lucas, and Kington 2000)—a robust indicator of the need for health

From: Brown, E. Richard, and Hongjian Yu. "Latinos' Access to Employment-Based Health Insurance." In *Latinos: Remaking America,* edited by Marcelo M. Suarez-Orozco and Mariela Paez, 236–48. Berkeley: University of California Press, 2002. Reprinted with permission.

services. This finding applies to the major Latino ethnic subgroups as well, although Puerto Ricans stand out as having particularly poor health status.

Although health status indicators demonstrate health care needs among Latinos that are at least equal to those of non-Latino whites, their use of health services demonstrates significant disparities. Among young children (aged 0 to 5 years), 8 percent of Latinos did not have a physician visit during the past year, compared to 5 percent of non-Latino whites; and among children aged 6 to 17, 16 percent of Latinos have not seen a physician in the past two years, compared to 7 percent of non-Latino whites. Among women who report being in fair or poor health, 13 percent of Latinas did not have a doctor visit in the past year—twice the rate of 6 percent for their non-Latino white counterparts (Brown et al. 2000).

Health insurance provides both children and adults an important degree of financial access to health services for acute and chronic conditions, as well as for preventive care (Freeman et al. 1990; Newacheck, Hughes, and Stoddard 1996; Halfon et al. 1997). Latinos' high uninsured rates thus adversely affect their access to health care. Uninsured Latino children aged 0 to 5 are two and one-half times as likely as those with any public or private health insurance not to have had even one doctor visit in the past year: 16 percent of uninsured children versus 6 percent of children with any coverage. Uninsured Latino children aged 6 to 17 are two to three times as likely as those with any private insurance or Medicaid not to have visited a doctor during the previous two years: 29 percent of the uninsured versus 12 percent of those with private insurance and 9 percent of those with Medicaid (Brown et al. 2000). Without physician visits, at these minimum frequencies, children cannot receive recommended monitoring of growth and development and essential preventive services. And this minimum standard does not take into account any additional health care needs for acute and chronic conditions.

Among uninsured adult Latinas in fair to poor health, one in four (24 percent) did not see a physician in a twelve-month period despite their health problems. This proportion is at least three times the rates of 7 percent and 8 percent, for those with private insurance and those with Medicaid, respectively, demonstrating the severe impact of not having insurance or access to health services. Among uninsured Latino men in fair or poor health, 40 percent did not have a doctor visit in the past year, twice the rate for those with coverage (19 percent).

The access barriers created by not having health insurance coverage undoubtedly contribute to the overall disparities in health status between Latinos and non-Latino whites. Health insurance coverage and access to medical care can ameliorate and improve Latinos' health problems and reduce disparities even if they cannot compensate fully for the powerful effects on health status of adverse economic and social factors (Brown et al. 2000). For persons in all ethnic groups, lack of health insurance coverage results in weak connections to the health care system and poor access to health services. Because insured persons in fair or poor health visit physicians more often than their uninsured counterparts, they are more likely to receive the care they need to manage their chronic conditions, such as diabetes or high blood pressure. Insured children and adults, whether in good or poor health, are more likely to receive preventive health services and care for acute conditions.

Latinos' Low Rates
of Employment-Based Health Insurance

. . . In the United States, employment is the most important source of health insurance coverage for the nonelderly population; 81 percent of those with any coverage receive it through their own or a family member's employment. Employers' payments for health benefits are the principal source of funding that helps make private health insurance affordable to many working families and individuals; employer contributions account for about three-fourths of all nongovernmental expenditures for private health insurance (U.S. Congressional Budget Office 1997). In the absence of employer contributions and group purchasing, health insurance coverage is less affordable for workers and their families, which diminishes their access to health services if other sources of coverage or services are not available. . . .

Compared to non-Latino white employees, Latinos—both overall and all subgroups—are far less likely to have employment-based health insurance (EBHI) and thus far more likely to be uninsured. Eight in ten (80.2 percent) non-Latino white employees receive EBHI through their own or a family member's employment, compared to just over half of Latinos (53.7 percent). . . . Among Latino subgroups, employees of Mexican and of Central and South American origin have the lowest rates of EBHI (52.3 percent and 51.0 percent, respectively), whereas Puerto Rican and Cuban-origin employees fare somewhat better (63.9 percent and 64.0 percent, respectively) but are still well below the rate for non-Latino whites.

Latino employees' lower EBHI rates result in an uninsured rate three times as high as that of non-Latino whites: 39.3 percent versus 13.3 percent. Because of their low EBHI rates, employees of Mexican origin and of Central and South American origin have the highest uninsured rates: 41.5 percent and 43.2 percent, respectively.

Employment is the most important source of private health insurance because the employer contribution and its tax exemption make health insurance more affordable for individuals and their families. However, an employee can access job-based insurance only through an employer or a union. Thus employees who work for an employer that does not offer health benefits at all, as well as those who are not eligible for the benefits that are offered, have no access to job-based insurance.

Not surprisingly, then, the ethnic-group disparity found in EBHI coverage is also found in "offer" rates. Three in ten Latino employees (30.5 percent) work for an employer that does not offer health benefits to any worker; this is twice the proportion of non-Latino white employees (13.6 percent). . . . To put it another way, only 69.5 percent of Latino employees work for an employer that offers health benefits, compared to 86.4 percent of non-Latino white employees. Again, employees of Central and South American origin (61.0 percent) and of Mexican origin (69.7 percent) are the most severely disadvantaged among Latinos, whereas Puerto Rican employees (83.4 percent) have an offer rate that approximates that of non-Latino whites. Thus Latino employees overall have less access to health insurance coverage through their employment, the primary way in which Americans get such coverage.

Overall, only about half (52.4 percent) of Latino employees, compared to two-thirds (66.3 percent) of non-Latino white employees, work for an employer that offers EBHI, are eligible for benefits, and accept them . . . . Employees of Central and South American origin and of Mexican origin are less likely than employees of Puerto Rican or Cuban origin to be in this advantageous situation. . . .

Latinos face many barriers to obtaining EBHI. Compared with non-Latino whites, Latino employees are more likely to be younger, to have lower educational attainment, to work full-time but for less than the full year, to work in a small firm, to work low-coverage industries and occupations, and to earn less and have lower family incomes . . . —all characteristics that are associated with lower levels of EBHI. These disadvantages are more characteristic of employees of Central and South American origin and of Mexican origin, who together make up about 86 percent of all Latino employees in the United States, than they are of Puerto Rican or Cuban-origin employees. . . .

CONCLUSION

Latinos' low rates of EBHI and their resulting high uninsured rates are due in large part to the low proportion of Latino employees whose employers offer job-based insurance. This disparity between Latinos and non-Latino whites prevails across most demographic, labor force, and income groups. Latinos, in general, experience lower "offer" rates than do non-Latino whites, regardless of whether they have less than a high school education or are college educated, whether they work full-time or part-time, whether they work in low-coverage or high-coverage industries and occupations, and whether they are low-wage or higher-wage employees. The problem is especially acute for employees of Mexican and Central and South American origin. Mexican-origin employees who are noncitizens are especially likely not to be offered EBHI, and more than half of those who are undocumented suffer this disadvantage.

This low proportion of Latino employees who work for an employer who offers health benefits suggests that Latinos experience a systematic disadvantage in the labor market—an apparent form of labor market inequity or abuse. A study of health insurance coverage in the nation's 85 largest urban areas provides further evidence of this inequity (Brown, Wyn, and Teleki 2000).

The inequities in health insurance coverage of Latinos have both ethical and public-policy importance. Inequities in access to health services offered by any group raise social justice concerns. In addition, diminished access to health insurance for any large group can have significant consequences for the nation, potentially increasing the amount of uncompensated care rendered by health care providers, decreasing the group's contributions to the economy, and boosting social tensions. Latinos are a large and growing population that represented 11.8 percent of the U.S. population in 2000 and is projected to reach 17.0 percent by 2020 (U.S. Bureau of the Census 2000); they represent an even larger share of the working age population. Their health and access to health services is [*sic*] an important concern of the nation.

Public policy could reduce the disparities in health insurance coverage experienced by Latinos. For the majority of uninsured persons of all ethnic groups, low incomes make insurance coverage unaffordable without substantial financial assistance. A national health care system that covered the entire population, or even a mandate that all employers cover those who work for them and those workers' dependents, would address the health insurance needs of this population. However, given the voluntary employment-based health insurance system that prevails in the United States, and given the limited prospects for a mandate that employers cover their employees, subsidies from government are the only feasible option.

In the absence of universal coverage, however, Medicaid or an alternative public program could provide more generous opportunities for working families and individuals to obtain subsidies for health insurance coverage. . . .

It is not hyperbole to suggest that the future of the nation and its economy depend on the well-being of Latinos. Effective public policies are needed and available to expand health insurance coverage and improve access to care for all population groups. . . .

REFERENCES

Abraaido-Lanza, A. F., B. P. Dohrenwend, D. S. Ng-Mak, and J. B. Turner. "The Latino Mortality Paradox: A Test of the 'Salmon Bias' and Healthy Migrant Hypothesis." *American Journal of Public Health* 1999;89(10) 543–48.

Brown, E. R., V. Ojeda, R. Wyn. and R. Levan, *Racial and Ethnic Disparities in Access to Health Insurance and Health Care,* Los Angeles and Menlo Park, CA: UCLA Center for Health Policy Research and Henry J. Kaiser Family Foundation, April 2000.

Brown, E. R., R. Wyn, and S. Teleki. *Disparities in Health Insurance and Access to Care for Residents across American Cities,* New York and Los Angeles: The Commonwealth Fund and the UCLA Center for Health Policy Research, August 2000.

Freeman, H. E., L. H., Aiken, R. J. Blendon, and C. R. Corey. "Uninsured Working-Age Adults: Characteristics and Consequences," *Health Services Research* 1990;24: 811–823.

Fuentes-Afflick, E., and P. Lurie. "Low Birth Weight and Latino Ethnicity: Examining the Epidemiologic Paradox," *Archives of Pediatrics and Adolescent Medicine,* 1997; 151(7):665–74.

Hajat, A., J. B. Lucas, and R. Kington, *Health Outcomes among Hispanic Subgroups: Data from the National Health Interview Survey, 199–95,* Advance Data from Vital and Health Statistics; no. 310. Hyattsville, MD: National Center for Health Statistics, 2000.

Halfon, N., D. L. Wood, R. B. Valdez, M. Pereyra, and N. Duan. "Medicaid Enrollment and Health Services Access by Latino Children in Inner-City Los Angeles," *Journal of American Medical Association* 1997; 277:636–41.

Newacheck, P. W., D. C. Hughes, and J. J. Stoddard. "Children's Access to Primary Care: Differences by Race, Income, and Insurance Status," *Pediatrics.* January 1996; 7(1)26–32.

U.S. Bureau of the Census, Population Division, Population Projections Branch. "Projections of the Resident Population by Race, Hispanic Origin, and Nation, Middle Series, 1999 to 2100," Washington, DC: U.S. Census Bureau, January 2000.

U.S. Congressional Budget Office, *Trends in Health Care Spending by the Private Sector* Washington, DC: U.S. Congressional Budget Office, April 1997.

Vega, W. A., and H. Amaro. "Latino Outlook: Good Health. Uncertain Program." *Annual Review of Public Health,* 1994; 15:39–67.

DISCUSSION QUESTIONS

1. What barriers do the authors identify as resulting in fewer Latinos being covered by health insurance?
2. Why are so many people in the United States not covered by health insurance? What social and cultural barriers prevent universal health care?

56

Beauty Myths and Realities and Their Impact on Women's Health

JANE SPRAGUE ZONES

The "beauty myth" defines a narrow range of social ideals for women's appearance. As Zones shows here, trying to attain such an ideal has consequences for women's physical and mental health.

Many women concur that personal beauty, or "looking good," is fostered from a very early age. It is probably true that the ways in which people assess physical beauty are not naturally determined but socially and culturally learned and therefore "in the eye of the beholder." However, we tend to discount the depth of our *common* perception of beauty, mistakenly assuming that

From: Zones, Jane Sprague. "Beauty Myths and Realities and Their Impact on Women's Health." In *Women's Health, Complexities and Differences,* edited by Sheryl Burt Ruzek, Virginia L. Oleson, and Adele E. Clark, 249–75. Columbus, OH: The Ohio State University, 1997. Reprinted with permission.

individuals largely set their own standards. At any period in history, within a given geographic and cultural territory, there are relatively uniform and widely understood models of how women "should" look. . . .

The preoccupation with appearance serves to control and contain women's ambitions and motivations to gain power in larger political contexts. To the degree that many females feel they must dedicate time, attention, and resources to maintaining and improving their looks, they neglect activities to improve social conditions for themselves or others. Conversely, as women become increasingly visible as powerful individuals in shaping events, their looks become targeted for irrelevant scrutiny and criticism in ways with which men in similar positions are not forced to contend. . . .

BEAUTY MYTHS AND THE EROSION OF SELF-WORTH

Perhaps the biggest toll the "beauty myth" takes is in terms of women's identity and self-esteem. Like members of other oppressed groups of which we may also be part, women internalize cultural stereotypes and expectations, perpetuating them by enforced acceptance and agreement. For women, this is intensified by the interaction of irrational social responses to physical appearance not only with gender but with other statuses as well—race, class, age, disability, and the like. Continuous questioning of the adequacy of one's looks drains attention from more worthwhile and confidence-building pursuits. . . .

QUANTIFYING BEAUTY: CONVENTIONALITY
AND COMPUTER ENHANCEMENT

The predominant, nearly universal standard for beauty in American society is to be slender, young, upper-class, and white without noticeable physical imperfections or disabilities. To the extent that a woman's racial or ethnic heritage, class background, age, or other social and physical characteristics do not conform to this ideal, assaults on opportunities and esteem increase. Physical appearance is at the core of racism and most other social oppressions, because it is generally what is used to classify individuals. . . .

BEAUTY AND THE CHALLENGE OF SOCIAL DIVERSITY

Although significant beauty ideals appear to transcend cultural subgroup boundaries, appearance standards do vary by reference group. Clothing preferred by adolescents, for example, which experiences quick fashion turnover, is considered inappropriate for older people. Body piercing, a current style for young white people in urban areas of the United States, is repellent to most older adults and some ethnic minorities in the same age group. Religious and political ideologies are often identified through appearance. Islamic fundamentalist women wear clothing that covers body and face, an expression of religious sequestering; Amish women wear conservative clothing and distinctive caps; orthodox Jewish women wear wigs or cover their hair; African American women for many years wore natural hairdos to show racial pride; and Native American women may wear tribal jewelry and distinctive clothes that indicate their respect for heritage. In recent years, the disability rights movement has encouraged personal visibility to accompany the tearing down of barriers to access, resulting in a greater variety of appliances (including elegant streamlined wheelchairs) and functional clothing.

Although there are varying and conflicting standards of good looks and appropriate appearance that are held simultaneously by social subgroups, the dominant ideals prevail and are legitimated most thoroughly in popular culture. . . .

One major way that dominant social forces have dealt with those who diverge is to remove these expressions from view—through ghettoization, anti-immigration policies, special education programs, retirement policies, and so on. The ultimate social insult is to render the oppressed invisible. Social barriers to visibility are expressed as well in pressures to avoid drawing attention to oneself. Those features that render us "different" are frequently the objects of harassment or unwanted attention. We learn to appear invisible. In the following sections, the gender effects of appearance in combination with other social statuses are described through personal accounts and social research. . . .

THE COMMERCIAL IMPERATIVE
IN THE QUEST FOR BEAUTY

. . . As new standards of beauty expectations are created, physical appearance becomes increasingly significant, and as the expression of alternative looks are legitimized, new products are developed and existing enterprises capitalize on the trends. Liposuction, developed relatively recently, has become the most popular of the cosmetic surgery techniques. Synthetic fats have been developed, and there is now a cream claimed to reduce thigh measurements.

Weight Loss

Regardless of the actual size of their bodies, more than half of American females between ages ten and thirty are dieting, and one out of every six college women is struggling with anorexia and bulimia (Iazzetto 1992). The quest to lose weight is not limited to white, middle-class women. Iazzetto cites studies that find this pervasive concern in black women, Native American girls (75 percent trying to lose weight), and high school students (63 percent dieting). However, there may be differences among adolescent women in different groups as to how rigid their concepts of beauty are and how flexible they are regarding body image and dieting (Parker et al. 1995). Studies of primary school girls show more than half of all young girls and close to 80 percent of ten- and eleven-year-olds on diets because they consider themselves "too fat" (Greenwood 1990; Seid 1989). Analyses of the origins, symbolic meanings, and impact of our culture's obsession with thinness (Chernin 1981; Freedman 1989; Iazzetto 1988; Seid 1989) occupy much of the body-image literature.

Concern about weight and routine dieting are so pervasive in the United States that the weight-loss industry grosses more than $33 billion each year. Over 80 percent of those in diet programs are women. These programs keep growing even in the face of 90 to 95 percent failure rates in providing and maintaining significant weight loss. Congressional hearings in the early 1990s presented evidence of fraud and high failure rates in the weight-loss industry, as well as indications of severe health consequences for rapid weight loss (Iazzetto 1992). The Food and Drug Administration (FDA) has reviewed documents submitted by major weight-loss programs and found evidence of safety and efficacy to be insufficient and unscientific. An expert panel urged consumers to consider program effectiveness in choosing a weight-loss method but acknowledged lack of scientific data for making informed decisions (Brody 1992).

Fitness

Whereas in the nineteenth century some physicians recommended sedentary lifestyle to preserve feminine beauty, in the past two decades of the twentieth century, interest in physical fitness has grown enormously. Nowhere is this change more apparent than in the gross receipts of some of the major fitness industries. In 1987, health clubs grossed $5 billion, exercise equipment $738 million (up from $5 million ten years earlier), diet foods $74 billion, and vitamin products $2.7 billion (Brand 1988). Glassner (1989) identifies several reasons for this surge of interest in fitness, including the aging of the "baby boom" cohort with its attendant desire to allay the effects of aging through exercise and diet, and the institution of "wellness" programs by corporations to reduce insurance, absentee, and inefficiency costs. A patina of health, well-toned but skinny robustness, has been folded into the dominant beauty ideal.

Clothing and Fashion

For most of us, first attempts to accomplish normative attractiveness included choosing clothing that enhanced our self-image. The oppressive effects of corsets, clothing that interfered with movement, tight shoes with high heels, and the like have been well documented (Banner 1983). Clothing represents the greatest monetary investment that women make in their appearance. Sales for *exercise* clothing alone in 1987 (including leotards, bodysuits, warm-up suits, sweats, and shoes) totaled $2.5 billion (Schefer 1988). To bolster sales, fashion leaders introduce new and different looks at regular intervals, impelling women to invest in what is currently in vogue. . . .

Cosmetics

The average person in North America uses more than twenty-five pounds of cosmetics, soaps, and toiletries each year (Decker 1983). The cosmetics industry produces over twenty thousand products containing thousands of chemicals, and it grosses over $20 billion annually (Becker 1991; Wolf 1991). Stock in cosmetics manufacturers has been rising 15 percent a year, in large part because of depressed petroleum prices. The oil derivative ethanol is the base for most products (Wolf 1991:82, 307). Profit margins for products are over 50 percent (McKnight 1989). Widespread false claims for cosmetics were virtually unchallenged for fifty years after the FDA became responsible for cosmetic industry oversight in 1938, and even now, the industry remains largely unregulated (Kaplan 1994). Various manufacturers assert that their goods can "retard aging," "repair the skin," or "restructure the cell." "Graphic evidence" of "visible improvement" when applying a "barrier" against "eroding effects" provides a pastiche of some familiar advertising catchphrases (Wolf 1991:109–10).

The FDA has no authority to require cosmetics firms to register their existence, to release their formulas, to report adverse reactions, or to show evidence of safety and effectiveness before marketing their products (Gilhooley 1978; Kaplan 1994). Authorizing and funding the FDA to regulate the cosmetics industry would allow some means of protecting consumers from the use of dangerous products.

Cosmetic Surgery

In interviews with cosmetic surgeons and users of their services, Dull and West (1991) found that the line between reconstructive plastic surgery (repair of deformities caused congenitally or by injury or disease) and aesthetic surgery has begun to blur. Doctors and their patients are viewing unimpaired features as defective and the desire to "correct" them as intrinsic to women's nature, rather than as a cultural imperative.

Because of an oversupply of plastic surgeons, the profession has made efforts to expand existing markets through advertising and by appeals to women of color. Articles encouraging "enhancement of ethnic beauty" have begun to appear, but they focus on westernizing Asian eyelids and chiseling African American noses. As Bordo (1993:25) points out, this technology serves to promote commonality rather than diversity.

Plastic surgery has been moving strongly in the direction of making appearance a bona fide medical problem. This has been played out dramatically in recent times in the controversy regarding silicone breast implants, which provides plastic surgeons with a substantial amount of income. Used for thirty years in hundreds of thousands of women (80 percent for cosmetic augmentation), the effects of breast implants have only recently begun to be studied to determine their health consequences over long periods (Zones 1992). . . .

HEALTH RISKS IN QUEST OF BEAUTY

Mental Health

For most women, not adhering to narrow, standardized appearance expectations causes insecurity and distraction, but for many, concerns about appearance can have serious emotional impact. Up until adolescence, boys and girls experience about the same rates of depression, but at around age twelve, girls' rates of depression begin to increase more rapidly. A study of over eight hundred high school students found that a prime factor in this disparity is girls' preoccupation with appearance. . . .

Physical Health

Perceived or actual variation from society's ideal takes a physical toll, too. High school and college-age females who were judged to be in the bottom half of their group in terms of attractiveness had significantly higher blood pressure than the young women in the top half. The relationship between appearance and blood pressure was not found for males in the same age group (Hansell, Sparacino, and Ronchi 1982). . . .

There are direct risks related to using commodities to alter appearance. According to the Consumer Products Safety Commission, more than 200,000 people visit emergency rooms each year as a result of cosmetics related health problems (Becker 1991). Clothing has its perils as well. In recent years, meralgia paresthetica, marked by sciatica, pain in the hip and thigh region, with tingling and itchy skin, has made an appearance among young women in the form of "skin-tight jean syndrome" (Gateless and Gilroy 1984). In earlier times, the same problems have arisen with the use of girdles, belts, and shoulder bags. . . .

Approximately 33 to 50 percent of all adult women have used hair coloring agents. Evidence over the past twenty-five years has shown that chemicals used in manufacturing hair dyes cause cancers in animals. Scientists at the National Cancer Institute (NCI) recently reported a significantly greater risk of cancers of the lymph system and of a form of cancer affecting bone marrow, multiple myeloma, in women who use hair coloring (Zahm et al. 1992). . . .

Because no cosmetic products require follow-up research for safety and effectiveness, virtually anything can be placed on the market without regard to potential health effects. Even devices implanted in the body, which were not regulated before 1978, can remain on the market for years without appropriate testing. During the decade of controversy over regulating silicone breast implants, the American Society of Plastic and Reconstructive Surgeons vehemently denied any need for controlled studies of the implant in terms of long-term safety. The society spent hundreds of thousands of dollars of its members' money in a public relations effort to avoid the imposition of requirements for such research to the detriment of investing in the expensive scientific follow-up needed (Zones 1992). . . .

Health consequences of beauty products extend beyond their impact on individuals. According to the San Francisco Bay Area Air Quality Management District, aerosols release 25 tons of pollution everyday. Almost half of that is from hairsprays. Although aerosols no longer use chlorofluorocarbons (CFCs), which are the greatest cause of depletion of the upper atmosphere ozone layer, aerosol hydrocarbons in hair sprays are a primary contributor to smog and ground pollution.

THE BEAUTY OF DIVERSITY

Both personal transformation and policy intervention will be necessary to allow women to present themselves freely. Governmental institutions, including courts and regulatory agencies, need to accord personal and product liability related to appearance products and services the attention they require to ensure public health and safety. The legal system must develop well-defined case law to assist the court in determining inequitable treatment based on appearance discrimination. . . .

The personal solution to individual self-doubt or even self-loathing of our physical being is to continuously make the decision to contradict the innumerable messages we are given that we are anything less than lovely as human beings. . . .

REFERENCES

Banner, Lois W. 1983. American Beauty. Chicago: University of Chicago Press.

Becker, Hilton. 1991. "Cosmetics: saving face at what price?" Annals of Plastic Surgery 26:171–73.

Bordo, Susan. 1993. Unbearable Weight: Feminism, Western Culture and the Body. Berkeley: University of California Press.

Brand, David. 1988. "A nation of health worrywarts?" Time, 25 July, 66.

Brody, Jane E. 1992. "Panel criticizes weight-loss programs." New York Times, 2 April, A10.

Chernin, Kim. 1981. The Obsession: Reflections on the Tyranny of Slenderness. New York: Harper Colophon Books.

Decker, Ruth. 1983. "The not-so-pretty risks of cosmetics." Medical Self-Care (Summer):25–31.

Dull, Diana, and Candace West. 1991. "Accounting for cosmetic surgery: the accomplishment of gender." Social Problems 38:54–70.

Freedman, Rita. 1986. Beauty Bound. Lexington, MA: Lexington Books.

———. 1989. Bodylove. New York: Harper and Row.

Gateless, Doreen, and John Gilroy. 1984. "Tight-jeans meralgia: hot or cold?" Journal of the American Medical Association 252:42–43.

Gilhooley, Margaret. 1978. "Federal regulation of cosmetics: an overview." Food Drug Cosmetic Law Journal 33:231–38.

Glassner, Barry. 1989. "Fitness and the postmodern self." Journal of Health and Social Behavior 30:180–91.

Greenwood, M. R. C. 1990. "The feminine ideal: a new perspective." UC Davis Magazine (July):8–11.

Hansell, Stephen, J. Sparacino, and D. Ronchi. 1982. "Physical attractiveness and blood pressure: sex and age differences." Personality and Social Psychology Bulletin 8:113–21.

Iazzetto, Demetria. 1988. "Women and body image: reflections in the fun house mirror." Pp. 34–53 in Carol J. Leppa and Connie Miller (eds.), Women's Health Perspectives: An Annual Review, Volcano, CA: Volcano Press.

———. 1992. "What's happening with women and body image?" National Women's Health Network News:1, 6, 7.

Kaplan, Sheila. 1994. "The ugly face of the cosmetics lobby." Ms. (Jan.–Feb.):88–89.

McKnight, Gerald. 1989. The Skin Game: The International Beauty Business Brutally Exposed. London: Sidgwick and Jackson.

Parker, Sheila, Mimi Nichter, Mark Nichter, Nancy Vuckovic, Colette Sims and Cheryl Ritenbaugh. 1995. "Body image and weight concerns among African Americans and white adolescent females: differences that make a difference." Human Organization 54(2):103–13.

Schefer, Dorothy. 1988. "Beauty: The real cost of looking good." Vogue (Nov.): 157–68.

Seid, Roberta Pollack. 1989. Never Too Thin: Why Women Are at War with Their Bodies. New York: Prentice-Hall.

Wolf, Naomi. 1991. The Beauty Myth: How Images of Beauty Are Used against Women. New York: William Morrow.

Zahm, Sheila Hoar, Dennis D. Weisenburger, Paula A. Babbitt, et al. 1992. "Use of hair coloring products and the risk of lymphoma, multiple myeloma, and chronic lymphocytic leukemia." American Journal of Public Health 82:990–97.

Zones, Jane Sprague. 1992. "The political and social context of silicone breast implant use in the United States." Journal of Long-Term Effects of Medical Implants 1:225–41.

DISCUSSION QUESTIONS

1. What health risks does Zones identify for women who are overly concerned with the beauty ideals established for them? What alternatives are there?

2. Much of the discussion of beauty and health has focused on women. How do ideals regarding appearance affect men's health?

57

Death and Social Structure

ROBERT BLAUNER

Death—a biological reality—is handled in all societies through rituals and within social institutions. Here Robert Blauner analyzes the particular ways that death is managed in modern societies. He finds that death is increasingly bureaucratized—that is, managed in complex organizations that are predictable and routinized, no matter how shocking the event of death itself.

Death is a biological and existential fact of life that affects every human society. Since mortality tends to disrupt the ongoing life of social groups and relationships, all societies must develop some forms of containing its impact. Mortuary institutions are addressed to the specific problems of the disposal of the dead and the rituals of transition from life to death. In addition, fertility practices, family and kinship systems, and religion take their shape partly in response to the pressure of mortality and serve to limit death's disorienting possibilities. . . . In particular, I hope to throw some light on the social and cultural consequences of modern society's organization of death. . . .

Modern societies control death through bureaucratization, our characteristic form of social structure. Max Weber has described how bureaucratization in the West proceeded by removing social functions from the family and the household and implanting them in specialized institutions autonomous of kinship considerations. Early manufacturing and entrepreneurship took place in or close to the home; modern industry and corporate bureaucracies are based on the separation of the workplace from the household. Similarly, only a few generations ago most people in the United States either died at home, or were brought into the home if they had died elsewhere. It was the responsibility of the family to lay out the corpse—that is, to prepare the body for the funeral. Today, of course, the hospital cares for the terminally ill and manages the crisis of dying; the mortuary industry (whose establishments are usually called "homes" in deference to past tradition) prepares the body for burial and makes many of the funeral arrangements. A study in Philadelphia found that about ninety percent of funerals started out from the funeral parlor, rather than from the home, as was customary in the past. This separation of the handling of illness and death from the family minimizes the average person's exposure to death and its disruption of the social process. When the dying are segregated among specialists for whom contact with death

From: Blauner, Robert. "Death and Social Structure." *Psychiatry: Journal for the Study of Interpersonal Process* 29 (November 1966): 378–94.

has become routine and even somewhat impersonal, neither their presence while alive nor as corpses interferes greatly with the mainstream of life.

Another principle of bureaucracy is the ordering of regularly occurring as well as extraordinary events into predictable and routinized procedures. In addition to treating the ill and isolating them from the rest of society, the modern hospital as an organization is committed to the routinization of the handling of death. Its distinctive competence is to contain through isolation, and reduce through orderly procedures, the disturbance and disruption that are associated with the death crisis. The decline in the authority of religion as well as shifts in the functions of the family underlies this fact. With the growth of the secular and rational outlook, hegemony in the affairs of death has been transferred from the church to science and its representatives, the medical profession and the rationally organized hospital.

Death in the modern hospital has been the subject of two recent sociological studies: Sudnow has focused on the handling of death and the dead in a county hospital catering to charity patients; and Glaser and Strauss have concentrated on the dying situation in a number of hospitals of varying status. The county hospital well illustrates various trends in modern death. Three-quarters of its patients are over 60 years old. Of the 250 deaths Sudnow observed, only a handful involved people younger than 40. This hospital is a setting for the concentration of death. There are 1,000 deaths a year; thus approximately three die daily, of the 330 patients typically in residence. But death is even more concentrated in the four wards of the critically ill; here roughly 75 percent of all mortality occurs, and one in 25 persons will die each day.

Hospitals are organized to hide the facts of dying and death from patients as well as visitors. Sudnow quotes a major text in hospital administration: "The hospital morgue is best located on the ground floor and placed in an area inaccessible to the general public. It is important that the unit have a suitable exit leading onto a private loading platform which is concealed from hospital patients and the public." Personnel in the high-mortality wards use a number of techniques to render death invisible. To protect relatives, bodies are not to be removed during visiting hours. To protect other inmates, the patient is moved to a private room when the end is foreseen. But some deaths are unexpected and may be noticed by roommates before the hospital staff is aware of them. These are considered troublesome because elaborate procedures are required to remove the corpse without offending the living.

The rationalization of death in the hospital takes place through standard procedures of covering the corpse, removing the body, identifying the deceased, informing relatives, and completing the death certificate and autopsy permit. Within the value hierarchy of the hospital, handling the corpse is "dirty work," and when possible attendants will leave a body to be processed by the next work shift. As with so many of the unpleasant jobs in our society, hospital morgue attendants and orderlies are often People of Color. Personnel become routinized to death and are easily able to pass from mention of the daily toll to other topics; new staff members stop counting after the first half-dozen deaths witnessed.

Standard operating procedures have even routinized the most charismatic and personal of relations, that between the priest and the dying patient. It is not that the church neglects charity patients. The chaplain at the county hospital daily goes through a file of the critically ill for the names of all known Catholic patients, then enters their rooms and administers extreme unction. After completing his round on each ward, he stamps the index card of the patient with a rubber stamp which reads: "Last Rites Administered. Date _____ Clergyman _____." Each day he consults the files to see if new patients have been admitted or put on the critical list. As Sudnow notes, this rubber stamp prevents him from performing the rites twice on the same patient. This example highlights the trend toward the depersonalization of modern death, and is certainly the antithesis of the historic Catholic notion of "the good death."

In the hospitals studied by Glaser and Strauss, depersonalization is less advanced. Fewer of the dying are comatose, and as paying patients with higher social status they are in a better position to negotiate certain aspects of their terminal situation. Yet nurses and doctors view death as an inconvenience, and manage interaction so as to minimize emotional reactions and fuss. They attempt to avoid announcing unexpected deaths because relatives break down too emotionally; they prefer to let the family members know that the patient has taken "a turn for the worse," so that they will be able to modulate their response in keeping with the hospital's need for order. And drugs are sometimes administered to a dying patient to minimize the disruptiveness of his passing—even when there is no reason for this in terms of treatment or the reduction of pain.

The dying patient in the hospital is subject to the kinds of alienation experienced by persons in other situations in bureaucratic organizations. Because doctors avoid the terminally ill, and nurses and relatives are rarely able to talk about death, he suffers psychic isolation. He experiences a sense of meaninglessness because he is typically kept unaware of the course of his disease and his impending fate, and is not in a position to understand the medical and other routines carried out in his behalf. He is powerless in that the medical staff and the hospital organization tend to program his death in keeping with their organizational and professional needs; control over one's death seems to be even more difficult to achieve than control over one's life in our society. Thus the modern hospital, devoted to the preservation of life and the reduction of pain, tends to become a "mass reduction" system, undermining the subjecthood of its dying patients.

The rationalization of modern death control cannot be fully achieved, however, because of an inevitable tension between death—as an event, a crisis, an experience laden with great emotionality—and bureaucracy, which must deal with routines rather than events and is committed to the smoothing out of affect and emotion. Although there was almost no interaction between dying patients and the staff in the county hospital studied by Sudnow, many nurses in the other hospitals became personally involved with their patients and experienced grief when they died. Despite these limits to the general trend, our society has gone far in containing the disruptive possibilities of mortality through its bureaucratized death control. . . .

DISCUSSION QUESTIONS

1. What does Blauner mean that death has a social structure? What evidence of this do you see in contemporary life?

2. Blauner argues that death is controlled in modern society through bureaucratic forms. Using hospitals and/or the funeral industry as examples, explain what he means.

http://infrotrac.thomsonlearning.com

InfoTrac College Edition

BONUS READING

Davies, Jon, et al. 2000. "Identifying Male College Students' Perceived Health Needs, Barriers to Seeking Help, and Recommendations to Help Men Adopt Healthier Lifestyles." *Journal of American College Health* 48 (May): 259–267.

This research examines several dimensions of men's greater likelihood of engaging in risky activities that are hazardous to their health. The authors find that gender stereotypes and men's compliance with them put men at risk for poor health. How are gender roles related to men's health and what differences might be found had women also been included in this study? How might these behaviors vary on different kinds of college campuses? What recommendations for a healthier student body would you suggest having reviewed this research?

SEARCH TERMS

access and health care
beauty myth
race, gender, and health
student health
universal health care

58

American Apartheid

DOUGLAS S. MASSEY AND NANCY A. DENTON

Douglas S. Massey and Nancy A. Denton argue that segregation, particularly residential segregation, is a fundamental dimension of race relations in the United States and is all too often ignored by policymakers and even scholars. It is a major cause of many of the ills of race relations in this country. They argue that it is the "missing link" in past attempts to understand the urban poor.

It is quite simple. As soon as there is a group area then all your uncertainties are removed and that is, after all, the primary purpose of this Bill [requiring racial segregation in housing].

Minister of the Interior,
Union of South Africa legislative debate on the the Group Areas Act of 1950

During the 1970s and 1980s a word disappeared from the American vocabulary. It was not in the speeches of politicians decrying the multiple ills besetting American cities. It was not spoken by government officials responsible for administering the nation's social programs. It was not mentioned by journalists reporting on the rising tide of homelessness, drugs, and violence in urban America. It was not discussed by foundation executives and think-tank experts proposing new programs for unemployed parents and unwed mothers. It was not articulated by civil rights leaders speaking out against the persistence of racial inequality; and it was nowhere to be found in the thousands of pages written by social scientists on the urban underclass. The word was segregation.

Most Americans vaguely realize that urban America is still a residentially segregated society, but few appreciate the depth of black segregation or the degree to which it is maintained by ongoing institutional arrangements and contemporary individual actions. They view segregation as an unfortunate holdover from a racist past, one that is fading progressively over time. If racial residential segregation persists, they reason, it is only because civil rights laws passed during the 1960s have not had enough time to work or because many blacks still prefer to live in black neighborhoods. The residential segregation of blacks is viewed charitably as a "natural" outcome of impersonal social and economic forces, the same forces that produced Italian and Polish neighborhoods in the past and that yield Mexican and Korean areas today.

From: "The Missing Link," reprinted by permission of the publisher from American Apartheid: Segregation and The Making of the Underclass by Douglas S. Massey and Nancy A. Denton, pp. 1–7. Cambridge, Mass.: Harvard University Press. Copyright © 1993 by the President and Fellows of Harvard College.

But black segregation is not comparable to the limited and transient segregation experienced by other racial and ethnic groups, now or in the past. No group in the history of the United States has ever experienced the sustained high level of residential segregation that has been imposed on blacks in large American cities for the past fifty years. This extreme racial isolation did not just happen; it was manufactured by whites through a series of self-conscious actions and purposeful institutional arrangements that continue today. Not only is the depth of black segregation unprecedented and utterly unique compared with that of other groups, but it shows little sign of change with the passage of time or improvements in socioeconomic status.

If policymakers, scholars, and the public have been reluctant to acknowledge segregation's persistence, they have likewise been blind to its consequences for American blacks. Residential segregation is not a neutral fact; it systematically undermines the social and economic well-being of blacks in the United States. Because of racial segregation, a significant share of black America is condemned to experience a social environment where poverty and joblessness are the norm, where a majority of children are born out of wedlock, where most families are on welfare, where educational failure prevails, and where social and physical deterioration abound. Through prolonged exposure to such an environment, black chances for social and economic success are drastically reduced.

Deleterious neighborhood conditions are built into the structure of the black community. They occur because segregation concentrates poverty to build a set of mutually reinforcing and self-feeding spirals of decline into black neighborhoods. When economic dislocations deprive a segregated group of employment and increase its rate of poverty, socioeconomic deprivation inevitably becomes more concentrated in neighborhoods where that group lives. The damaging social consequences that follow from increased poverty are spatially concentrated as well, creating uniquely disadvantaged environments that become progressively isolated—geographically, socially, and economically—from the rest of society.

The effect of segregation on black well-being is structural, not individual. Residential segregation lies beyond the ability of any individual to change; it constrains black life chances irrespective of personal traits, individual motivations, or private achievements. For the past twenty years this fundamental fact has been swept under the rug by policymakers, scholars, and theorists of the urban underclass. Segregation is the missing link in prior attempts to understand the plight of the urban poor. As long as blacks continue to be segregated in American cities, the United States cannot be called a race-blind society.

THE FORGOTTEN FACTOR

The present myopia regarding segregation is all the more startling because it once figured prominently in theories of racial inequality. Indeed, the ghetto was once seen as central to black subjugation in the United States. In 1944 Gunnar Myrdal wrote in *An American Dilemma* that residential segregation "is basic in a mechanical sense. It exerts its influence in an indirect and impersonal way: because Negro people do not live near white people, they cannot . . . associate with each other in the many activities founded on common neighborhood. Residential segregation . . . becomes reflected in uni-racial schools, hospitals, and other institutions" and creates "an artificial city . . . that permits any prejudice on the part of public officials to be freely vented on Negroes without hurting whites."

Kenneth B. Clark, who worked with Gunnar Myrdal as a student and later applied his research skills in the landmark *Brown v. Topeka* school integration case, placed residential segregation at the heart of the U.S. system of racial oppression. In *Dark Ghetto,* written in 1965, he argued that "the dark ghetto's invisible walls have been erected by the white society, by those who have power, both to confine those who have no power and to perpetuate their powerlessness. The dark ghettos are social, political, educational, and—above all—economic colonies. Their inhabitants are subject peoples, victims of the greed, cruelty, insensitivity, guilt, and fear of their masters."

Public recognition of segregation's role in perpetuating racial inequality was galvanized in the late 1960s by the riots that erupted in the nation's ghettos. In their aftermath, President Lyndon B. Johnson appointed a commission chaired by Governor Otto Kerner of Illinois to identify the causes of the violence and to propose policies to prevent its recurrence. The Kerner Commission released its report in March 1968 with the shocking admonition that the United States was "moving toward two societies, one black, one white—separate and unequal." Prominent among the causes that the commission identified for this growing racial inequality was residential segregation.

In stark, blunt language, the Kerner Commission informed white Americans that "discrimination and segregation have long permeated much of American life; they now threaten the future of every American." "Segregation and poverty have created in the racial ghetto a destructive environment totally unknown to most white Americans. What white Americans have never fully understood—but what the Negro can never forget—is that white society is deeply implicated in the ghetto. White institutions created it, white institutions maintain it, and white society condones it."

The report argued that to continue present policies was "to make permanent the division of our country into two societies; one, largely Negro and poor, located in the central cities; the other, predominantly white and affluent, located in the suburbs." Commission members rejected a strategy of ghetto enrichment coupled with abandonment of efforts to integrate, an approach they saw "as another way of choosing a permanently divided country." Rather, they insisted that the only reasonable choice for America was "a policy which combines ghetto enrichment with programs designed to encourage integration of substantial numbers of Negroes into the society outside the ghetto."

America chose differently. Following the passage of the Fair Housing Act in 1968, the problem of housing discrimination was declared solved, and residential segregation dropped off the national agenda. Civil rights leaders stopped pressing for the enforcement of open housing, political leaders increasingly debated employment and educational policies rather than housing integration, and academicians focused their theoretical scrutiny on everything from culture to family structure, to institutional racism, to federal welfare systems. Few people spoke of racial segregation as a problem or acknowledged its persisting consequences. By the end of the 1970s residential segregation became the forgotten factor in American race relations.

While public discourse on race and poverty became more acrimonious and more focused on divisive issues such as school busing, racial quotas, welfare, and affirmative action, conditions in the nation's ghettos steadily deteriorated. By the end of the 1970s, the image of poor minority families mired in an endless cycle of unemployment, unwed childbearing, illiteracy, and dependency had coalesced into a compelling and powerful concept: the urban underclass. In the view of many middle-class whites, inner cities had come to house a large population of poorly educated single mothers and jobless men—mostly black and Puerto Rican—who were unlikely to exit poverty and become self-sufficient. In the ensuing national debate on the causes for this persistent poverty, four theoretical explanations gradually emerged: culture, racism, economics, and welfare.

Cultural explanations for the underclass can be traced to the work of Oscar Lewis, who identified a "culture of poverty" that he felt promoted patterns of behavior inconsistent with socioeconomic advancement. According to Lewis, this culture originated in endemic unemployment and chronic social immobility, and provided an ideology that allowed poor people to cope with feelings of hopelessness and despair that arose because their chances for socioeconomic success were remote. In individuals, this culture was typified by a lack of impulse control, a strong present-time orientation, and little ability to defer gratification. Among families, it yielded an absence of childhood, an early initiation into sex, a prevalence of free marital unions, and a high incidence of abandonment of mothers and children.

Although Lewis explicitly connected the emergence of these cultural patterns to structural conditions in society, he argued that once the culture of poverty was established, it became an independent cause of persistent poverty. This idea was further elaborated in 1965 by the Harvard sociologist and then Assistant Secretary of Labor Daniel Patrick Moynihan, who in a confidential report to the President focused on the relationship between male unemployment, family instability, and the intergenerational transmission of poverty, a process he labeled a "tangle of pathology." He warned that because of the structural absence of employment in the ghetto, the black family was disintegrating in a way that threatened the fabric of community life.

When these ideas were transmitted through the press, both popular and scholarly, the connection between culture and economic structure was somehow lost, and the argument was popularly perceived to be that "people were poor because they had a defective culture." This position was later explicitly adopted by the conservative theorist Edward Banfield, who argued that lower-class culture—with its limited time horizon, impulsive need for gratification, and psychological self-doubt—was primarily responsible for persistent urban poverty. He believed that these cultural traits were largely imported, arising primarily because cities attracted lower-class migrants.

The culture-of-poverty argument was strongly criticized by liberal theorists as a self-serving ideology that "blamed the victim." In the ensuing wave of reaction, black families were viewed not as weak but, on the contrary, as resilient and well adapted survivors in an oppressive and racially prejudiced society. Black disadvantages were attributed not to a defective culture but to the persistence of institutional racism in the United States. According to theorists of the underclass such as Douglas Glasgow and Alphonso Pinkney, the black urban underclass came about because deeply imbedded racist practices within American institutions—particularly schools and the economy—effectively kept blacks poor and dependent.

As the debate on culture versus racism ground to a halt during the late 1970s, conservative theorists increasingly captured public attention by focusing on a third possible cause of poverty: government welfare policy. According to Charles Murray, the creation of the underclass was rooted in the liberal welfare state. Federal antipoverty programs altered the incentives governing the behavior of poor men and women, reducing the desirability of marriage, increasing the benefits of unwed childbearing, lowering the attractiveness of menial labor, and ultimately resulted in greater poverty.

A slightly different attack on the welfare state was launched by Lawrence Mead, who argued that it was not the generosity but the permissiveness of the U.S. welfare system that was at fault. Jobless men and unwed mothers should be required to display "good citizenship" before being supported by the state. By not requiring anything of the poor, Mead argued, the welfare state undermined their independence and competence, thereby perpetuating their poverty.

This conservative reasoning was subsequently attacked by liberal social scientists, led principally by the sociologist William Julius Wilson, who had long been arguing for the increasing importance of class over race in understanding the social and economic problems facing blacks. In his 1987 book *The Truly Disadvantaged,* Wilson argued that persistent urban poverty stemmed primarily from the structural transformation of the inner-city economy. The decline of manufacturing, the suburbanization of employment, and the rise of a low-wage service sector dramatically reduced the number of city jobs that paid wages sufficient to support a family, which led to high rates of joblessness among minorities and a shrinking pool of "marriageable" men (those financially able to support a family). Marriage thus became less attractive to poor women, unwed childbearing increased, and female-headed families proliferated. Blacks suffered disproportionately from these trends because, owing to past discrimination, they were concentrated in locations and occupations particularly affected by economic restructuring.

Wilson argued that these economic changes were accompanied by an increase in the spatial concentration of poverty within black neighborhoods. This new geography of poverty, he felt, was enabled by the civil rights revolution of the 1960s, which provided middle-class blacks with new opportunities outside the ghetto. The out-migration of middle-class families from ghetto areas left behind a destitute community lacking the institutions, resources, and values necessary for success in postindustrial society. The urban underclass thus arose from a complex interplay of civil rights policy, economic restructuring, and a historical legacy of discrimination.

Theoretical concepts such as the culture of poverty, institutional racism, welfare disincentives, and structural economic change have all been widely debated. None of these explanations, however, considers residential segregation to be an important contributing cause of urban poverty and the underclass. In their principal works, Murray and Mead do not mention segregation at all and Wilson refers to racial segregation only as a historical legacy from the past, not as an outcome that is institutionally supported and actively created today. Although Lewis mentions segregation sporadically in his writings, it is not assigned a central role in the set of structural factors responsible for the culture of poverty, and Banfield ignores it entirely. Glasgow, Pinkney, and other theorists of institutional racism mention the ghetto frequently, but generally call not for residential desegregation but for race-specific policies to combat the effects of discrimination in the schools and labor markets. In general, then, contemporary theorists of urban poverty do not see high levels of black-white segregation as particularly relevant to understanding the underclass or alleviating urban poverty.

The purpose of this [argument] is to redirect the focus of public debate back to issues of race and racial segregation and to suggest that they should be fundamental to thinking about the status of black Americans and the origins of the urban underclass. Our quarrel is less with any of the prevailing theories of urban poverty than with their systematic failure to consider the important role that segregation has played in mediating, exacerbating, and ultimately amplifying the harmful social and economic processes they treat.

We join earlier scholars in rejecting the view that poor urban blacks have an autonomous "culture of poverty" that explains their failure to achieve socioeconomic success in American society. We argue instead that residential segregation has been instrumental in creating a structural niche within which a deleterious set of attitudes and behaviors—a culture of segregation—has arisen and flourished. Segregation created the structural conditions for the emergence of an oppositional culture that devalues work, schooling, and marriage and that stresses attitudes and behaviors that are antithetical and often hostile to success in the larger economy. Although poor black neighborhoods still contain many people who lead conventional, productive lives, their example has been overshadowed in recent years by a growing concentration of poor, welfare-dependent families that is an inevitable result of residential segregation.

We readily agree with Douglas, Pinkney, and others that racial discrimination is widespread and may even be institutionalized within large sectors of American society, including the labor market, the educational system, and the welfare bureaucracy. We argue, however, that this view of black subjugation is incomplete without understanding the special role that residential segregation plays in enabling all other forms of racial oppression. Residential segregation is the institutional apparatus that supports other racially discriminatory processes and binds them together into a coherent and uniquely effective system of racial subordination. Until the black ghetto is dismantled as a basic institution of American urban life, progress ameliorating racial inequality in other arenas will be slow, fitful, and incomplete.

We also agree with William Wilson's basic argument that the structural transformation of the urban economy undermined economic supports for the black community during the 1970s and 1980s. We argue, however, that in the absence of segregation, these structural changes would not have produced the disastrous social and economic outcomes observed in inner cities during these decades. Although rates of black poverty were driven up by the economic dislocations Wilson identifies, it was segregation that confined the increased deprivation to a small number of densely settled, tightly packed, and geographically isolated areas.

Wilson also argues that concentrated poverty arose because the civil rights revolution allowed middle-class blacks to move out of the ghetto. Although we remain open to the possibility that class-selective migration did occur, we argue that concentrated poverty would have happened during the 1970s with or without black middle-class migration. Our principal objection to Wilson's focus on middle-class out-migration is not that it did not occur, but that it is misdirected: focusing on the flight of the black middle class deflects attention from the real issue, which is the limitation of black residential options through segregation.

Middle-class households—whether they are black, Mexican, Italian, Jewish, or Polish—always try to escape the poor. But only blacks must attempt their escape within a highly segregated, racially segmented housing market. Because of segregation, middle-class blacks are less able to escape than other groups, and as a result are exposed to more poverty. At the same time, because of segregation no one will move into a poor black neighborhood except other poor blacks. Thus both middle-class blacks and poor blacks lose compared with the poor and middle class of other groups: poor blacks live under unrivaled concentrations of poverty and affluent blacks live in neighborhoods that are far less advantageous than those experienced by the middle class of other groups.

Finally, we concede Murray's general point that federal welfare policies are linked to the rise of the urban underclass, but we disagree with his specific hypothesis that generous welfare payments, by themselves, discouraged employment, encouraged unwed childbearing, undermined the strength of the family, and thereby caused persistent poverty. We argue instead that welfare payments were only harmful to the socioeconomic well-being of groups that were residentially segregated. As poverty rates rose among blacks in response to the economic dislocations of the 1970s and 1980s, so did the use of welfare programs. Because of racial segregation, however, the higher levels of welfare receipt were confined to a small number of isolated, all-black neighborhoods. By promoting the spatial concentration of welfare use, therefore, segregation created a residential environment within which welfare dependency was the norm, leading to the intergenerational transmission and broader perpetuation of urban poverty. . . .

Our fundamental argument is that racial segregation—and its characteristic institutional form, the black ghetto—are the key structural factors responsible for the perpetuation of black poverty in the United States. Residential segregation is the principal organizational feature of American society that is responsible for the creation of the urban underclass. . . .

DISCUSSION QUESTIONS

1. Regardless of your race or ethnicity, did you grow up in a racially segregated environment? How central in your life was this fact? What consequences did it have? If you know anyone who did, what in your estimation were the effects on their life?

2. What is the "culture of poverty" view? Do you agree with it? What do Massey and Denton have to say about it?

59

Black, Brown, Red, and Poisoned

REGINA AUSTIN AND MICHAEL SCHILL

The principle of environmental racism states that race, more so than class (socioeconomic status), explains the unfortunate residential closeness of people of color to toxic waste dumps, chemical plants, oil refineries, incinerators, and other toxic sources. Here Regina Austin and Michael Schill review the issue of environmental racism as well as activist strategies to combat it.

People of color throughout the United States are receiving more than their fair share of the poisonous fruits of industrial production. They live cheek by jowl with waste dumps, incinerators, landfills, smelters, factories, chemical plants, and oil refineries whose operations make them sick and kill them young. They are poisoned by the air they breathe, the water they drink, the fish they catch, the vegetables they grow, and, in the case of children, the very ground they play on. Even the residents of some of the most remote rural hamlets of the South and Southwest suffer from the ill effects of toxins.

This [essay] examines some of the reasons why communities of color bear a disparate burden of pollution. It also brings into focus the commonality of their struggles and some strategies that are useful in overcoming environmental injustice.

From: Austin, Regina, and Michael Schill. "Black, Red, Brown, and Poisoned." In *Unequal Protection: Environmental Justice and Communities of Color*, ed. Robert D. Bullard, 53–73. San Francisco: Sierra Club Books, 1994. Reprinted with permission.

THE PATH OF LEAST RESISTANCE

The disproportionate location of sources of toxic pollution in communities of color is the result of various development patterns. In some cases, the residential communities where people of color now live were originally the homes of whites who worked in the facilities that generate toxic emissions. The housing and the industry sprang up roughly simultaneously. Whites vacated the housing (but not necessarily the jobs) for better shelter as their socioeconomic status improved, and poorer black and brown folks who enjoy much less residential mobility took their place. In other cases, housing for African Americans and Latino Americans was built in the vicinity of existing industrial operations because the land was cheap and the people were poor. For example, Richmond, California, was developed downwind from a Chevron oil refinery when African Americans migrated to the area to work in shipyards during World War II.

In yet a third pattern, sources of toxic pollution were placed in existing minority communities. The explanations for such sitings are numerous; some reflect the impact of racial and ethnic discrimination. The impact, of course, may be attenuated and less than obvious. The most neutral basis for a siting choice is probably the natural characteristics of the land, such as mineral content of the soil. . . . Low population density would appear to be a similar criterion. It has been argued, however, that in the South, a sparse concentration of inhabitants is correlated with poverty, which is in turn correlated with race. "It follows that criteria for siting hazardous waste facilities which include density of population will have the effect of targeting rural black communities that have high rates of poverty."

Likewise, the compatibility of pollution with preexisting uses might conceivably make some sites more suitable than others for polluting operations. Pollution tends to attract other sources of pollutants, particularly those associated with toxic disposal. For example, Chemical Waste Management, Inc. (Chem Waste) has proposed the construction of a toxic waste incinerator outside of Kettleman City, California, a community composed largely of Latino farm workers. Chem Waste also has proposed to build a hazardous waste incinerator in Emelle, a predominantly African American community located in the heart of Alabama's "black belt." The company already has hazardous waste landfills in Emelle and Kettleman City.

According to the company's spokeswoman, Chem Waste placed the landfill in Kettleman City "because of the area's geological features. Because the landfill handles toxic waste, . . . it is an ideal spot for the incinerator"; the tons of toxic ash that the incinerator will generate can be "contained and disposed of at the installation's landfill." Residents of Kettleman City face a "triple whammy" of threats from pesticides in the fields, the nearby hazardous waste landfill, and a proposed hazardous waste incinerator. This case is not unique.

After reviewing the literature on hazardous waste incineration, one commentator has concluded that "[m]inority communities represent a 'least cost' option for waste incineration . . . because much of the waste to be incinerated is already in these communities." Despite its apparent neutrality, then, siting based on compatibility may be related to racial and ethnic discrimination, particularly if such discrimination influenced the siting of preexisting sources of pollution.

Polluters know that communities of low-income and working-class people with no more than a high school education are not as effective at marshaling opposition as communities of middle- or upper-income people. People of color in the United States have traditionally had less clout with which to check legislative and executive abuse or to challenge regulatory laxity. Private corporations, moreover, can have a powerful effect on the behavior of public officials. Poor minority people wind up the losers to them both.

People of color are more likely than whites to be economically impoverished, and economic vulnerability makes impoverished communities of color prime targets for "risky" technologies. Historically, these communities are more likely than others to tolerate pollution-generating commercial development in the hope that economic benefits will inure to the community in the form of jobs, increased taxes, and civic improvements. Once the benefits start to flow, the community may be reluctant to forgo them even when they are accompanied by poisonous spills or emissions. This was said to be the case in Emelle, in Sumter County, Alabama, site of the nation's largest hazardous waste landfill.

Sumter County's population is roughly 70 percent African American, and 30 percent of its inhabitants fall below the poverty line. Although the landfill was apparently leaking, it was difficult to rally support against the plant among African American politicians because its operations contributed an estimated $15.9 million to the local economy in the form of wages, local purchases of goods and services, and per-ton landfill user fees.

Of course, benefits do not always materialize after the polluter begins operations. For example, West Harlem was supposed to receive, as a trade-off for accepting New York City's largest sewage treatment plant, an elaborate state park to be built on the roof of the facility. The plant is functioning, fouling the air with emissions of hydrogen sulfide and promoting an infestation of rats and mosquitoes. The park, however, has yet to be completed, the tennis courts have been removed from the plan completely, and the "first-rate" restaurant has been scaled down to a pizza parlor.

In other cases, there is no net profit to distribute among the people. New jobs created by the poisonous enterprises are "filled by highly skilled labor from outside the community," while the increased tax revenues go not to "social services or other community development projects, but . . . toward expanding the infrastructure to better serve the industry."

Once a polluter has begun operations, the victims' options are limited. Mobilizing a community against an existing polluter is more difficult than organizing opposition to a proposed toxic waste-producing activity. Resignation sets in, and the resources for attacking ongoing pollution are not as numerous, and the tactics not as potent, as those available during the proposal stage. Furthermore, though some individuals are able to escape toxic poisoning by moving out of the area, the flight of others will be blocked by limited incomes, housing discrimination, and restrictive land use regulations.

THREAT TO BARRIOS, GHETTOS, AND RESERVATIONS

Pollution is no longer accepted as an unalterable consequence of living in the "bottom" (the least pleasant, poorest area minorities can occupy) by those on the bottom of the status hierarchy. Like anybody else, people of color are distressed by accidental toxic spills, explosions, and inexplicable patterns of miscarriages and cancers, and they are beginning to fight back, from Maine to Alaska.

To be sure, people of color face some fairly high barriers to effective mobilization against toxic threats, such as limited time and money; lack of access to technical, medical, and legal expertise; relatively weak influence in political and media circles; and ideological conflicts that pit jobs against the environment. Limited fluency in English and fear of immigration authorities will keep some of those affected, especially Latinos, quiescent. Yet despite the odds, poor minority people are responding to their poisoning with a grass-roots movement of their own.

Activist groups of color are waging grass-roots environmental campaigns all over the country. Although they are only informally connected, these campaigns reflect certain shared characteristics and goals. The activity of activists of color is indicative of a grassroots movement that occupies a distinctive position relative to both the mainstream movement and the white grass-roots environmental movement. The environmental justice movement is antielitist and antiracist. It capitalizes on the social and cultural differences of people of color as it cautiously builds alliances with whites and persons of the middle class. It is both fiercely environmental *and* conscious of the need for economic development in economically disenfranchised communities. Most distinctive of all, this movement has been extremely outspoken in challenging the integrity and bona fides of mainstream establishment environmental organizations.

People of color have not been mobilized to join grass-roots environmental campaigns because of their general concern for the environment. Characterizing a problem as being "environmental" may carry weight in some circles, but it has much less impact among poor minority people. It is not that people of color are uninterested in the environment—a suggestion the grass-roots activists find insulting. In fact they are more likely to be concerned about pollution than are people who are wealthier and white. Rather, in the view of many people of color, environmentalism is associated with the preservation of wildlife and wilderness, which simply is not more important than the survival of people and the communities in which they live; thus, the mainstream movement has its priorities skewed. . . .

CAPITALIZING ON THE RESOURCES
OF COMMON CULTURE

For people of color, social and cultural differences such as language are not handicaps but the communal resources that facilitate mobilization around issues like toxic poisoning. As members of the same race, ethnicity, gender, and even age cadre, would-be participants share cultural traditions, modes, and mores that encourage cooperation and unity. People of color may be more responsive to organizing efforts than whites because they already have experience with collective action through community groups and institutions such as churches, parent-teacher associations, and town watches or informal social networks. Shared criticisms of racism, a distrust of corporate power, and little expectation that government will be responsive to their complaints are common sentiments in communities of color and support the call to action around environmental concerns.

Grass-roots environmentalism is also fostered by notions that might be considered feminist or womanist. Acting on a realization that toxic poisoning is a threat to home and family, poor minority women have moved into the public realm to confront corporate and government officials whose modes of analysis reflect patriarchy, white supremacy, and class and scientific elitism. There are numerous examples of women of color whose strengths and talents have made them leaders of grass-roots environmental efforts.

The organization Mothers of East Los Angeles (MELA) illustrates the link between group culture and mobilization in the people of color grass-roots environmental movement. Persistent efforts by MELA defeated proposals for constructing a state prison and a toxic waste incinerator in the group's mostly Latino American neighborhood in East Los Angeles.

Similarly, the Lumbee Indians of Robeson County, North Carolina, who attach spiritual significance to a river that would have been polluted by a hazardous waste facility proposed by the GSX Corporation, waged a campaign against the facility on the ground of cultural genocide. Throughout the campaign, "Native American dance, music, and regalia were used at every major public hearing. Local Lumbee churches provided convenient meeting locations for GSX planning sessions. Leaflet distribution at these churches reached significant minority populations in every pocket of the county's nearly 1,000 square miles."

Concerned Citizens of Choctaw defeated a plan to locate a hazardous waste facility on their lands in Philadelphia, Mississippi. The Good Road Coalition, a grass-roots Native American group based on the Rosebud Reservation in South Dakota, defeated plans by a Connecticut-based company to build a 6,000-acre garbage landfill on the Rosebud. Local residents initiated a recall election, defeating several tribal council leaders and the landfill proposal. The project, dubbed "dances with garbage," typifies the lengths that the Lakota people and other Native Americans will go to preserve their land—which is an essential part of their religion and culture.

Consider, finally, the Toxic Avengers of El Puente, a group of environmental organizers based in the Williamsburg section of Brooklyn, New York. The name is taken from the title of a horror movie. The group attacks not only environmental racism but also adultism and adult superiority and privilege. The members, whose ages range from nine to twenty-eight, combine their activism with programs to educate themselves and others about the science of toxic hazards.

The importance of culture in the environmental justice movement seems not to have produced the kind of distrust and misgivings that might impede interaction with white working-class and middle-class groups engaged in grass-roots environmental activism. There are numerous examples of ethnic-based associations working in coalitions with one another, with majority group associations, and with organizations from the mainstream. There are also localities in which the antagonism and suspicion that are the legacy of white racism have kept whites and African Americans from uniting against a common toxic enemy. The link between the minority groups and the majority groups seems grounded in material exchange, not ideological fellowship. The white groups attacking toxins at the grass-roots level have been useful sources of financial assistance and information about tactics and goals. . . .

BRIDGING THE JUSTICE-ENVIRONMENT GAP

At the same time that environmental justice activists are battling polluters, some are engaged on another front in a struggle against elitism and racism that exist within the mainstream environmental movement. There are several substantive points of disagreement between grass-roots groups of color and mainstream environmental organizations. First, communities of color are tired of shouldering the fallout from environmental regulation. A letter sent to ten of the establishment environmental organizations by the Southwest Organizing Project and numerous activists of color engaged in the grass-roots environmental struggle illustrates the level of exasperation:

> Your organizations continue to support and promote policies which emphasize the clean-up and preservation of the environment on the backs of working people in general and people of color in particular. In the name of eliminating environmental hazards at any cost, across the country industrial and other economic activities which employ us are being shut down, curtailed or prevented while our survival needs and cultures are ignored. We suffer the end results of these actions, but are never full participants in the decision-making which leads to them. . . .

Another threat to communities of color is the growing popularity of NIMBY (not in my backyard) groups. People of color have much to fear from these groups because their communities are the ones most likely to lose the contests to keep the toxins out. The grass-roots environmentalists argue that rather than trying to bar polluters, who will simply locate elsewhere, energies should be directed at bringing the amount of pollution down to zero. In lieu of NIMBY, mainstream environmentalists should be preaching NIABY (not in anyone's backyard).

Finally, conservation organizations are making "debt-for-nature" swaps throughout the so-called Third World. Through swaps, conservation organizations procure ownership of foreign indebtedness (either by gift or by purchase at a reduced rate) and negotiate with foreign governments for reduction of the debt in exchange for land. Grass-roots environmental activists of color complain that these deals, which turn conservation organizations into creditors of so-called Third World peoples, legitimate the debt and the exploitation on which it is based. . . .

DISCUSSION QUESTIONS

1. In your opinion, is environmental racism (the closeness of people of color to toxic waste sites) mostly because of race, class (socioeconomic status), or both, or some other reason all together? Discuss.

2. Assume that you are asked to organize a group to combat environmental racism in or near your own hometown or city. What kinds of issues would you focus on and how might you begin to go about organizing such a group?

60

Mobilizing Minority Communities

Social Capital and Participation in Urban Neighborhoods

KENT E. PORTNEY AND JEFFREY M. BERRY

In a study of five U.S. cities, the authors show how important neighborhood associations can be for building a sense of community among urban residents. They also examine how such participation varies according to the racial makeup of neighborhoods.

An unusually strong consensus has emerged among academics as to the problem: Americans are disengaged from civic life. Increasingly we live in splendid isolation, relating to friends but paying little attention to our community (Barber, 1984; Putnam, 1995). We no longer live in Tocqueville's America but in a society where the town square has been replaced by the mall, cable television, and the Internet. Yet Americans claim they want to be more involved in working together with their neighbors on civic problems, that they want to have more of a voice in the political process, and that we ought to have smaller government and more voluntarism by rank-and-file citizens (Bellah, Madden, Sullivan, Swidler, & Tipton, 1985; Kettering Foundation, 1991).

From: Portney, Kent E., and Jeffrey M. Berry. "Mobilizing Minority Communities: Social Capital and Participation in Urban Neighborhoods." *American Behavioral Scientist* 40 (March–April 1997): 632–44. Reprinted with permission of Sage Publications, Inc.

These sentiments are easily articulated to pollsters; actually taking the time to work with one's neighbors is another thing entirely.

Although there is widespread agreement as to the indictment, there is not much consensus as to the underlying causes. There is even less agreement as to the cure. Whatever the causes and potential solutions, it is clear that a central issue in determining the public's involvement in community life is how the opportunities to participate are structured. If people want to participate but are failing to do so, it may well be that they do not find avenues for participation accessible or inviting. In this article, we ask what kinds of political organizations are most effective in mobilizing minorities in city politics. We look at people of minority status because the debate about social capital and civic engagement largely concentrates on White, middle-class America. Virtually none of the debate and, as far as we can find, no empirical analysis considers whether poor people, people of minority racial or ethnic status, and people in inner cities have also experienced the trends in civic engagement. Indeed, there is an assumption that whatever the state of engagement is in general, it must be worse for African Americans and Hispanics, who are certainly thought to be less connected, less civically engaged, and less well equipped to compete in mainstream political processes.

We emphasize neighborhoods because neighborhoods are where the bonds of community are built. They are the wellsprings of social capital. People's sense of community, their sense of belonging to a neighborhood, their caring about the people who live there, and their belief that people who live there care about them are critical attitudes that can nurture or discourage participation.

Drawing on data gathered from five American cities, we examine the impact of structures of strong democracy on community building. By structures of strong democracy, we mean institutions that give all residents an opportunity to participate at all stages of the policy-making process and allow citizens the authority to determine the final outcomes of the policy matters considered in this participatory process. By community building, we mean those beliefs and types of political behavior that contribute to positive attitudes by residents about their neighborhood and encourage a willingness to work cooperatively on its behalf.

If our goal is to understand civic engagement, civil society, and the creation of social capital, neighborhood associations seem to be an ideal focus. They offer citizens an opportunity to discuss community problems with other neighbors. They are forums in which city officials come to respond to the questions and demands of ordinary citizens. Successful neighborhood associations should promote cooperation, encourage voluntarism, and enhance feelings of community.

FIVE AMERICAN CITIES

To further explore these issues of participation, race, and community, we conducted research in five cities—Birmingham (Alabama), Dayton (Ohio), Portland (Oregon), St. Paul (Minnesota), and San Antonio (Texas)—where participatory democracy is taken seriously. Elsewhere we explain how these cities were chosen, but in general terms they were selected because they were thought to have the most impressive citizen participation systems in the country (Berry, Portney, & Thomson, 1993).

Four of these cities have citywide systems of neighborhood associations. In each and every neighborhood there is a neighborhood association that has substantial authority over the community and is run cooperatively by residents. These neighborhood associations have real powers and are not merely planning or advisory boards. Their strongest authority lies in the area of zoning and land use. In St. Paul, for example, a resident wanting a zoning permit or a variance must go to his or her neighborhood association rather than to city hall. For all intents and purposes, there is no appealing a zoning decision made by one of the city's neighborhood associations. This applies to businesspeople and developers as well as homeowners. Thus project proposals for new businesses or housing must gain the support of the neighborhood association before they can be carried out. . . .

RACE AND NEIGHBORHOODS

Why are cities that are unusual in their commitment to participatory democracy good laboratories for studying race and political behavior? What we can find out about race in urban settings is certainly limited by focusing only on structures of participatory democracy. What is valuable about using these cities and the data we have collected is that the underlying theory of participatory democracy is built on the notion of community. The extent to which people might be said to possess a sense of community—the extent to which they feel a sense of belonging to a neighborhood and care about the people who live there, believing that others who live in their neighborhoods care about them—may well be critical elements underlying minority empowerment.

Neighborhood associations are one way of trying to build community. These organizations encourage neighbors to come together and talk to each other about their community and the problems that it faces. In face-to-face meetings, men and women can learn from each other, reason with one another, and search for common interests. Ideally, bonds of friendship and community are forged as neighbors look for solutions to the issues before them. Political participation becomes an educative device rather than an occasionally exercised civic obligation. . . .

Within the five cities, there is substantial diversity in their racial and ethnic makeup. Birmingham is 55.6% Black, and Dayton 36.9% Black. Portland (7.6%) and St. Paul (4.9%) have considerably smaller Black populations. None of these four cities has a significant number of Hispanic residents. San Antonio is 53.7% Hispanic and 7.3% black. The surveys we took indicated that race was not a dominant issue in any city at the time of the research. Race, however, is not far beneath the surface in these cities either. In San Antonio, issues often divide Anglos and Hispanics. In Birmingham, the relatively new majority status of the African American population was a cause of some uneasiness on the part of the city's White minority. In Dayton, White flight is a continuing concern, and in

Dayton and St. Paul, racial quotas in city hiring were an issue during the time of our fieldwork.

Surveys of about 1,100 residents in each of the five cities were completed from 1986 to 1987. Around the same time, fieldwork was conducted in each city, and elite interviews were done with city councillors, administrators, interest group leaders, and activists in the neighborhood associations. . . .

. . . First, we look at participation and mobilization. What is the relationship between the racial composition of neighborhoods and the racial composition of activists in the neighborhood associations? Is there a racial bias in the patterns of participation? Second, we ask about neighborhood and sense of community. Do neighborhood associations designed around principles of strong democracy build community more successfully than other kinds of political organizations?

PARTICIPATION AND MOBILIZATION

One central element underlying the idea of strong democracy is what we call face-to-face political participation—communal activities that seem capable of helping to rebuild American democracy. In this article, we focus on four specific types of activity, including some of the most frequently practiced forms of face-to-face activity in American cities. We concentrate on participation in the neighborhood associations, independent issue-based citizen organizations, neighborhood crime watch organizations, and social, civic, self-help, and service organizations. In any major U.S. city, significant numbers of residents are likely to be active in these kinds of organizations. . . .

. . . The participation rate of low socioeconomic status (SES) residents of predominantly African American neighborhoods is almost twice that of low SES residents of low minority population neighborhoods. The same general pattern appears for people at each socioeconomic level. This pattern stands in stark contrast to the findings for independent issue-based citizen organizations. Although citizen group participation among residents of low minority population neighborhoods is comparable to that for neighborhood associations, these independent organizations do not seem to be as effective in reaching residents of racially diverse or predominantly Black neighborhoods.

When the participation rates are broken down by the race of the respondent . . . instead of socioeconomic status, the patterns lead to similar conclusions. Participation among African Americans in neighborhood associations is consistently higher than that for issue-based organizations, regardless of the racial composition of the neighborhood. African Americans and Whites participate in social, service, and self-help organizations at relatively high rates, but only in predominantly non-African American neighborhoods. In predominantly Black neighborhoods, only neighborhood associations stand out as . . . providing clearly superior opportunities for African Americans to become civically engaged. . . .

Overall, if judged solely on the basis of how well they mobilize people in racially diverse and predominantly Black neighborhoods, the neighborhood associations appear to stand out as being the most effective among the four forms of face-to-face participation. But, of course, we are also interested in attitudes about the community, resulting from some of the consequences of participation.

COMMUNITY

. . . Measuring a sense of community is difficult because it is an elusive concept, seemingly encompassing all that is positive about neighborhood life. . . .

Despite the variety in the way the concept is treated by social scientists and philosophers, two fundamental concerns seem to be at the core of this idea. The first is the belief that there is an identifiable community and that one feels a sense of belonging to it (Crenson, 1983). The second is a sense of common purpose. People may believe that they and their neighbors share goals and that there is a willingness to work together to achieve them (Fowler, 1991). The question wording used in these surveys should tap either or both of these sentiments.

. . . Two findings stand out. First, the neighborhood . . . associations appear superior to other types of organizations in nurturing a strong sense of community. In all three racial composition groupings, neighborhood associations achieve the highest response rate for those who indicate they have a strong sense of community.

The second pattern is that in the neighborhoods that have a majority of African American residents, respondents' sense of community tends to be higher across the board. The one exception to this is that for social and service organizations, there is no significant difference between the largely Black neighborhoods and those with greater racial diversity. This pattern is not an unexpected finding. It seems logical that members of a group that has minority status in the broader society might feel more a part of a neighborhood where they are in the majority. Our earlier analysis also showed that sense of community does not vary by class. Poor, middle-class, and wealthy residents are indistinguishable in their feelings of attachment with their neighborhoods (Berry et al., 1993). . . .

If sense of community is an essential building block for creating a participatory democracy and transforming the way people behave in the political process, then the cities with citywide systems of neighborhood associations have a valuable resource. These neighborhood associations are more effective than other types of organizations in nurturing feelings of identity and shared purpose with one's neighbors.

CONCLUSION

Political theorists have long argued that participatory democracy will lead to more citizen involvement in community life. If people talk to each other and work with each other to solve problems, this will lead to a community where people care about each other. Individuals will discover common purposes and develop stronger ties to their neighborhoods and cities. If America is becoming more individualistic and less community oriented, then participatory democracy would appear to be an appealing solution. . . . Are structures of strong democracy of particular help to minority groups as they compete in urban political systems?

It appears so. Although neighborhood associations do not increase the overall number of people who participate in city politics, minority participation rates tend to increase as their percentage of the neighborhood population increases. It seems clear that neighborhood associations are comfortable places for residents in racially mixed and predominantly African American areas. Furthermore, in comparison to other kinds of organizations, participation in the neighborhood associations is more strongly associated with a high sense of community.

The neighborhood associations in Birmingham, Dayton, Portland, and St. Paul are not just another kind of political organization. Although they are independent of government because they are run by volunteers, they are also an official part of city government. They have meaningful and autonomous powers, and city officials have strong incentives to be responsive to the neighborhood associations' requests and preferences. Neighborhood residents understand that these organizations have clout within city government. In stark contrast to these successful citywide systems is San Antonio, where only one section of town is organized. On a host of attitudinal and behavioral measures, both Hispanics and Whites in San Antonio demonstrate less support and less involvement in the governmental process.

Poor Black neighborhoods are often drawn in stereotypical terms as communities where social and political institutions have badly deteriorated and where antisocial behavior is all too prevalent (Wilson, 1987). In the cities we studied, poor Black neighborhoods and Black neighborhoods of all economic stripes demonstrate relatively high levels of political participation in neighborhood associations. These neighborhood associations are effective in mobilizing African Americans and cultivating attitudes that are supportive of the community. In neighborhoods with significant numbers of Black residents, structures of strong democracy work.

REFERENCES

Barber, B. R. (1984). Strong democracy. Berkeley: University of California Press.

Bellah, R. N., Madden, R., Sullivan, W. M., Swidler, A., & Tipton, S. M. (1985). Habits of the heart: Individualism and commitment in American life. Berkeley: University of California Press.

Berry, J. M., Portney, K. E., & Thomson, K. (1993), The rebirth of urban democracy. Washington, DC: The Brookings Institution.

Crenson, M. A. (1983). Neighborhood politics. Cambridge, MA: Harvard University Press.

Fowler, R. B. (1991). The dance with community. Lawrence: University Press of Kansas.

Kettering Foundation. (1991). Citizens and politics. Dayton, OH: Kettering Foundation.

Putnam, R. D. (1995). Bowling alone: America's declining social capital. Journal of Democracy, 6, 65–78.

Wilson, W. J. (1987). The truly disadvantaged. Chicago: University of Chicago Press.

DISCUSSION QUESTIONS

1. What kinds of organizations do the authors find most enhance a sense of civic engagement and community belonging?

2. What is unique about some of the cities studied in terms of city support for neighborhoods and what does this suggest for social policies designed to enhance Americans' sense of community belonging?

http://infotrac.thomsonlearning.com

InfoTrac College Edition

BONUS READING

Muller, Peter O. "The Suburban Transformation of the Globalizing American City." *The Annals of the American Academy of Political and Social Science* 551 (May 1997): 44–58.

Although many people who study urbanization have focused on center cities and how they are affected by contemporary social change, here the author argues that social change, especially that occurring on a global level, is also affecting suburbanization. What changes, in particular, are transforming both cities and suburbs and what implications do these changes have for different groups in the United States?

SEARCH TERMS

environmental justice/environmental racism
hypersegregation
residential segregation
suburbanization
urban and social capital

61

Generations X, Y, and Z:
Are They Changing America?

DUANE F. ALWIN

What creates social change? Is it the values of young people who bring new perspectives and new issues to society? Or is it the influence of major historical events? In this article, Duane Alwin examines generational sources of social change, arguing that historical events and shifts in the individual lives because of aging are both sources of social change.

The Greatest Generation saved the world from fascism. The Dr. Spock Generation gave us rebellion and free love. Generation X made cynicism and slacking off the hallmarks of the end of the 20th century. In the media, generation is a popular and all-purpose explanation for change in America. Each new generation replaces an older one's zeitgeist with its own.

Generational succession is increasingly a popular explanation among scholars, too. Recently, political scientist Robert Putnam argued in *Bowling Alone* that civic engagement has declined in America even though individual Americans have not necessarily become less civic minded. Instead, he argues, older engaged citizens are dying off and being replaced by younger, more alienated Americans who are less tied to institutions such as the church, lodge, political party and bowling league.

Next to characteristics like social class, race, and religion, generation is probably the most common explanatory tool used by social scientists to account for differences among people. The difficulties in proving such explanations, however, are not always apparent and are often overshadowed by the seductiveness of the idea. Generational arguments do not always hold the same allure once they are given closer scrutiny.

Changes in the worldviews of Americans result not only from the progression of generations but also from historical events and patterns of aging. For example, generational replacement seems to explain why fewer Americans now than 30 years ago say they trust other people, but historical events seem to explain why fewer say they trust government. Similarly, historical events in interaction with aging (or life cycle change) may explain lifetime changes in church attendance and political partisanship better than generational shifts.

From: Alwin, Duane F. "Generations X, Y, and Z: Are They Changing America?"
Contexts 1 (Winter 2002): 42–50.

EXPLAINING SOCIAL CHANGE

Some rather massive changes over the past 50 years in Americans' attitudes need explaining. Consider this short list of examples:

- In 1977, 66 percent said that it is better if the man works and the woman stays home; in 2000, only 35 percent did.

- In 1972, 48 percent said that sex before marriage is wrong; in 2000, 36 percent did.

- In 1972, 39 percent said that there should be a law against interracial marriage; in 2000, 12 percent did.

- In 1958, 78 percent said that one could trust the government in Washington to do right; in 2000 only 44 percent did.

Do changes in beliefs and behaviors reflect the experiences of specific generations, do they occur when Americans of all ages change their orientations, or do they result from something else? Although the idea of generational succession is promising and useful, it also has problems that may limit it as an all-purpose explanation of social change.

SOME PRELIMINARIES

Before we begin to deconstruct the idea of generational replacement we need to clarify a few issues. The first is that when sociologists use the term *generation* it can refer to one of three quite different things:

1. All people born at the same time.
2. A unique position within a family's line of descent (as in the second generation of Bush presidents).
3. A group of people self-consciously defined, by themselves and by others, as part of an historically based social movement (as in the "hippie" generation).

There are many examples in the social science literature of all three uses, and this can create a great deal of confusion. Here I refer mainly to the first use, measuring generation by year of birth. Demographers prefer to use the term cohort. Either way the reference is to the historical period in which people grow to maturity. I use the terms cohort and generation more or less interchangeably.

When sociologists are discussing social change in less precise terms, they may refer to generations in a somewhat more nuanced, cultural sense. Generations in this usage do not necessarily map neatly to birth years. Rather, the distinction between generations is a matter of quality, not degree, and their exact time boundaries cannot always be easily identified. It is also clear that statistically there is no way to identify cohort or generation effects unequivocally. The interpretation of generational differences depends entirely on one's ability or willingness to make some rather hefty assumptions about other processes, such as how aging affects attitudes, but as we shall see, we can nonetheless develop reasonable conclusions.

COHORT REPLACEMENT

Cohort replacement is a fact of social life. Earlier-born cohorts die off and are replaced by those born more recently. The question is: do the unique formative experiences of cohorts become distinctively imprinted onto members' worldviews, making them distinct generations over the course of their lifetimes, or do people of all cohorts adapt to change, remaining pliable in their beliefs throughout their lives?

When historical events mainly affect the young, we have the makings of a generation. Such an effect—labeled a cohort effect—refers to the outcomes attributable to having been born in a particular historical period. When, for example, people describe the Depression generation as particularly thrifty, they imply that the experience of growing up under privation permanently changed the economic beliefs and style of life of people who grew to maturity during that time.

Unique events that happen during youth are no doubt powerful. Certainly, some eras and social movements, like the Civil Rights era and the women's movement, or some new ideologies (e.g. Roosevelt's New Deal) provide distinctive experiences for youth during particular times. As Norman Ryder put it, "the potential for change is concentrated in the cohorts of young adults who are old enough to participate directly in the movements impelled by change, but not old enough to have become committed to an occupation, a residence, a family of procreation or a way of life."

To some observers, today's younger generations—Generation X and its younger counterpart—display a distinctive lack of social commitment. The goals of individualism and the good life have replaced an earlier generation's involvement in social movements and organizations. Is this outlook simply part of being young, or is it characteristic of a particular generation?

Each generation resolves issues of identity in its own way. In the words of analyst Erik Erikson, "No longer is it merely for the old to teach the young the meaning of life . . . it is the young who, by their responses and actions, tell the old whether life as represented by the old and presented to the young has meaning; and it is the young who carry in them the power to confirm those who confirm them and, joining the issues, to renew and to regenerate or to reform and to rebel." . . .

Before we accept this way of understanding change, however, we should consider other possibilities. One is that people change as they get older, which we call an effect of aging. The older people get, for example, the more intensely they may hold to their views. America, as a whole, may be becoming more politically partisan because the population is getting older—an age effect. Another possibility is that people change in response to specific historical events, what sociologists call period effects. The Civil Rights movement, for example, may have changed many Americans' ideas about race, not just the views of the generation growing up in the 1960s. The events of September 11, 2001, likely had an effect on the entire nation, not just those in the most impressionable years of youth.

A third possibility is that the change is located in only one segment of society. Members of the Roman Catholic faith, for example, may be the most responsive to the current turmoil over the sexual exploitation of youth by some priests in ways that hardly touch the lives of Protestants. Let us weigh these possibilities more closely, looking at the issues raised by Putnam in *Bowling Alone.*

CHANGES IN SOCIAL CONNECTEDNESS AND TRUST

It is often relatively easy to construct a picture of generational differences by comparing data from different age groups in social surveys and polls, but determining what produced the data is considerably more complex.

Take, for example, one of the key empirical findings of Putnam's analysis: the responses people give to the question of whether they trust their fellow human beings. The General Social Survey (administered regularly to a nationwide, representative sample of American residents since 1972) asks the following question: "Do you think most people would try to take advantage of you if they got the chance, or would they try to be fair?" Figure 1 . . . presents the percentage of respondents in each set of cohorts who responded that people would try to be fair. The results show that birth cohorts were consistently different from one another, the recent ones being more cynical about human nature. There has been little change in this outcome over the years except insofar as new generations replaced older ones. These results reinforce the Putnam thesis, that the degree of social connectedness in the formative years of people's generation shapes their sense of trust.

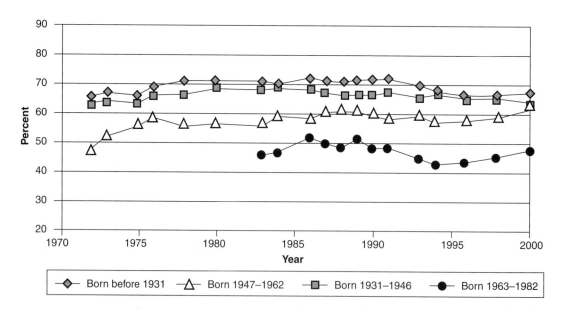

FIGURE 1 Percent of U.S. Population Who Believe "Most People Try to Be Fair"

Still, I would note some problems with these conclusions. First, generational experiences are not the only factors that differentiate these four groups, they also differ by age. Second, these data do not depict the young lives of the cohorts born before 1930 (who were 42 years of age or older in 1972), so we have little purchase on their beliefs before 1972. Third, there is remarkable growth in trust among the Baby Boom cohorts—those born from 1947 to 1962—over their midlife period, and in 2000 they had achieved a level of trust on a par with earlier cohorts. Finally, even the most recent cohorts (the lowest line in the figure) show some tendency to gain trust in recent years. The point is that while the data appear to show a pattern of generational differences—less trust among more recent cohorts—age might be just as plausible an explanation of the differences: trust goes up as people mature.

There may be more than one way to explain changes in Americans' trust of people, but generations do not explain changes in Americans' trust of government. In 1958 the National Election Studies (NES) began using the following question: "How much of the time do you think you can trust the government in Washington to do what is right—just about always, most of the time, or only some of the time?"

There are two important things to note about Figure 2. First, there are hardly any differences among birth cohorts who say most of the time or always in their responses to this question; the lines are virtually identical. Thus, generational replacement explains none of the very dramatic decline of trust in government. That decline may be better explained by historical events that affected all cohorts—the Vietnam War, the feminist movement, or the Watergate and White-water scandals—and there is little basis for arguing that more recent cohorts are more alienated from government than those born earlier. (Note that affirmations of trust in government rose dramatically right after 9/11.) . . .

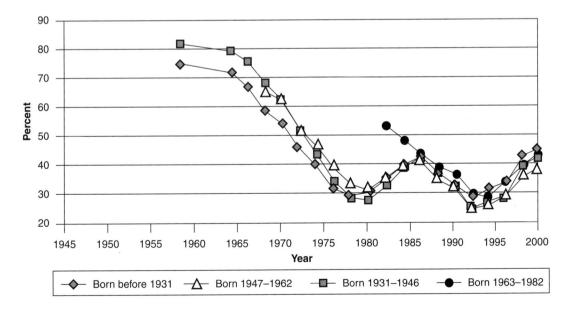

FIGURE 2 Percent of U.S. Population Who Believe "You Can Trust the Government in Washington to Do What Is Right"

GENERATIONS AND SOCIAL CHANGE

Society reflects, at any given time, the sum of its generations. Where one set of cohorts is especially large—like the Baby Boomers—its lifestyle dominates the society as it passes through the life course. Baby Boomers' taste in music and clothes, for example, disproportionately influence the whole culture. However, in cases where there are no major differences among generations (as in the example of trust in government), then generational succession cannot explain social change.

Where generations persistently differ, however, their succession will produce social change. Certainly, if the more recent generations have less affiliation and involvement with traditional religious groups, this will lead to social change, at least until they develop their own form of religiosity.

Because of the Baby Boomer generation's sheer size, its liberal positions on political and social issues will probably shape beliefs and behavior well into the new century, as Boomers replace the generations that came before. But even here, the Baby Boomers' distinctiveness may wane under the influence of historical events and processes of aging. Baby Boomers, for example, may be growing more conservative with age. This argues in favor of an alternative to the generational view: Generations do not necessarily differ in the same ways over time; individuals are not particularly consistent over their lives; and social change results as much from shifts in individual lives due either to aging or historical events. . . .

The existence of generation effects may depend very much on when one takes the snapshot of generational differences, and how generations differ may depend on which groups in society one examines. All fair warnings for the next essay you read on Generations X, Y or Z.

DISCUSSION QUESTIONS

1. What is a cohort effect and how is it significant to the process of social change? What social changes are the result of cohort effects among your age generation?

2. As you imagine the future, what social changes do you think will result from the various sources of generational change that Alwin identifies?

62

Jihad vs. McWorld

BENJAMIN R. BARBER

The author compares and contrasts two countervailing sources of social change: Jihad and McWorld. Jihad is a traditionalist, anti-modern worldview (now especially familiar because of the post-9/11 social and political context). McWorld, on the other hand, is a world increasingly driven by market values and the commercialization of all forms of life. Though they seem like opposites, Barber shows how they produce each other and result from common forces of social change.

. . . [A]nyone who reads the daily papers carefully, taking in the front page accounts of civil carnage as well as the business page stories on the mechanics of the information superhighway and the economics of communication mergers, anyone who turns deliberately to take in the whole 360-degree horizon, knows that our world and our lives are caught between what William Butler Yeats called the two eternities of race and soul: that of race reflecting the tribal past, that of soul anticipating the cosmopolitan future. Our secular eternities are corrupted, however, race reduced to an insignia of resentment, and soul sized down to fit the demanding body by which it now measures its needs. Neither race nor soul offers us a future that is other than bleak, neither promises a polity that is remotely democratic.

The first scenario rooted in race holds out the grim prospect of a retribalization of large swaths of humankind by war and bloodshed: a threatened balkanization of nation-states in which culture is pitted against culture, people against people, tribe against tribe, a Jihad in the name of a hundred narrowly conceived faiths against every kind of interdependence, every kind of artificial social cooperation and mutuality: against technology, against pop culture, and against integrated markets; against modernity itself as well as the future in which modernity issues. The second paints that future in shimmering pastels, a busy portrait of onrushing economic, technological, and ecological forces that demand integration and uniformity and that mesmerize peoples everywhere with fast music, fast computers, and fast food—MTV, Macintosh, and McDonald's—pressing nations into one homogenous global theme park, one McWorld tied together by communications, information, entertainment, and commerce. Caught between Babel and Disneyland, the planet is falling precipitously apart and coming reluctantly together at the very same moment.

Some stunned observers notice only Babel, complaining about the thousand newly sundered "peoples" who prefer to address their neighbors with sniper rifles and mortars; others—zealots in Disneyland—seize on futurological platitudes and the promise of virtuality, exclaiming "It's a small world after all!" Both are right, but how can that be?

We are compelled to choose between what passes as "the twilight of sovereignty" and an entropic end of all history; or a return to the past's most fractious and demoralizing discord; to "the menace of global anarchy," to Milton's capital of hell, Pandemonium; to a world totally "out of control."

The apparent truth, which speaks to the paradox at the core of this argument, is that the tendencies of both Jihad *and* McWorld are at work, both visible sometimes in the same country at the very same instant. Iranian zealots keep one ear tuned to the mullahs urging holy war and the other cocked to Rupert Murdoch's Star television beaming in *Dynasty, Donahue* and *The Simpsons* from hovering satellites. Chinese entrepreneurs vie for the attention of party cadres in Beijing and simultaneously pursue KFC franchises in cities like Nanjing, Hangzhou, and Xian where twenty-eight outlets serve over 100,000 customers a day. The Russian Orthodox church, even as it struggles to renew the ancient faith, has entered a joint venture with California businessmen to bottle and sell natural waters under the rubric Saint Springs Water Company. Serbian assassins wear Adidas sneakers and listen to Madonna on Walkman headphones as they take aim through their gunscopes at scurrying Sarajevo civilians looking to fill family watercans. Orthodox Hasids and brooding neo-Nazis have both turned to rock music to get their traditional messages out to the new generation, while fundamentalists plot virtual conspiracies on the Internet.

. . . It is not Jihad and McWorld but the relationship between them that most interests me. For, squeezed between their opposing forces, the world has been sent spinning out of control. Can it be that what Jihad and McWorld have in common is anarchy: the absence of common will and that conscious and collective human control under the guidance of law we call democracy?

Progress moves in steps that sometimes lurch backwards; in history's twisting maze, Jihad not only revolts against but abets McWorld, while McWorld not only imperils but re-creates and reinforces Jihad. They produce their contraries and need one another. . . .

What then does it mean in concrete terms to view Jihad and McWorld dialectically when the tendencies of the two sets of forces initially appear so intractably antithetical? After all, Jihad and McWorld operate with equal strength in opposite directions, the one driven by parochial hatreds, the other by universalizing markets, the one re-creating ancient subnational and ethnic borders from within, the other making national borders porous from without. Yet Jihad and McWorld have this in common: they both make war on the sovereign nation-state and thus undermine the nation-state's democratic institutions. Each eschews civil society and belittles democratic citizenship, neither seeks alternative democratic institutions. Their common thread is indifference to civil liberty. Jihad forges communities of blood rooted in exclusion and hatred, communities that slight democracy in favor of tyrannical paternalism or consensual tribalism.

McWorld forges global markets rooted in consumption and profit, leaving to an untrustworthy, if not altogether fictitious, invisible hand issues of public interest and common good that once might have been nurtured by democratic citizenries and their watchful governments. Such governments, intimidated by market ideology, are actually pulling back at the very moment they ought to be aggressively intervening. What was once understood as protecting the public interest is now excoriated as heavy-handed regulatory browbeating. Justice yields to markets, even though, as Felix Rohatyn has bluntly confessed, "there is a brutal Darwinian logic to these markets. They are nervous and greedy. They look for stability and transparency, but what they reward is not always our preferred form of democracy." If the traditional conservators of freedom were democratic constitutions and Bills of Rights, "the new temples to liberty," George Steiner suggests, "will be McDonald's and Kentucky Fried Chicken.". . .

Jihad is, I recognize, a strong term. In its mildest form, it betokens religious struggle on behalf of faith, a kind of Islamic zeal. In its strongest political manifestation, it means bloody holy war on behalf of partisan identity that is metaphysically defined and fanatically defended. Thus, while for many Muslims it may signify only ardor in the name of a religion that can properly be regarded as universalizing (if not quite ecumenical), I borrow its meaning from those militants who make the slaughter of the "other" a higher duty. I use the term in its militant construction to suggest dogmatic and violent particularism of a kind known to Christians no less than Muslims, to Germans and Hindis as well as to Arabs. The phenomena to which I apply the phrase have innocent enough beginnings: identity politics and multicultural diversity can represent strategies of a free society trying to give expression to its diversity. What ends as Jihad may begin as a simple search for a local identity, some set of common personal attributes to hold out against the numbing and neutering uniformities of industrial modernization and the colonizing culture of McWorld. . . .

. . . Jihad is then a rabid response to colonialism and imperialism and their economic children, capitalism and modernity; it is diversity run amok, multiculturalism turned cancerous so that the cells keep dividing long after their division has ceased to serve the healthy corpus.

Even traditionally homogenous integral nations have reason to feel anxious about the prospect of Jihad. The rising economic and communications interdependence of the world means that such nations, however unified internally, must nonetheless operate in an increasingly multicultural global environment. Ironically, a world that is coming together pop culturally and commercially is a world whose discrete subnational ethnic and religious and racial parts are also far more in evidence, in no small part as a reaction to McWorld. . . .

THE SMALLING WORLD OF MCWORLD

. . . Every demarcated national economy and every kind of public good is today vulnerable to the inroads of transnational commerce. Markets abhor frontiers as nature abhors a vacuum. Within their expansive and permeable domains, interests are private, trade is free, currencies are convertible, access to banking is open, contracts are enforceable (the state's sole legitimate economic function), and the laws of production and consumption are sovereign, trumping the laws of legislatures and courts. In Europe, Asia, and the Americas such markets have already eroded national sovereignty and given birth to a new class of institutions—international banks, trade associations, transnational lobbies like OPEC, world news services like CNN and the BBC, and multinational corporations—institutions that lack distinctive national identities and neither reflect nor respect nationhood as an organizing or a regulative principle. . . .

McWorld is a product of popular culture driven by expansionist commerce. Its template is American, its form style. Its goods are as much images as material, an aesthetic as well as a product line. It is about culture as commodity, apparel as ideology. Its symbols are Harley-Davidson motorcycles and Cadillac motorcars hoisted from the roadways, where they once represented a mode of transportation, to the marquees of global market cafés like Harley-Davidson's and the Hard Rock where they become icons of lifestyle. You don't drive them, you feel their vibes and rock to the images they conjure up from old movies and new celebrities, whose personal appearances are the key to the wildly popular international café chain Planet Hollywood. Music, video, theater, books, and theme parks—the new churches of a commercial civilization in which malls are the public squares and suburbs the neighborless neighborhoods—are all constructed as image exports creating a common world taste around common logos, advertising slogans, stars, songs, brand names, jingles, and trademarks. Hard power yields to soft, while ideology is transmuted into a kind of videology that works through sound bites and film clips. Videology is fuzzier and less dogmatic than traditional political ideology: it may as a consequence be far more successful in instilling the novel values required for global markets to succeed. . . .

The dynamics of the Jihad-McWorld linkage are deeply dialectical. Japan has, for example, become more culturally insistent on its own traditions in recent years even as its people seek an ever greater purchase on McWorld. In 1992, the number-one restaurant in Japan measured by volume of customers was McDonald's, followed in the number-two spot by the Colonel's Kentucky Fried Chicken. . . .

In Russia, in India, in Bosnia, in Japan, and in France too, modern history then leans both ways: toward the meretricious inevitability of McWorld, but also into Jihad's stiff winds, heaving to and fro and giving heart both to the Panglossians and the Pandoras, sometimes for the very same reasons. The Panglossians bank on Euro-Disney and Microsoft, while the Pandoras await nihilism and a world in Pandemonium. Yet McWorld and Jihad do not really force a choice between such polarized scenarios. Together, they are likely to produce some stifling amalgam of the two suspended in chaos. Antithetical in every detail, Jihad and McWorld nonetheless conspire to undermine our hard-won (if only half-won) civil liberties and the possibility of a global democratic future. In the short run the forces of Jihad, noisier and more obviously nihilistic than those of McWorld, are likely to dominate the near future, etching small stories of local tragedy and regional genocide on the face of our times and creating a climate of instability marked by multimicrowars inimical to global integration. But in the long run, the forces of McWorld are the forces underlying the slow certain thrust of Western civilization and as such may be unstoppable. Jihad's microwars will hold the headlines well into the next century, making predictions of the end of history look terminally dumb. But McWorld's homogenization is likely to establish a macropeace that favors the triumph of commerce and its markets and to give to those who control information, communication, and entertainment ultimate (if inadvertent) control over human destiny. Unless we can offer an alternative to the struggle between Jihad and McWorld, the epoch on whose threshold we stand—postcommunist, postindustrial, postnational, yet sectarian, fearful, and bigoted—is likely also to be terminally postdemocratic.

DISCUSSION QUESTIONS

1. What does Barber mean when he says that what Jihad and McWorld may have in common is anarchy?
2. Why does Barber conclude that modern history leans both toward Jihad and McWorld and what does he see as the potential implications of this?

63

The Genius of the Civil Rights Movement

Can It Happen Again?

ALDON MORRIS

The civil rights movement was arguably the most influential movement in the United States during the twentieth century. Aldon Morris reviews the development of the civil rights movements and notes the products of this movement, including the mobilization of other national and international movements and the transformations in academic scholarship that the movement generated. By identifying the particular historical and social

From: Morris, Aldon. "The Genius of the Civil Rights Movement: Can It Happen Again?" 2001. Northwestern University. Reprinted by permission of author.

circumstances in which the civil rights movement developed, he also asks whether such a movement is possible again.

It is important for African Americans, as well as all Americans, to take a look backward and forward as we approach the turn of a new century, indeed a new millennium. When a panoramic view of the entire history of African Americans is taken into account, it becomes crystal clear that African American social protest has been crucial to Black liberation. In fact, African American protest has been critical to the freedom struggles of people of color around the globe and to progressive people throughout the world.

The purpose of this essay is: (1) to revisit the profound changes that the modern Black freedom struggle has achieved in terms of American race relations; (2) to assess how this movement has affected the rise of other liberation movements both nationally and internationally; (3) to focus on how this movement has transformed how scholars think about social movements; (4) to discuss the lessons that can be learned from this groundbreaking movement pertaining to future African American struggles for freedom in the next century.

It is hard to imagine how pervasive Black inequality would be today in America if it had not been constantly challenged by Black protests throughout each century since the beginning of slavery. The historical record is clear that slave resistance and slave rebellions and protest in the context of the Abolitionist movement were crucial to the overthrow of the powerful slave regime.

The establishment of the Jim Crow regime was one of the great tragedies of the late nineteenth and early twentieth centuries. The overthrow of slavery represented one of those rare historical moments where a nation had the opportunity to embrace a democratic future or to do business as usual by reinstalling undemocratic practices. In terms of African Americans, the White North and South chose to embark along undemocratic lines.

For Black people, the emergence of the Jim Crow regime was one of the greatest betrayals that could be visited upon a people who had hungered for freedom so long; what made it even worse for them is that the betrayal emerged from the bosom of a nation declaring to all the world that it was the beacon of democracy.

The triumph of Jim Crow ensured that African Americans would live in a modern form of slavery that would endure well into the second half of the twentieth century. The nature and consequences of the Jim Crow system are well known. It was successful in politically disenfranchising the Black population and in creating economic relationships that ensured Black economic subordination. Work on wealth by sociologists Melvin Oliver and Thomas Shapiro (1995), as well as Dalton Conley (1999), are making clear that wealth inequality is the most drastic form of inequality between Blacks and Whites. It was the slave and Jim Crow regimes that prevented Blacks from acquiring wealth that could have been passed down to succeeding generations. Finally, the Jim Crow regime consisted of a comprehensive set of laws that stamped a badge of inferiority on Black people and denied them basic citizenship rights.

The Jim Crow regime was backed by the iron fist of southern state power, the United States Supreme Court, and white terrorist organizations. Jim Crow was also held in place by white racist attitudes. As Larry Bobo has pointed out, "The available survey data suggests that anti-Black attitudes associated with Jim Crow were once widely accepted . . . [such attitudes were] expressly premised on the notion that Blacks were the innately intellectual, cultural, and temperamental inferior group relative to Whites (Bobo, 1997:35)." Thus, as the twentieth century opened, African Americans were confronted with a powerful social order designed to keep them subordinate. As long as the Jim Crow order remained intact, the Black masses could breathe neither freely nor safely. Thus, nothing less than the overthrow of a social order was the daunting task that faced African Americans during the early decades of the twentieth century.

The voluminous research on the modern civil rights movement has reached a consensus: That movement was the central force that toppled the Jim Crow regime. To be sure, there were other factors that assisted in the overthrow including the advent of the television age, the competition for Northern Black votes between the two major parties, and the independence movement in Africa which sought to overthrow European domination. Yet it was the Civil Rights movement itself that targeted the Jim Crow regime and generated the great mass mobilizations that would bring it down.

What was the genius of the Civil Rights movement that made it so effective in fighting a powerful and vicious opposition? The genius of the Civil Rights movement was that its leaders and participants recognized that change could occur if they were able to generate massive crises within the Jim Crow order—crises of such magnitude that the authorities of oppression must yield to the demands of the movement to restore social order. Max Weber defined power as the ability to realize one's will despite resistance. Mass disruption generated power. That was the strategy of nonviolent direct action. By utilizing tactics of disruption, implemented by thousands of disciplined demonstrators who had been mobilized through their churches, schools, and voluntary associations, the Civil Rights movement was able to generate the necessary power to overcome the Jim Crow regime. The famous crises created in places like Birmingham and Selma, Alabama, coupled with the important less visible crises that mushroomed throughout the nation, caused social breakdown in Southern business and commerce, created unpredictability in all spheres of social life, and strained the resources and credibility of Southern state governments while forcing white terrorist groups to act on a visible stage where the whole world could watch. At the national level, the demonstrations and repressive measures used against them generated foreign policy nightmares because they were covered by foreign media in Europe, the Soviet Union, and Africa. Therefore what gave the mass-based sit-ins, boycotts, marches, and jailings their power was their ability to generate disorder.

As a result, within ten years—1955 to 1965—the Civil Rights movement had toppled the Jim Crow order. The 1964 Civil Rights Bill and the 1965 Voting Rights Act brought the regime of formal Jim Crow to a close.

The Civil Rights movement unleashed an important social product. It taught that a mass-based grass roots social movement that is sufficiently organized, sustained, and disruptive is capable of generating fundamental social change. In other words, it showed that human agency could flow from a relatively powerless and despised group that was thought to be backward, incapable of producing great leaders.

Other oppressed groups in America and around the world took notice. They reasoned that if American Blacks could generate such agency they should be able to do likewise. Thus the Civil Rights movement exposed the agency available to oppressed groups. By agency I refer to the empowering beliefs and action of individuals and groups that enable them to make a difference in their own lives and in the social structures in which they are embedded.

Because such agency was made visible by the Civil Rights movement, disadvantaged groups in America sought to discover and interject their agency into their own movements for social change. Indeed, movements as diverse as the Student movement, the Women's movement, the Farm Worker's movement, the Native American movement, the Gay and Lesbian movement, the Environmental movement, and the Disability Rights movement all drew important lessons and inspiration from the Civil Rights movement. From that movement other groups discovered how to organize, how to build social movement organizations, how to mobilize large numbers of people, how to devise appropriate tactics and strategies, how to infuse their movement activities with cultural creativity, how to confront and defeat authorities, and how to unleash the kind of agency that generates social change.

For similar reasons, the Black freedom struggle was able to effect freedom struggles internationally. For example, nonviolent direct action has inspired oppressed groups as diverse as Black South Africans, Arabs of the Middle East, and pro-democracy demonstrators in China to engage in collective actions. The sit-in tactic made famous by the Civil Rights movement, has been used in liberation movements throughout the third world, in Europe, and in many other foreign countries. The Civil Rights movement's national anthem "We Shall Overcome" has been interjected into hundreds of liberation movements both nationally and internationally. Because the Civil Rights movement has been so important to international struggles, activists from around the world have invited civil rights participants to travel abroad. Thus early in Poland's Solidarity movement Bayard Rustin was summoned to Poland by that movement. As he taught the lessons of the Civil Rights movement, he explained that "I am struck by the complete attentiveness of the predominantly young audience, which sits patiently, awaiting the translations of my words." (Rustin, undated)

Therefore, as we seek to understand the importance of the Black Freedom Struggle, we must conclude the following: the Black Freedom Struggle had provided a model and impetus for social movements that have exploded on the American and international landscapes. This impact has been especially pronounced in the second half of the twentieth century.

What is less obvious is the tremendous impact that the Black Freedom Struggle has had on the scholarly study of social movements. Indeed, the Black freedom struggle has helped trigger a shift in the study of social movements and collective action. The Black movement has provided scholars with profound empirical and theoretical puzzles because it has been so rich organizationally and tactically and because it has generated unprecedented levels of mobilization. Moreover, this movement has been characterized by a complex leadership base, diverse gender roles, and it has revealed the tremendous amount of human agency that usually lies dormant within oppressed groups. The empirical realities of the Civil Rights movement did not square with the theories used by scholars to explain social movements prior to the 1960s.

Previous theories did not focus on the organized nature of social movements, the social movement organizations that mobilize them, the tactical and strategic choices that make them effective, nor the rationally planned action of leaders and participants who guide them. In the final analysis, theories of social movements lacked a theory that incorporated human agency at the core of their conceptual apparatuses. Those theories conceptualized social movements as spontaneous, largely unstructured, and discontinuous with institutional and organizational behavior. Movement participants were viewed as reacting to various forms of strain and doing so in a non-rational manner. In these frameworks, human agency was conceptualized as reactive, created by uprooted individuals seeking to reestablish a modicum of personal and social stability. In short, social movement theories prior to the Civil Rights movement operated with a vague, weak vision of agency to explain phenomena that are driven by human action.

The predictions and analytical focus of social movement theories prior to the 1970s stood in sharp contrast to the kind of theories that would be needed to capture the basic dynamics that drove the Civil Rights movement. It became apparent to social movement scholars that if they were to understand the Civil Rights movement and the multiple movements it spun, the existing theoretical landscape would have to undergo a radical process of reconceptualization.

As a result, the field of social movements has been reconceptualized and this retheoritization will affect research well into the new millennium. To be credible in the current period any theory of social movements must grapple conceptually with the role of rational planning and strategic action, the role of movement leadership, and the nature of the mobilization process. How movements are gendered, how movement dynamics are bathed in cultural creativity, and how the interactions between movements and their opposition determine movement outcomes are important questions. At the center of this entire matrix of factors must be an analysis of the central role that human agency plays in social movements and in the generation of social change.

Thanks, in large part, to the Black freedom struggle, theories of social movements that grapple with real dynamics in concrete social movements are being elaborated. Intellectual work in the next century will determine how successful scholars will be in unraveling the new empirical and theoretical puzzles thrust forth by the Black freedom movement. Although it was not their goal, Black demonstrators of the Civil Rights movement changed an academic discipline.

A remaining question is: Will Black protest continue to be vigorous in the twenty-first century, capable of pushing forward the Black freedom agenda? It is not obvious that Black protest will be as sustainable and as paramount as it has been in previous centuries. To address this issue we need to examine the factors important to past protests and examine how they are situated in the current context.

Social movements are more effective when they can identify a clear-cut enemy. Who or what is the clear-cut enemy of African Americans of the twenty-first century? Is it racism, and if so, who embodies it? Is it capitalism, and if so, how is this enemy to be loosened from its abstract perch and concretized? In fact, we do not currently have a robust concept that grasps the modern form of domination that Blacks currently face. Because the modern enemy has become opaque, slippery, illusive, and covert, the launching of Black protest has become more difficult because of conceptual fuzziness.

Second, during the closing decades of the twentieth century the Black class structure has become more highly differentiated and it is no longer firmly anchored in the Black community. There is some danger, therefore, that the cross fertilization between different strata within the Black class structure so important to previous protest movements may have become eroded to the extent that it is no longer fully capable of launching and sustaining future Black protest movements.

Third, will the Black community of the twenty-first century possess the institutional strength required for sustaining Black protest? Black colleges have been weakened because of the racial integration of previously all white institutions of higher learning and because many Black colleges are being forced to integrate. The degree of institutional strength of the church has eroded because some of them have migrated to the suburbs in an attempt to attract affluent Blacks. In other instances, the Black Church has been unable to attract young people of the inner city who find more affinity with gangs and the underground economy. Moreover, a great potential power of the Black church is not being realized because its male clergy refuse to empower Black women as preachers and pastors. The key question is whether the Black church remains as close to the Black masses—especially to poor and working classes—as it once was. That closeness determines its strength to facilitate Black protest.

In short, research has shown conclusively that the Black church, Black colleges and other Black community organizations were critical vehicles through which social protest was organized, mobilized and sustained. A truncated class structure was also instrumental to Black protest. It is unclear whether during the twenty-first century these vehicles will continue to be effective tools of Black protest or whether new forces capable of generating protest will step into the vacuum.

In conclusion, I foresee no reason why Black protest should play a lesser role for Black people in the twenty-first century. Social inequality between the races will continue and may even worsen especially for poorer segments of the Black communities. Racism will continue to effect the lives of all people of color. If future changes are to materialize, protest will be required. In 1898 as Du Bois glanced toward the dawn of the twentieth century, he declared that in order for Blacks to achieve freedom they would have to protest continuously and energetically. This will become increasingly true for the twenty-first century. The question is whether organizationally, institutionally, and intellectually the Black community will have the wherewithal to engage in the kind of widespread and effective social protest that African Americans have utilized so magnificently. If previous centuries are our guide, then major surprises on the protest front should be expected early in the new millennium.

REFERENCES

Bobo, L. 1997. "The Color Line, the Dilemma, and the Dream: Race Relations in America at the Close of the Twentieth Century." In *Civil Rights and Social Wrongs: Black-White Relations since World War II,* edited by J. Higham, pp. 31–55. University Park, PA: Penn State University Press.

Conley, Dalton. 1999. *Being Black, Living in the Red: Race, Wealth, and Social Policy in America.* Berkeley: University of California Press.

Oliver, Melvin, and Thomas E. Shapiro. 1995. *Black Wealth/White Wealth: A New Perspective on Racial Inequality.* New York: Routledge.

Rustin, Bayard. no date. *Report on Poland.* New York: A. Philip Randolph Institute.

DISCUSSION QUESTIONS

1. What does Morris mean by "the genius of the Civil Rights movement?" Can you imagine such a strategy being an effective means of combating the oppression of racial groups today? If so, how; if not, why not?

2. What does Morris identify as the products of the civil rights movement? What does this teach you about the connections between contemporary social movements and the civil rights movement?

http://infotrac.thomsonlearning.com

InfoTrac College Edition

BONUS READING

Edwords, Fred. "How Biotechnology Is Transforming What We Believe In and How We Live." *The Humanist* 59 (September 1999): 23–29.

Not many years ago, biotechnology, the cloning of human beings, and the genetic alteration of food would have seemed like science fiction. Now such scientific possibilities are no longer imaginary. These new scientific technologies can potentially seriously alter how we live. What potential changes are identified by the author and what new ethical issues are raised by the possibility of these changes?

SEARCH TERMS

Black power movement
generational change
social activism
social change
world future

Glossary

achieved status a status attained by effort

age cohort an aggregate group of people born during the same time period

alienation the feeling of powerlessness and separation from one's group or society

anomie a social structural condition existing when social regulations (norms) in a society breakdown

anti-Semitism hostility toward or discrimination against Jewish people

ascribed status a status determined by birth

assimilation process by which a minority becomes socially, economically, and culturally absorbed within the dominant society

authority power that is perceived by others as legitimate

beliefs shared ideas held collectively by people within a given culture

bigotry the state of mind of people intolerantly devoted to their own opinions and prejudices

block grant a large amount of money allocated by the federal government to a state government for a specific purpose

boundary maintenance the act(s) of preserving something that indicates or fixes a limit into group membership

bourgeoisie term used to loosely describe either the ruling or middle class in a capitalist society

bureaucracy a type of formal organization characterized by an authority hierarchy, with a clear division of labor, explicit rules, and impersonality

capital accumulated goods devoted to the production of other goods

capitalism an economic system based on the principles of market competition, private property, and the pursuit of profit

capitalist class those persons who own the means of production in a society

caste system a system of stratification (characterized by low social mobility) in which one's place in the stratification system is determined and fixed by birth

chauvinism an attitude of superiority toward members of the opposite sex; also, behavior expressive of such an attitude

chi-square   a test statistic used to determine the statistical significance of a relationship between two variables

class   see social class

classism   prejudice or discrimination based on social class

clique   a narrow, exclusive circle or group of persons, especially one held together by common interests, views, or purposes

cohort effect   the outcomes attributable to having been born in a particular historical period

collective consciousness   the body of beliefs that are common to a community or society and that give people a sense of belonging

coming out process   the process of defining oneself in terms of some prior latent or "secret" role, such as publicly defining oneself as gay or lesbian

commodity chain   the network of production and labor processes by which a product becomes a finished commodity

communism   an economic system where the state is the sole owner of the systems of production

conflict theory   a theoretical perspective that emphasizes the role of power and coercion in producing social order

consumerism   a preoccupation with and inclination toward the buying of consumer goods

contingent worker   a temporary worker

correlation   a statistical technique that analyzes patterns of association between pairs of sociological variables

counterculture   subculture created as a reaction against the values of the dominant culture

cult   a religious group devoted to a specific cause or charismatic leader

cultural imperialism   forcing all groups in a society to accept the dominant culture as their own

culture   the complex system of meaning and behavior that defines the way of life for a given group or society

culture lag   the delay in cultural adjustments to changing social conditions

culture of poverty   the argument that poverty is a way of life and, like other cultures, is passed on from generation to generation

cyberspace interaction   interaction occurring when two or more persons share a virtual reality experience via communication and interaction with each other

data   the systematic information that is used to investigate research questions

data set   a large collection of systematic information used in sociological research

decentralization   the redistribution of population and industry from urban centers to outlying areas

de facto   by practice or in fact, even if not in law

de jure   by rule, by policy, or by law

deindustrialization   term referring to the structural transformation of a society from a manufacturing-based economy to a service-based economy

deterrence   a punishment intended to prevent or discourage the behavior being punished

deviance   behavior that is recognized as violating expected rules and norms

discrimination   overt negative and unequal treatment of the members of some social group or stratum solely because of their membership in that group or stratum

division of labor   the systematic interrelation of different tasks that develops in complex societies

dominant culture   the culture of the most powerful group in society

double consciousness   having the realization that one is being viewed as the "other" in a social situation, and also being able to perceive and understand the dominant group

downsizing   term referring to action by companies to eliminate job positions in order to cut the firm's operating costs

dramatic metaphor   a perspective used to suggest that social interaction has a likeness to dramas presented on a stage

dual labor market   theoretical description of the occupational system defining it as divided into two major segments: the primary labor market and the secondary labor market

dyad a group consisting of two people

economic restructuring contemporary transformations in the basic structure of work that are permanently altering the workplace, including the changing composition of the workplace, deindustrialization, the use of enhanced technology, and the development of a global economy

economy the system on which the production, distribution, and consumption of goods and services is based

egalitarian societies or groups where men and women share power

emotional labor (or management) work intended to produce a desired emotional effect on a client

endogamy the practice of selecting mates from within one's group

Enlightenment the period in eighteenth- and nineteenth-century Europe characterized by faith in the ability of human reason to solve society's problems

environmental racism phrase describing the disproportionate location of sources of toxic pollution in or very near communities of color

epistemology the division of philosophy that investigates the nature and origin of knowledge

ethnic enclave a regional or neighborhood location with people of distinct cultural origins

ethnic group a social category of people who share a common culture, such as a common language or dialect, a common religion, and common norms, practices, and customs

ethnography a descriptive account of social life and culture in a particular social system based on observations of what people actually do

ethnomethodology a technique for studying human interaction by deliberately disrupting social norms and observing how individuals attempt to restore normalcy

exchange theory an approach that explains social interaction on the basis of rewards, punishments, and the exchange of valued resources with others

exogamy the practice of selecting mates from outside one's group

export processing zones an area in a developing country where little bureaucratic regulation is enforced on import or exports

false consciousness the idea that subordinated classes internalize the view of the ruling class

family a primary group of people— usually related by ancestry, marriage, or adoption—who form a cooperative economic unit to care for any offspring (and each other) and who are committed to maintaining the group over time

feeling rules situational guidelines for emotional displays appropriate to a specific situation

feminism beliefs, actions, and theories that attempt to bring justice, fairness, and equity to all women, regardless of their race, age, class, sexual orientation, or other characteristics

feminization of poverty the trend whereby a growing proportion of the poor are women and children

field research the process of gathering data in a naturally occurring social setting

folkways the general standards of behavior adhered to by a group

frequency the number of individuals in a single class or category

functionalism a theoretical perspective that interprets each part of society in terms of how it contributes to the stability of the whole society

game stage the stage in childhood when children become capable of taking a multitude of roles at the same time

gemeinschaft German for community; a state characterized by a sense of fellow feeling among the members of a society, including strong personal ties, sturdy primary group memberships, and a sense of personal loyalty to one another; associated with rural life

gender socially learned expectations and behaviors associated with members of each sex

gender identity one's definition of self as a woman or man

gender role the learned expectations associated with being a man or a woman

gendered institution total pattern of gender relationships embedded in social institutions

generalized other the abstract composite of social roles and social expectations

genocide the deliberate and systematic destruction of a racial, political, or cultural group

gesellschaft German for society; a form of social organization characterized by a high division of labor, less prominence of personal ties, the lack of a sense of community among the members, and the absence of a feeling of belonging; associated with urban life

global care chain a social network where a series of personal links between mothers across the globe transfer caring (paid or unpaid) to children other than their own

global economy term used to refer to the fact that all dimensions of the economy now cross national borders

globalization increased economic, political, and social interconnectedness and interdependence among societies in the world

group a collection of individuals who interact and communicate, share goals and norms, and who have a subjective awareness as "we"

hegemony ascendancy or dominance of one social group or culture over another

heterogeneous consisting of dissimilar or diverse ingredients or constituents

heterosexism the institutionalization of heterosexuality as the only socially legitimate sexual orientation

homogamy the pattern by which people select mates with similar social characteristics to their own

homogeneous of the same or similar kind or composition

homophobia the fear and hatred of homosexuality

hypothesis a statement about what one expects to find when one does research

ideology a belief system that tries to explain and justify the status quo

impression management a process by which people control how others perceive them

indicator something that represents an abstract concept

individualism a doctrine that states that the interests of the individual are and ought to be paramount

industrialization term referring to sustained economic growth following the application of raw materials and other more intellectual resources to mechanized production

in-group a group with which one feels a sense of solidarity or community of interests

initiation rite ceremonies, ordeals, or instructions by which one is made a member of a sect or society or is invested with a particular function or status

institution see social institution

institutional privileges institutionalized benefits given to those of the dominant group that seem to "naturally" afford this group greater opportunity

instrumental emotionally neutral, task-oriented (goal-oriented) behavior

interlocking directorate organizational linkages created when the same people sit on the boards of directors of a number of different corporations

issues problems that affect large numbers of people and have their origins in the institutional arrangements and history of a society

kinship system the pattern that defines people's family relationships to one another

labeling theory a theory that interprets the responses (or "label") of others as most significant in determining the behavior of people

labor market the available supply of jobs

lesbian feminism a feminist theoretical perspective and movement philosophy that posits that women's relationship with each other, sexual and otherwise, should be valued and supported

liberal feminism a feminist theoretical perspective asserting that the origin of women's inequality is in barriers blocking women's advancement

life chances   the opportunities that people have in common by virtue of belonging to a particular class

longitudinal study   a research design dealing with change within a specific group over a period of time

looking-glass self   the idea that people's conception of self arises through reflection about their relationship to others

matrilineal kinship system   kinship systems in which family lineage (or ancestry) is traced through the mother

"me"   the social being we see as ourself in any social situation

means of production   the system by which goods are produced and distributed

medicalization of deviance   a social process through which a norm-violating behavior is culturally defined as a disease and is treated as a medical condition

meritocracy   a system (as an educational system) presuming that the most talented are chosen and moved ahead on the basis of their achievement

methodology   term referring to the practices and techniques used to gather, process, and interpret theories about social life

minority group   any distinct group in society that shares common group characteristics and is forced to occupy low status in society because of prejudice and discrimination

miscegenation   the mixing of the races through marriage

model minority   a minority group used as an example to suggest that social mobility is possible for minority groups (ignoring the fact that the presumed model minority has only achieved partial success)

modernization   term referring to the social process by which traditional societies achieve industrial capitalist economies

mores   strict norms that control moral and ethical behavior

mythopoetic men's movement organized group of men inspired by

Robert Blye's philosophies of recapturing a form of masculinity represented in Western mythology to promote a positive male identity that some believe was allegedly lost in the feminist movement

norms   the specific cultural expectations for how to act in a given situation

nuclear family   social unit comprised of a man and a woman living together with their children

objectification   process of treating a person or group of people as inanimate objects, devoid of their humanity

objectivity   the absence of bias in making or interpreting observations

occupational segregation   the pattern by which workers are separated into different occupations on the basis of social characteristics such as race and gender

oppression   systematic, institutionalized mistreatment of one social group by another social group

organization   see social organization

out-group   a group that is distinct from one's own and is usually an object of hostility or dislike

outsourcing   hiring temporary workers outside of a firm to do certain jobs; such temporary workers seldom receive employee benefits

pacifist   a person opposed to war or violence of any kind

paradigm   a framework for understanding a phenomenon based on a particular and identifiable set of assumptions

participant observation   a method whereby the sociologist becomes both a participant in the group being studied and a scientific observer of the group

patriarchy   a society or group where men have power over women

political economy   term referring to the interdependent workings and interests of governing and material wealth production and distribution systems

positivism   a system of thought in which accurate observation and description is considered the highest form of knowledge

postmodernism   a theoretical perspective based on the idea that society is not an objective thing but is found in the words and images—or discourses—that people use to represent and describe behavior and ideas

poverty line   the figure established by the government to indicate the amount of money needed to support the basic needs of a household

power   a person or group's ability to exercise influence and control over others

power elite model   a theoretical model of power positing a strong link between government and business

prejudice   the negative evaluation of a social group, and individuals within that group, based upon conceptions about that social group that are held despite facts that contradict it

presentation of self   Goffman's phrase referring to how people display themselves during social interaction

prestige   the subjective value with which different groups or people are judged

primary group   a group characterized by intimate, face-to-face interaction and relatively long-lasting relationships

proletarianization   (1) the process by which parts of the middle class become effectively absorbed into the working class; (2) the process by which an occupational category is downgraded in occupational status and is more closely akin to working-class jobs

proletarians   the laboring class; especially the class of industrial workers who do not own means of production and hence sell their labor to live

proportion   the ratio of a part to the total

Protestant ethic   belief that hard work and self-denial lead to salvation and success

race   a social category or social construction based on certain characteristics, some biological, that have been assigned social importance in the society

racial formation   process by which groups come to be defined as a "race" through social institutions such as the law and the schools

racial project   an organized effort to interpret and represent resources along particular racial lines

racism   the perception and treatment of a racial or ethnic group, or member of that group, as intellectually, socially, and culturally inferior to the dominant group

radical feminism   feminist theoretical perspective that interprets patriarchy as the primary cause of women's oppression

redlining   practice by which bank officials excluded racial groups from purchasing homes in particular residential areas to preserve the all-white composition of such neighborhoods

reductionism   explaining a complex system's working with a simplistic explanation

religiosity   the intensity and consistency of practice of a person's (or group's) faith

religious socialization   the process by which one learns a particular religious faith

ritual   symbolic activities that express a group's spiritual convictions

role   the expected behavior associated with a given status in society

role conflict   two or more roles associated with contradictory expectations

role negotiation   working out for oneself the expectations of a specific role based on one's own beliefs and the social beliefs associated with the role

role strain   conflicting expectations within the same role

sample   any subset of units from a population that a researcher studies

scientific method   the steps in a research process, including observation, hypothesis testing, analysis of data, and generalization

script   a learned performance of a social role

secular   the ordinary beliefs of daily life that are specifically not religious

seder   a Jewish home or community service including a ceremonial dinner held on the first evening of Passover in commemoration of freedom of Jews from slavery and their exodus from Egypt.

self   the relatively stable set of perceptions of who we are in relation to ourselves, others, and the social system

self-fulfilling prophecy   the process by which merely applying a label changes behavior and thus tends to justify the label

service sector the part of the labor market that makes up the nonmanual, nonagricultural jobs

sex biological identity as male or female

sex segregation the distribution of men and women in any social group or society

sexism a system of practices and beliefs through which women are controlled and exploited because of the significance given to differences between the sexes

sexual orientation the manner in which individuals experience sexual arousal and pleasure

social class the social structural position that groups hold relative to the economic, social, political, and cultural resources of society

social construction of reality the process by which what we perceive as real is given objective meaning through a process of social interaction

social control a process by which groups and individuals within those groups are brought into conformity with dominant social expectations

social facts social patterns that are external to individuals

social identity the role or status that distinguishes a person within a social context

social institution an established and organized system of social behavior with a recognized purpose

social interaction behavior between two or more people that is given meaning

social mobility a person's movement over time from one social class to another

social movement a group that acts with some continuity and organization to promote or resist change in society

social network a set of links of social interaction between individuals or other social units such as groups or organizations

social order the stable social organization of the social world

social organization the order established in social groups

social sanctions mechanisms of social control that enforce norms

social speedup the phenomenon of having more to do in the same amount of time than was once the case

social stratification a relatively fixed hierarchical arrangement in society by which groups have different access to resources, power, and perceived social worth; a system of structured social inequality

social structure the patterns of social relationships and social institutions that comprise society

socialist feminism a feminist theoretical perspective that interprets the origins of women's oppression as lying in the system of capitalism

society a system of social interactions that includes both culture and social organization

socioeconomic status (SES) a measure of class standing, typically indicated by income, occupational prestige, and educational attainment

sociological imagination the ability to see the societal patterns that influence individual and group life

sociology the study of human behavior in society

state the organized system of power and authority in society

statistically significant term describing the relationship between variables that is larger than would be expected by chance alone

status an established position in a social structure that carries with it a degree of prestige

status hierarchy the power and prestige order of the group

steering the practice of some real estate agents by which ethnic and racial minorities are influenced to buy houses that are not located in predominantly white communities in order to preserve the racial makeup of white communities

stereotype an oversimplified set of beliefs about the members of a social group or social stratum that is used to categorize individuals of that group

stigma an attribute that is socially devalued and discredited

subculture the culture of groups whose values and norms of behavior are somewhat different from those of the dominant culture

subjectivity belonging to the thinking subject rather than outside of the subject

symbolic interaction theory a theoretical perspective claiming that people act toward things because of the meaning things have for them

taboo a prohibition imposed by social custom

troubles privately felt problems that come from events or feelings in one individual's life

underemployment a term used to describe being employed at a level below what would be expected, given a person's level of training or education

unemployment rate the percentage of those not working but officially defined as looking for work

urban underclass a grouping of people, largely minority and poor, who live at the absolute bottom of the socioeconomic ladder in urban areas

values the abstract standards in a society or group that define ideal principles

wealth the monetary value of what someone actually owns, calculated by adding all financial assets (stocks, bonds, property, insurance, value of investments, etc.) and subtracting debts; also called net worth

welfare system the public benefit system of a society designed to provide for the needs of those who cannot fully provide for themselves or their families

working class those persons who do not own the means of production and therefore must sell their labor in order to earn a living

working poor employed people whose wages are too low to bring their standard of living above the poverty level

Index

"Abiding Faith" (Chaves and Hagaman),
  277–81
Abusive treatment, of migrant workers, 180
Accumulation of wealth, 143–48
Achieved status, 137–38
Acquaintance rape. *See* Rape culture
Addiction, 49
Adler, Patricia and Peter Adler, "Clique
  Dynamics," 95–102
Adolescence, 15
Advertising, 43–49
  affluent and, 44–45
  alcohol and, 47
  body image and, 41
  children and, 46–47
  color blindness and, 190
  editorial content and, 47–48
  expenditures on, 43
  gays and lesbians and, 45–46
  mass media and, 43–44
  minorities and, 45
  newspaper dating services, 85
  poor and, 44–45
  population culture and, 48–49
  smoking and, 48
  Super Bowl and, 43–44
  women and, 47–48
  See also Internet dating

Advertising Age, 44–47
Advertorials, 47
Aerosols, 359
AFDC, 17
Affluent, advertising and, 44–45
African Americans
  advertising and, 45
  alienation and, 80
  Black churches, 402
  Black colleges, 402
  Black-White model, 178–83
  Civil Rights Movement and,
    397–403
  code for white man, 81
  double consciousness, 176
  economic marginality and, 81–83
  employment and, 79, 81–82
  family and, 81–83
  girls' self-esteem, 291
  incarceration of, 83
  male family members, 83
  males, 203, 205–6
  middle class, 142, 370, 372
  poor children, 80
  residential segregation and, 366–73
  street codes of the inner city, 78–83
  suicide rates, 11
  teen pregnancy, 13–21

African Americans, *continued*
  wealth and, 143–48
  women, 142
African-American women
  mother-daughter relationships, 208–18
    background, 209–10
    community, 214–15
    concrete learnings, 212–13
    double consciousness, 213–14
  See also African Americans
Age effect, 389, 392
Ageism, 70–73
  dependence and, 70–72
  inequalities for women, 71–73
  inequality and, 71
  institutionalization of retirement (Social
    Security), 70–72
  men, 72–73
  old bodies and, 72–73
  "passing" for younger ages, 71
Air Jordan Nike, 174
Alcohol
  advertising and, 47–48
  and college sex, 225, 228
Alcoholics Anonymous, 124
Alcoholism, 121, 124
Alger, Horatio, Jr., 337–38
Allport, Floyd, 225
Alternative Center, 16, 17
Alwin, Duane, "Generations X, Y,
    and Z: Are They Changing America?,"
    387–92
Ambivalence, of college-bound students,
    51–52
American Apartheid (Massey and Denton),
    366–73
American Association of University Women
    (AAUW), *Shortchanging Girls,
    Shortchanging America,* 290–92
American cultural icons, 33–39
American Dilemma, An (Myrdal), 367–68
American Dream, 14, 136, 190
American economy (WWII–2000), 136–37
American Express, 45
Americanization of global media, 171
Anarchy, 394
Andersen, Margaret L., "Social Construction
    of Gender, The," 197–202
Anderson, Elijah, "Code of the Street,"
    78–83
Anholt, Simon, 48
Antabuse, 123
Anti-Americanism, 174
  in Asia, 173
Anti-modern worldview (Jihad), 393

"Arbondale" welfare office, 149
"Are American Jews Vanishing Again?"
    (Goldscheider), 281–86
Arendell, Terry, "Divorce and
    Remarriage," 253–62
Armstrong, Louis, 48
Arnold Communications of Boston,
    43–44
Asceticism, 275
Ascribed status, 137–38
Ashdlaa'o (Navajo), 63
Asian Americans, 181–82
  model minority myth, 182–83
Asian NICs, 156, 160
Asian/Pacific Islanders, 181
Astaire, Fred, 172
Astrology, 78
"Attitudes toward Violence against Women:
    A Cross-Nation Study" (Nayak,
    Byrne, Martin, and Abraham), 271
Aurelius, Marcus, 76
Austin, Regina and Michael Schill, "Black,
    Brown, Red, and Poisoned," 373–79
Avila, Ernestine, 164

Babel, 393–94
Baby Boomers, 392
Backstreet Boys, 48
Balkanization, 393
Banfield, Edward, 369
Barber, Benjamin R., "Jihad vs. McWorld,"
    393–97
"Barbie Doll Culture and the American
    Waistland" (Cunningham), 39–43
Barbie Dolls, 48
  culture and, 39–43
Bars, date rape and, 106
Basketball, globalization of, 170–74
Bathing, 31
Baywatch, 48
BBC, 396
Beauty
  beauty of diversity, 359
  beauty myths and self-worth, 354
  clothing and fashion, 356–57
  conventional, 354
  cosmetic surgery, 357
  cosmetics, 357
  fitness and, 356
  health risks
    mental health, 358
    physical health, 358–59
  quantifying, 354
  social diversity and, 354–55
  weight loss and, 355–56

"Beauty Myths and Realities and Their Impact on Women's Health" (Zones), 353–60

Beer Connoisseur, 44

Behavior
private vs. public, 225
See also Pluralistic ignorance

Bellah, Robert, *Habits of the Heart,* 51

Berdaches (Navajo third gender), 199

Best, Joel, "Promoting Bad Statistics," 27

Bible, 277

Bin Laden, Osama, 39

Biographies, history and, 2–3

Biomedicine, 122–24

Biotechnology, 404

Birmingham, AL, 380–84

Bitsj' yishtlizh (Navajo), 63

"Black, Brown, Red, and Poisoned" (Austin and Schill), 373–79

Black Freedom Struggle, 400

Black liberation, 398
See also Civil Rights Movement

Black Newspaper Network, 47

Black Wealth/White Wealth (Oliver and Shapiro), 147

Black-White model, 178–83

Blaming the victim, 370

Blauner, Robert, "Death and Social Structure," 361–64

Blended families. *See* Remarriage; Stepfamilies

Blue jeans, 172

Boarding schools, 65

Body image, 39–43

"Body Ritual among the Nacirema" (Miner), 28–32

Bonacich, Edna, Lucie Cheng, Norma Chinchilla, Nora Hamilton, and Paul Ong, "Garment Industry in the Restructuring Global Economy, The," 155–62

Boswell, A. Ayres and Joan Z. Spade, "Fraternities and Collegiate Rape Culture: Why Are Some Fraternities More Dangerous Places for Women?," 102–11

Bourgeois, 132–36

Bowling Alone (Putnam), 279–80, 387, 390

Boycotts, 399

Boys, sports and, 74

"Breadwinner father," 241

Break-up patterns (relationships), 234–37

Brown, E. Richard and Hongjian Yu, "Latinos' Access to Employment-Based Health Insurance," 348–53

Brown v. Board of Education, 292–93, 297–99, 368

Bureaucratization, of death, 361–64

Burger King, 36

"Buy This 24–Year-Old and Get All His Friends Absolutely Free" (Kilbourne), 43–49

Capital, vs. culture, 173

Capitalism, 7, 336
global, 170–74
Marxism and, 132–36
"McWorld," 393–97
Protestant ethic and, 272–76

"Captive Labor: America's Prisoners as Corporate Workforce" (Lafer), 331

Careers, vs. jobs, 306–7

"Caring for Our Young: Child Care in Europe and The United States" (Clawson and Gerstel), 263–70

Carothers, Suzanne C., "Catching Sense: Learning from Our Mothers to Be Black and Female," 208–18

Carter, Susan, 16

"Catching Sense: Learning from Our Mothers to Be Black and Female" (Carothers), 208–18

Cathedrals of consumption, 33, 34, 37–38

Catholic Welfare Agency, 186

"Challenges for Middle Eastern Women" (Fernea), 218–23

Channeling, 278

Charms, 29

Charter schools, 301

Chat rooms, 92

Chauffeur image, 232–33

Chaves, Mark and Dianne Hagaman, "Abiding Faith," 277–81

Chemical Waste Management, Inc., 374

Chicago Bulls, 171

Chicago Tribune, 44

Child-care
declining value of, 169
family and, 166–67
fathers and, 169
federal subsidies, 150–51
globalization of, 162–70
See also Nannies

Child support, 256–57

Children
advertising and, 46–47
African Americans, 80
divorce and, 253–62
of divorce and remarriage, 253–62

Children, *continued*
  mothers and work and, 239–45
  raising Jewish, 284–85
Chlorofluorocarbons, 359
Chopping down the cherry tree myth, 28
Christian Science Monitor, The, "When
    Agencies Sleep," 116
Church, Black community and, 402
Church attendance, 279
Cigar Aficionado, 44
Civic engagement, 379–92
Civil Rights Bill (1964), 399
Civil Rights Movement, 136, 389, 397–403
Civil Rights Project, 296
Clark, Kenneth B., *Dark Ghetto,* 368
Class
  college students and, 51–52
  defined, 139
  gender and, 200–201
  old age and, 73
  social stratification and, 139–40
  urban underclass, 366–73
Class struggle, 132–36
Class system, defined, 139
Clawson, Dan and Naomi Gerstel, "Caring
    for Our Young: Child Care in Europe
    and The United States," 263–70
"Clique Dynamics" (Adler and Adler),
    95–102
Cliques in elementary school, 95–102
  exclusion techniques
    compliance, 100
    expulsion, 101
    in-group subjugation, 99–100
    out-group subjugation, 99
    stigmatization, 100–1
  inclusion techniques
    application, 97
    friendship realignment, 97
    ingratiation, 98
    recruitment, 96
Clothing, beauty and, 356–57
CNN (Cable News Network), 171, 396
Coastal Living, 44
Coca-Cola, 47, 172
"Code of the Street" (Anderson), 78–83
Codes, street codes of African Americans,
    78–83
Cohabitation, 254
Cohort effect, 389
Cohort replacement, 388–90
Collateral, 36
Collective consciousness, crime and, 118
Collective identities, Navajo women,
    58–69

Collective solutions, 8
College
  Black colleges, 402
  leaving home for, 51–58
    ambivalence about, 51–52
    change and, 53
    family and, 55–56
    identity changes, 52–55
    independence, 51
    upper-middle class survey
        participants, 51
  pluralistic ignorance and hooking up,
      225–29
  sex and, 225–29
Colonized mentality, 180
Color-blind myth, 189–95
"Color-Blind Privilege: The Social and
    Political Functions of Erasing the
    Color Line in Post Race America"
    (Gallagher), 189–95
Coltrane, Scott, "Family Rituals and the
    Construction of Reality," 245–52
Coming-of-age, 51
Commitment, 235, 236
Commodity chains, 156
Common ground, class and, 139
Communication, breaking up and,
    235–37
"Communist Manifesto, The" (Marx and
    Engels), 132–36
Communities
  African-American women and,
      214–15
  study of 5 minority, 379–85
Community, African-American women
    and, 214–15
Comparative data, 25–26
Concerned Citizens of Choctaw, 377
Concrete learnings, 212–13
Congressional Research Service, 256
Conley, Dalton M., 398
  "Wealth Matters," 143–48
Conrad, Peter and Joseph W. Schneider,
    "Medicalization of Deviance, The,"
    120–26
Conservative theorists, government welfare
    policy and, 370
Conservatives, 14
Consumerism, *See also* Advertising
Consumption, 275
  See also Cathedrals of consumption
Consumption culture, 33–39
Consumption sites, 37–38
  See also Malls; Wal-Mart
Conventional beauty, 354

Conversion
    Jewish, 283–84
    Jews, 283–84
Corporations, 6, 9–10, 333
Cosmetic surgery, 357
Cosmetics, 357
Cosmopolitan, 47
Cost of living, 146
Coyote, 185
Crawford, Cindy, 48
Credit cards, 33, 34, 36–37
Crime
    collective consciousness and, 118–19
    functions of, 117–20
    inner city, 78–80
    medicalization of deviance, 120–26
    necessity of, 119
    system designed to fail, 129–30
Crimes rates, 127
Criminal justice policy, 126–30
Criminality, progression of, 117
Criminals
    designing a "class" of, 127–30
    traditional beliefs about, 129–30
    as "work of poor," 128–29
Critical modernist view, of global care
        chains, 167–69
Critical understandings, 212–13
Cuba, immigration example, 186
Cultural basis of gender, 197
Cultural icons, Barbie, 39–43
Cultural norms, 39
Cultural practices, the Nacirema, 28–32
Culture
    vs. capital, 173
    vs. racism, 370
"Culture of poverty," 369–71
Culture of segregation, 371
Cunningham, Kamy, "Barbie Doll Culture
        and the American Waistland,"
        39–43
Current Population Survey, 313
Custody arrangements, 255–56
Cyberselves, vs. real selves, 89
Cynical performers, 75–76

Dark Ghetto (Clark), 368
Darwinism, 395
Date rape. *See* Rape culture
Dating cites. *See* Internet dating
Davies, Jon, et al., "Identifying Male
        College Students' Perceived Health
        Needs, Barriers to Seeking Help, and
        Recommendations to Help Men
        Adopt Healthier Lifestyles," 365

Day care
    European models of, 266–68
    in France, 263–65, 267
    in the U.S., 263–64, 267–69
Dayton, OH, 380–84
Death
    bureaucratization of, 361–64
    hospitals and, 362–63
"Death and Social Structure" (Blauner),
        361–64
"Debt-for-nature," 378
Decent people, vs. street people, 79–81, 83
Deindustrialization, 81, 137, 155
    See also Restructuring
Democratic party, diversity and power elite
        and, 339
Denmark, day care in, 266–67
Dependence, age and, 70–72
Desegregation, 297–303
    alternative choices
        home schooling, 301–2
        magnet schools, 299–300
        privatization and charter schools, 301
        vouchers, 300–1
    background, 297–99
    See also Residential segregation;
        Segregation
Desipio, Louis and Rodolfo O. de la Garza,
        "Forever Seen as New: Latino
        Participation in American Elections,"
        342–46
Developed countries, 155
Deviance
    changing definitions, 121
    medicalization of, 120–26
Deviance designation, 124–25
Diario Las Americas, 47
Diaz, Vicky, 162, 169
Direct foreign investment, 156
Disability, gender and, 94
Disability Rights movement, 400
Discounters, 37–38
Discrimination, age and, 70–73
Disneyland, 393–94
Displacement theory, 164
"Dissed" (disrespected), 80
Disulfiram (Antabuse), 123
Diverse work teams, 6
Diversity, 6
    beauty and, 354–55
    beauty of, 359
    in networking, 111–15
    power elite and, 337–42
"Diversity in the Power Elite"
        (Zweigenhaft and Domhoff), 337–42

Division of labor, gender and, 141
Division of Labor in Society, The
    (Durkheim), 121
Divisions of labor, family, 248–50
Divorce, 253–62
    child outcomes, 257–58
    custody arrangements, 255–56
    demographic patterns in divorce and
        remarriage, 253–54
    economic support of children, 256–57
    and nonstandard working hours, 316
    postdivorce parenting, 254–55
    remarriage, 258–59
    stepfamilies, 253–54, 258–59
"Divorce and Remarriage" (Arendell),
    253–62
Domestic violence, 150
Domestic workers, globalization of,
    162–70
Donlevy, Jim, "Educational Reforms and
    High-Stakes Testing: Are Public
    Schools Still for the Public?," 304
Double consciousness, 176
    African-American mother-daughter
        relationships and, 213–14
Downey, Robert, Jr., 246
Downsizing, 137
Dr. Spock Generation, 387
Dramaturgical model of interaction, 54
"Drapetomania," 124
Drug offenders, punishing, 131
Drugs, 80
Du Bois, W. E. B., *Souls of Black Folk*,
    176–77, 402
Dualism, 183
Durkheim, Emile
    "Functions of Crime, The," 117–20
    The Division of Labor in Society, 121

Eating disorders, 40
Ebony, 47
Écoles maternelles (French day care), 263
Economic marginality, of African
    Americans, 81–83
Economic Policy Institute, 319
Economy, 5
    American (WWII–2000), 136–37
    power elite and, 332–42
Edin, Kathryn, "Few Good Men: Why
    Poor Women Don't Remarry," 154
Edin, Kathryn and Laura Lein, *Making
    Ends Meet: How Single Mothers
    Survive Welfare and Low-Wage
    Work*, 330
Editorial content, advertising and, 47–48

Education
    color-blindness and, 193
    desegregation, 297–303
    in Middle East, 220
    minorities and, 180
    of Navajo women, 60–61
    social inequality and, 141–42
    See also College
Educational Council for Foreign Medical
    Graduates, 186
"Educational Reforms and High-Stakes
    Testing: Are Public Schools Still for
    the Public?" (Donlevy), 304
Edwords, Fred, "How Biotechnology Is
    Transforming What We Believe In and
    How We Live," 404
Efficiency, restructuring and, 157–59
Egypt the Sphinx, 177
Ehrenreich, Barbara, "Nickel-and-Dimed:
    On (Not) Getting by in America,"
    317–30
Elderly
    care of, 166–67
    See also Ageism
Elementary school cliques. *See* Cliques in
    elementary school
Eligibility criteria, for aid, 146
Eliot, T. S., 15
Elite
    Latino, 345
    power elite, 332–42
Ellington, Duke, 172
Ellis Island, 188
Emelle, AL, 374, 375
Emotional labor, 306
Employment
    African Americans and, 79, 81–82
    diverse networking and, 112–13
    of Navajo women, 60–61
    nonstandard work schedules, 313–17
    service industry, 305–12
    See also Headings under Work
Employment-based health insurance
    (EBHI), Latinos and, 348–53
Enlightenment, 7
Environmental injustice, 373–79
Environmental justice movement, 376–78
Environmental movement, 400
Environmental racism, 373–79
Environmental standards, 161
EPZs. *See* Export processing zones
Erickson, Bonnie, "Social Networks: The
    Value of Variety," 111–15
Erikson, Erik, 389
Escape, men and, 233

Ethiopia the Shadow, 177
Ethnic group, defined, 140
Ethnicity, 140–41
    defined, 140
    networking and, 115
    power elite and, 338
Ethnographic approach, 19
European colonization, 179–80
European models of day care, 263–68
Evangelical Christians, 278
Exclusion
    cliques and, 98–101
    men and, 233
Execretory functions, 31–32
Exploitation, 133–35
Export processing zones (EPZs), 156
Export-led development, 159
    See also Restructuring
Export-led industrialization, 156
Extended families, Navajos, 60

Face mutilation, 30
"Faggot," 230
Fair Housing Act, 368
Family, 9
    African American mother-daughter
        relationships, 208–18
    African Americans and, 81–83
    child-care and, 166–67
    and college-bound students, 55–56
    "culture of poverty" and, 369–71
    divorce and remarriage, 253–62
    effects of nonstandard workweek, 315–16
    extended, 60
    feminism and, 221
    home ownership vs. renting, 143–48
    Jews, 281–86
    men and work, 240–42
    Navajos and, 60–61
    rituals, 245–52
    stepfamilies, 253–54, 258–59
    work and mothering, 239–45
    work roles, 248–50
Family leave, 268
"Family man," 240
Family and Medical Leave Act (FMLA), 268
Family oriented, vs. work oriented, 240–45
"Family Rituals and the Construction of
    Reality" (Coltrane), 245–52
Fanjul family, 340
Farm Worker's movement, 400
Fashion, 36
    beauty and, 356–57
Fashion Play Barbie, 40
Fast food, 172

Fast food industry, 307
Fast food restaurants, 33–36
Fathers
    child-care and, 169
    custody and, 255–56
    divorce and, 145
    mothers' and fathers' working images,
        240–42
    and nonstandard working hours, 315–16
    teen mothers and, 17
Fear, men and, 202–3, 230, 233
Federal child-care subsidies, 150–51
Federal Reserve system, 173
Federal surveys, 23
Feminism, defining masculinity, 232
Feminists, division of labor and, 141
Fernea, Elizabeth, "Challenges for Middle
    Eastern Women," 218–23
Feudalism, 132
"Few Good Men: Why Poor Women
    Don't Remarry" (Edin), 154
Filipina immigrants, 165–66
Fitness, beauty and, 356
Flaming, 87, 91
*Flat Broke With Children: Women in the Age
    of Welfare Reform* (Hays), 149
Fleetwood Enterprises, 340
Flexible-use workers, 308–9
FMLA. *See* Family and Medical Leave Act
Focus groups, 190–91
Food stamps, 146
"Football Versus Barbies: Childhood Play
    Activities as Predictors of Sport
    Participation by Women" (Giuliano,
    Popp, Knight), 74
Forbid/allow experiment, 24–25
Fordist production, 137
"Forest and the Trees, The" (Johnson), 6–13
"Forever Seen as New: Latino Participation
    in American Elections" (Desipio and
    de la Garza), 342–46
Fortune telling, 278
Foster, Jodie, 245
France, day care in, 263–65, 267
Franchises, 35
Frankenberg, Erica and Chungmei Lee,
    "Race in American Public Schools:
    Rapidly Resegregating School
    Districts," 292–96
Franklin, Benjamin, 181
Fraternities, rape and. *See* Rape culture
"Fraternities and Collegiate Rape Culture:
    Why Are Some Fraternities More
    Dangerous Places for Women?"
    (Boswell and Spade), 102–11

Frederick's of Hollywood, 40
Free trade, 157–59
Freud, Anna, 52
Freud, Sigmund, 7
    displacement theory, 164
Front, 76
Fuentes, Thomas, 340
Fugger, Jacob, 273
Functionalist theory, crime, 117–20
"Functions of Crime, The" (Durkheim),
    117–20

Gallagher, Charles A., "Color-Blind
    Privilege: The Social and Political
    Functions of Erasing the Color Line in
    Post Race America," 189–95
Gallup, George, 22
Gap, 38
Garey, Anita, "Weaving Work and
    Motherhood," 239–45
"Garment Industry in the Restructuring
    Global Economy, The" (Bonacich,
    Cheng, Chinchilla, Hamilton, and
    Ong), 155–62
Gay and lesbian movement, 400
Gay marriage, 246
Gays, 206
    advertising and, 45
    Internet dating and, 91
    See also Lesbians
Gender
    ageism and, 70–73
    cultural basis, 197
    date rape and, 106–8
    defined, 198, 200
    disability and, 94
    hierarchical division of labor, 141
    institutional basis of, 199–201
    networking and, 114
    power elite and, 337–42
    power relations, 204
    service industry and, 307–11
    social construction of, 197–98
    social stratification and, 138, 141
    socialization of girls in school, 288–92
    third genders, 199
"Gender and Aging" (Slevin and Calasanti),
    70–73
    See also Ageism
Gender bending, 199
Gender boundaries, 231
Gender norms, 15
Gender police, 231
Gender roles, 200
    Thanksgiving example, 245–52

Gendered, 200
General Social Survey (trust survey), 390–91
Generation X, 387
Generations, 387–92
    changes in social connectedness and trust,
        390–91
    cohort replacement, 389–90
    definitions, 388–89
    explaining social change, 388
    and social change, 392
"Generations X, Y, and Z: Are They
    Changing America?" (Alwin), 387–92
Genetic engineering, 123
"Genius of the Civil Rights Movement:
    Can It Happen Again?" (Morris),
    397–403
Genocide, 396
Gerbner, George, 49
Gerschick, Thomas J., "Toward a Theory
    of Disability and Gender," 94
Gershwin, George, 172
Getty, Ann, 16
Ghetto, 79, 367–72
    out-migration, 370–72
Girls
    socialization in school, 288–92
    sports and, 74
"Girls' Low Self-Esteem Slows Their
    Progress" (San Francisco Examiner), 290
"Girls' Self-Esteem Is Lost on the Way to
    Adolescence" (New York Times), 290
Glasgow, Douglas, 370
Glenn, Evelyn Nakano, 164
Global care chains, 162–70
    critical modernist view, 167–69
    end of, 166–67
    primordialist view, 167–68
    sunshine modernists view, 167–68
"Global cities," 156
Global media, 171
Globalization, 38
    of love, 162–70
    Marxism and, 133
    of media, 171
    of nannies, 162–70
    restructuring and, 155–62
    of sports (basketball), 170–74
God, 277–78
Goddess of the Empty Woman, 42
Goffman, Erving, "Presentation of Self in
    Everyday Life," 75–77
Golden arches, 36
Goldscheider, Calvin, "Are American Jews
    Vanishing Again?," 281–86
Good Road Coalition, 377

Government
  power elite and, 332–42
  trust in, 388, 391
Government welfare policy, conservative
    theorists and, 370
Graham, Billy, 337
Graham, Martha, 172
Grass-roots environmental campaigns, 376–78
Grass-roots social movement, 399–400
Great American Celebration, 338
Great Depression, 338
"Great Divides" (Shapiro), 136–43
Greatest Generation, 387
Greenspan, Alan, 173
Group identities, Navajo women, 58–69
GSX Corporation, 377
Guiliano, Traci, Kathryn Popp, and Jennifer
    Knight, "Football Versus Barbies:
    Childhood Play Activities as Predictors
    of Sport Participation by Women," 74
Guns, 80
Gurin, Patricia Y., Eric L. Dey, and Gerald
    Gurin, "How Does Racial/Ethnic
    Diversity Promote Education?," 196
Gwich'in tribe, 49

Habits of the Heart (Bellah), 51
Harassment, on-line, 87
Hard power, 396
Hardey, Michael, "Life Beyond the Screen:
    Embodiment and Identity through the
    Internet," 84–93
Harlem, 14
Harris, Diane, 16–17
Hathaway Shirt Company, 340
Hays, Sharon
    Flat Broke With Children: Women in the
      Age of Welfare Reform, 149
    "Studying the Quagmire of Welfare
      Reform," 148–53
Head baking, 30
Head Start, 146
Health, diverse networking and, 113–14
Health insurance, Latinos and, 348–53
Health risks, in the quest of beauty, 358–59
Healthcare, college men and, 365
Heaven, 277
Hell, 277
Hemingway, Ernest, 48
Herbalists, 29
Hierarchical social structure, 137
    See also Social stratification
Hierarchy, in power elite, 333
Higgins, Mary, 16
Hijras (Indian third gender), 199

Hines, Revathi, "Silent Voices: 2000
    Presidential Election and the Minority
    Vote in Florida, The," 347
Hispanic Network, 45
Hispanics, 142
    advertising and, 45
    environmental activism, 377
    immigration examples, 184–85, 186
    See also Latinos
Historical events, power elite and, 333–34
History, 1
    biography and, 2–3
    society and, 2
    transformative power of, 3
Hochschild, Arlie Russell
    "Nanny Chain, The," 162–70
    Second Shift, The, 167
    Time Bind, The, 167
Holidays, Thanksgiving and gender roles,
    245–52
Hollywood, advertising and, 48–49
Holy war, 394
"Holy-mouth-men," 29–30
Home for the Holidays (film), 245–50
Home owners, vs. renters, 143–48
Home schooling, 301–2
Homogeneous teams, 6
Homogenization, 397
Homophobia, masculinity as, 230–34
Homosexuality, 199
    See also Gays; Lesbians
Hondagneu-Sotelo, Pierette, 164
Hooking up, 225–29
Hope, Bob, 337
Horatio Alger awards, 337
Hospitals, death and, 362–63
Household work, 212–13
"How Biotechnology Is Transforming
    What We Believe In and How We
    Live" (Edwords), 404
"How Does Racial/Ethnic Diversity
    Promote Education?" (Gurin, Dey,
    and Hurtado), 196
"How to Judge Globalism" (Sen), 175
Hsun Tzu We, 76
Human agency, 400
Human nature, 3
Hunter, Holly, 245
Hyperconsumerism, 36, 38

"Identifying Male College Students'
    Perceived Health Needs, Barriers to
    Seeking Help, and Recommendations
    to Help Men Adopt Healthier
    Lifestyles" (Davies, et al., 365

Identity
    changing in college, 52–55
    masking true self, 76–77
    Navajo women, 58–69
IMF. *See* International Monetary Fund
*Immigrant America: Who They Are and Why
        They Come* (Portes and Rumbaut),
        184–89
Immigration, 184–89
    Cuban example, 186
    domestic workers and, 163–70
    from developing countries, 157–60
    Hispanic example, 184–85
    Indian example, 186–87
    Latino, 343–44
    new vs. old, 184, 187–88
    statistics, 187
    Vietnamese example, 185–86
Imperialism, 395
Impression management, 75–77
Incarceration, of African Americans, 83
Inclusion, cliques and, 96–98
Income disparities, between men and
        women, 71
Independence, leaving for college and, 51
India, *hijras* (third gender), 199
Indians. *See* Native Americans; Navajos
Individual character flaws, 7
Individualism, 6–9, 341, 389
    background of, 7
Individualistic models, 10–13
Individuals, vs. systems, 12–13
Indo-American Cultural Institute, 186
Inequality, 6
    age and, 71–73
    See also Social inequality; Social
        stratification
Inner city, African Americans in, 78–83
Inner-city schools, 82
Insider interviewer, 19–20
Institutional basis of gender, 199–201
Institutional contradiction, 1
Institutional racism, residential segregation
        and, 366–73
Integration, 292–93
    See also Desegregation; Residential
        segregation; Segregation
Interactive service work, 306
Interlocking directorate, 335
Intermarriage, Jews and, 281–86
International economics, 7
International Labor Organization, 166, 169
International Monetary Fund (IMF), 157
International movements, Civil Rights
        Movement as precursor to, 398, 400

International Organization for Migration, 163
Internet advertising, 44, 46
Internet dating, 84–93
    advertising nature of, 86–87
    form and content of sites, 85
    from virtual to copresence, 90
    negotiating relationships, 88–90
    pure relationship concept, 84, 86
    researching, 85–86
        questionnaires and participants, 86
    rules and rituals, 88–89
    vs. real world dating, 84–85
Interracial marriage, 388
Intimidation, on-line, 87
Iron Cages (Takaki), 181
Islam, 287, 395
Issues, 4

James, William, 7
Japanese NICs, 156
Jenkins, Evie, 16–17
Jews
    in America, 281–86
    future of, 285–86
    intermarriage, 281–86
    population, 281
Jihad, 393–97
    defined, 393, 395
"Jihad vs. McWorld" (Barber), 393–97
Jim Crow regime, 398–99
Jobs, vs. careers, 306
Johnson, Allan G., "The Forest and the
        Trees," 6–13
Johnson, Lyndon B., 368
Jordan, Michael, 170–74
Judeo-Christian beliefs, U.S., 277–81
"Just Say No," 14

Kaplan, Elaine Bell, "Not Our Kind of
        Girl," 13–21
Karp, David, Lynda Lytle Holmstrom, and
        Paul S. Gray, "Leaving Home for
        College: Expectations for Selective
        Reconstruction of Self," 51–58
Ken Dolls, 48
Kentucky Fried Chicken, 36
Kerner Commission, 368
Kerner, Otto, 368
Kettleman City, CA, 374
Keynes, John Maynard, 49
Kilbourne, Jean, "Buy This 24–Year-Old
        and Get All His Friends Absolutely
        Free," 43–49
Kimmel, Michael S., "Masculinity as
        Homophobia," 230–34

Kittrie, Nicholas, 123
Knight, Phil, 172
Korean Americans, 142

Labor
    exploitation of (Marxism), 133–35
    globalization of domestic workers, 162–70
    restructuring and, 155–62
    service industry, 305–12
    women in U.S., 166–67
    women's occupations in the Middle East, 219–20
    working mothers, 239–45
Labor market, social networks and, 112–13
Labor unions, 136–37
Lafeber, Walter, "Michael Jordan and the New Global Capitalism," 170–74
Lafer, Gordon, "Captive Labor: America's Prisoners as Corporate Workforce," 331
Laissez-faire racism, 194
Lambert, Tracy A., Arnold S. Kahn, and Kevin J. Apple, "Pluralistic Ignorance and Hooking Up," 225–29
Language, 9
Las Vegas showgirls, 41–42
Latina, 45
Latinas, girls' self-esteem, 291–92
Latinos
    Black-White model and, 178–83
    elites, 345
    employment-based health insurance and, 348–53
    geographic concentration of and voting, 344–45
    health status and access to health care, 348–49
    naturalized citizens, 343–44
    non-naturalized immigrants, 344
    statistics, 342–43
    voting participation, 342–46
    See also Hispanics
"Latinos' Access to Employment-Based Health Insurance" (Brown and Yu), 348–53
Latipso (temple) ceremonies, 30–31
Lazarus, Emma, 188
Leaders, women, 222
"Leaving Home for College: Expectations for Selective Reconstruction of Self" (Karp, Holmstrom, and Gray), 51–58
Leeds, Shana, 16–17
Lesbians, advertising and, 45
Less-developed countries, 155, 157
Level playing field, 190
Lewis, Oscar, 369

Liberal individualism, 341
Liberal theorists, culture of poverty and, 370
Liberation movements, Civil Rights Movement as precursor to, 398, 400
"Life Beyond the Screen: Embodiment and Identity through the Internet" (Hardey), 84–93
Linton, Ralph, 28
"Listener," 31
Literary Digest poll, 22, 24
Little, Dana, 16
"Long Goodbye, The" (Vaughan), 234–37
Look, "The Vanishing American Jew," 281
Los Angeles, Mexican population, 188
Low-wage workers, 317–30, 370
Loyalty, African-American women and, 215

MacDonald, Cameron Lynne and Carmen Sirianni, "Service Society and the Changing Experience of Work, The," 305–12
Macropeace, 397
McDonald's, 33, 35–36, 172
McWorld vs. Jihad, 393–97
Madonna, 40
Magazines, advertising and, 44
Magical beliefs, 28
    See also "Body Ritual among the Nacirema"
Magical potions, 29
Magnet schools, 299–300
Making Ends Meet: How Single Mothers Survive Welfare and Low-Wage Work (Edin and Lein), 330
Male privilege, 202–4
Malls, 33, 36–38, 49
Man, 199
Management, in service sector, 323–24
Managerial work, 306
"Manic reformism," 124
Manufacturing jobs, 305
Marches, 399
Markets, restructuring and, 157–59
Marriage, 4
Martinez, Elizabeth, "Seeing More than Black & White," 178–83
Marx, Karl, 140
Marx, Karl and Engels, Friedrich, "Communist Manifesto, The," 132–36
Masculinities
    costs of, 205
    definitions of, 203
    feminist definition of, 232
    as homophobia, 230–34
    See also Men

"Masculinity as Homophobia" (Kimmel), 230–34
Masochism, 30
Mass marketing, 45
Mass media, advertising and, 44
Massey, Douglas S. and Nancy A. Denton, *American Apartheid,* 366–73
MasterCard, 37
Masturbation, 124
Matching process, 138
Mauer, Marc, "Social Cost of America's Race to Incarcerate, The," 131
Mead, Lawrence, 370, 371
Medicaid, 146
Medical model, 123–25
Medicalization of deviance, 120–26
　examples of, 121
"Medicalization of Deviance, The" (Conrad and Schneider), 120–26
Medicine men, 29–31
Meeks, Loretta F., Wendell A. Meeks, and Claudia A. Warren, "Racial Desegregation: Magnet Schools, Vouchers, Privatization, and Home Schooling," 297–303
MELA. *See* Mothers of East Los Angeles
Melting pot, 58
Men
　age and, 72–73
　differences and inequalities among, 205–6
　families and work, 240–42
　fear and, 202–3, 230, 233
　gay men, 206
　husbands of immigrant domestic workers, 165–66
　labor in Marxism, 134–35
　male privilege, 202–4
　Middle Eastern, 218–19
　networking and, 114
　remarriages, 253–54
　suicide rates, 11
Men's movement, chauffeur image, 232–33
Mental health risks, and quest of beauty, 358
Mentally ill, crime and, 120–26
Messner, Michael A., *Politics of Masculinities, The,* 202–7
Methadone, 123
Metropolis, 4–5
Mexican *ranchero* elite, 181
Mexicans, 181
　in Los Angeles, 188
"Michael Jordan and the New Global Capitalism" (Lafeber), 170–74
Microwars, 396–97

Middle class
　African-American, 142, 211
　white, 142
Middle Eastern women, changing role of, 218–23
Migra, 185
Migrant workers
　abuse of, 180
　Middle Eastern, 219
Military establishment, power elite and, 332–42
Mills, C. Wright
　Power Elite, The, 332–36, 338
　"Sociological Imagination, The," 1–5
Milwaukee Parental Choice Program, 300
Miner, Horace, "Body Ritual among the Nacirema," 28–32
Minorities, segregation study, 293–96
Minority groups, old age and, 73
"Mobilizing Minority Communities: Social Capital and Participation in Urban Neighborhoods" (Portney and Berry), 379–85
Model minority myth, 182–83
Monroe, Marilyn, 40
Moral neutrality, 124
Morris, Aldon, "Genius of the Civil Rights Movement: Can It Happen Again?," 397–403
Morrow, Ronald G. and Diane L. Gill, "Perceptions of Homophobia and Heterosexism in Physical Education," 238
Mos Burger, 35
"Most of Us," 228
Mother-substitute model of child care, 267
Mothers of East Los Angeles (MELA), 377
Mouth rituals, 29–30
Moynihan, Patrick, 369
Ms. Teenage America Pageant, 40
MTV, 46
Muller, Peter O., "Suburban Transformation of the Globalizing American City, The," 386
Multinational chains, 49
Multiple identities, 65
Murdock, George P., 28
Murray, Charles, 370, 371, 372
Muslims, 395
Myrdal, Gunnar, *American Dilemma, An,* 367–68

Nacirema, 28–32
　bathing, 31
　excretory functions, 31

face and head rites, 30
geographically, 28
"holy-mouth-men," 29–30
latipso ceremonies, 30–31
"listener," 31
market economy of, 28
medicine men, 29–31
mouth-rites, 29–30
reproductive functions, 31–32
teeth rites, 30
Nannies, immigrant, 162–70
"Nanny Chain, The" (Hochschild),
    162–70
Nation-state system, 5
National Electoral Studies (NES), 391
National Jewish Population Study
    (2000–01), 281
Native American movement, 400
Native Americans, 181
    environmental activism, 377
    Navajo women survey, 58–69
Naturalized citizens, Latinos, 343–44
Navaho Indians, *berdaches* (third
    gender), 199
"Navajo Women and the Politics of
    Identity" (Schulz), 58–69
Navajos
    generational differences among women,
        58–69
    See also Surveys
Nayak, Madhabika B., Christina A. Byrne,
    Mutsumi K. Martin, and Anna George
    Abraham, "Attitudes toward Violence
    against Women: A Cross-Nation
    Study," 271
Negro Problem, 177
Neighborhood associations, 379–85
Neighborhoods, race and, 381–82
Neo-colonialism, 179–80
NES. *See* National Electoral Studies
Network variety, 111–15
    ethnicity and, 115
    gender and, 114
    health and, 113–14
    jobs and, 112–13
    security industry study, 112–13
Networking. *See* Network variety
NIABY (not in anyone's backyard), 378
"Nickel-and-Dimed: On (Not) Getting by
    in America" (Ehrenreich), 317–30
NICs, 156
Nike, 38, 171–72
NIMBY (not in my backyard), 378
Nirulas, 35
Non-Hispanic Whites, 181

Norms, sex and, 225–26
"Not Our Kind of Girl" (Kaplan), 13–21
Notgnihsaw, 28

O'Connor, Sandra Day, 340
Off-line world, 91
"Off-shore sourcing," 156
Oliver, Melvin, 398
Oliver, Melvin and Thomas Shapiro, *Black
    Wealth/White Wealth,* 147
Omissions, surveys and, 23–24
One-night stands, 225
O'Neill, Eugene, 48
OPEC, 396
Oppositional culture, 82
Oppression, 6–7
    hierarchy of, 178–79
Orenstein, Peggy, "School Girls," 288–92
Organization for Economic Cooperation
    and Development, 266
Organized religion, 279
Orientation model of work and family,
    239–45
Oscars, advertising and, 44

Pa-To-Mac, 28
Palestinian Authority, 221
Pandemonium, 394
Pandoras, 396
Panglossians, 396
"Paranoia with counterrevolutionary
    delusions," 124
Parens patriae, 123
Parenting
    African-Americans and, 210
    custody arrangements, 255–56
    postdivorce, 254–55
    See also Divorce; Remarriage;
        Stepfamilies
Parks, Terry, 16, 19
Parrenas, Rhacel, 162
    Servants of Globalization, 163
Parrenas, Rhacel, 164
Participant-observation, 78
Participation, 9–10
Participatory democracy, neighborhoods
    and, 379–85
Party, 140
Passing, for younger ages, 70–71
Patriarchy, 132, 141, 222
    in workforce, 160
Patterson, Lois, 16
"Peaceful Faith, A Fanatic Few"
    (Woodward), 287
Pentagon, 33

People's laws, 78
"Perceptions of Homophobia and
    Heterosexism in Physical Education"
    (Morrow and Gill), 238
Perelman, Max, 171
Period effect, 389
Periphery workers, 305–12
Personal front, 77
Personal Responsibility Act (1996), 148–53
Philadelphia, MS, 377
Physical appearance, old age and gender,
    72–73
Physical health, and quest of beauty, 358–59
Pinkney, Alphonso, 370
Pluralism, 338
Pluralistic ignorance, 225–29
"Pluralistic Ignorance and Hooking Up"
    (Lambert, Kahn, and Apple), 225–29
Poe, Edgar Allan, 48
Police, 126–27
    inner city Blacks and, 78, 80
Politicians, power elite and, 332–42
Politics, and Middle Eastern women,
    221–22
Politics of exclusion, 233
Politics of Masculinities, The (Messner),
    202–7
Pollution, environmental racism and,
    373–79
Poor
    advertising and, 44–45
    criminals and, 128–29
Popular culture, 393–97
    advertising and, 48–49
    technological innovations and, 172
Population size, 23
Populations, surveys and, 26
Portes, Alejandro and Rubén Rumbaut,
    *Immigrant America: Who They Are and
    Why They Come,* 184–89
Portland, OR, 380–84
Portney, Kent E. and Jeffrey M. Berry,
    "Mobilizing Minority Communities:
    Social Capital and Participation in
    Urban Neighborhoods," 379–85
Poverty, 15
    individualism and, 8
    inner city, 78–83
    residential segregation and, 366–73, 367
Poverty line, 144, 146
    welfare mothers and, 152
Powell, Colin, 191
Power
    men and, 232–33
    Weber's definition, 399

Power elite, 332–42, 915
    diversity and, 337–42
    economy, politics, and military
        components, 332–42
    hierarchy of, 333
Power Elite, The (Mills), 332–36, 338
Powerlessness, men and, 232–33
Pregnancy, African-American teens, 13–21
"Presentation of Self in Everyday Life"
    (Goffman), 75–77
Presser, Harriet B., "Toward a 24–Hour
    Economy," 313–17
Primordialist view, of global care chains,
    167–68
Principle of form-resistant correlations, 25
Prison experience, 127
Prison population, 126, 131
Private gain, 76
Privatization, 301
Privileged groups, 6–7
PRIZM, 46
Probability sampling theory, 22–23
Production, Marxism and, 133–135
Proletarianization, 156, 160
Proletarians, 134–35
"Promoting Bad Statistics" (Best), 27
Property ownership, importance of, 143–48
Proposition, 187, 180
"Props" (proper due), 79
Protestant ethic, 272–76
*Protestant Ethic and the Spirit of Capitalism,
    The* (Weber), 272–76
Proudhon, Pierre Joseph, 144
Psychoactive medications, 123
Psychology, 7
Psychosurgery, 121, 123
Psychotechnology, examples of, 123
Psychotherapy, 8
Public policy, African Americans and, 82
Punishment, revising, 120
Pure relationship concept, 84, 86, 91
Puritan ethic, 274–75
Putnam, Robert, *Bowling Alone,* 279–80,
    387, 390

Quasi-stepparents, 254

Race
    defined, 140
    gender and, 200–1
    inequality and wealth, 143–48
    neighborhoods and, 381–82
    power elite and, 337–42
    and service industry work, 307–11
    social stratification and, 140–41

"Race in American Public Schools: Rapidly Resegregating School Districts" (Frankenberg and Lee), 292–96
Race based privilege, 191
   See also Color-blind myth
Race and ethnicity, social stratification and, 138, 140–41
"Racial Desegregation: Magnet Schools, Vouchers, Privatization, and Home Schooling," (Meeks, Meeks, Warren), 297–303
Racial-formation theory, 140
Racism, 402
   Black-White model, 178–83
   color-blindness, 189–95
   environmental, 373–79
   institutional racism and residential segregation, 366–73
   laissez-faire, 194
   teen pregnancy and, 15
   vs. culture, 370
Racist rejection, 80
Rape culture
   college and, 226
   college statistics, 102
   and fraternities, 102–11
      See also Studies
Rational mind, 7
Rationalism, 273, 275
Raw materials, 133
Reagan, Ronald, 14, 337, 340
Real selves, vs. cyberselves, 89
Recidivism, 127
"Red Oxen," 171
Redlining, 147
Reformation, 274
Reiman, Jeffrey H., "Rich Get Richer and the Poor Get Prison, The," 126–30
Reincarnation, 278
Reiss, Steven, "Why America Loves Reality TV," 50
Relationships, break-up patterns, 234–37
Religion
   church attendance in U.S. and other countries, 279
   participation, 279–80
   in the U.S., 277–81
Religious fanatics, 287
Religious movements, 280
Remarriage, 253–54, 258–59
Renters, vs. home owners, 143–48
Reproductive functions, 31
Research
   on cliques, 95–102
   Internet dating, 85–86

minority communities (5 cities), 379–85
   See also Studies; Surveys
Resegregation, 292–96
Residential segregation, 366–73
   See also Desegregation; Segregation
Respect
   African-American women and, 215–16
   codes of the street and, 78, 79–80
Restructuring, 137
   critical view of, 159–61
   globalization and, 155–62
   positive view of, 157–59
Retirement, institutionalization of (Social Security), 70–72
Revolutionary class, 135
Rice, Condelleeza, 191
"Rich Get Richer and the Poor Get Prison, The" (Reiman), 126–30
Richmond Youth Service Agency, 16
Riots, ghetto, 368
Rituals
   among the Nacirema, 28–32
   family, 245–52
Ritzer, George, "September 11, 2001: Mass Murder and Its Roots in the Symbolism of American Consumer Culture," 33–39
Roediger, David, 194
Romero, Mary, 164
Roosevelt, Theodore, 22
Rosebud Reservation, SD, 377
Rugg, Donald, 24
Rural population, 133
Russkoye Bistro, 35
Rustin, Bayard, 400

Sadism, 30
St. Paul, MN, 380–84
Sakioka family, 340
Sample size, 22–23
Sampling, 22
San Antonio, TX, 380–84
San Francisco Bay Area Air Quality Management District, 359
Sanctions, against welfare mothers, 150–51
Savant, 177
Scholasticism, 274
"School Girls" (Orenstein), 288–92
Schulz, Amy J., "Navajo Women and the Politics of Identity," 58–69
Schuman, Howard, "Sense and Nonsense about Surveys," 21–26
Search engines, Internet dating, 85
Second Shift, The (Hochschild), 167
Secrecy, breaking up and, 235–37

Secularism, 277
"Seeing More than Black & White"
     (Martinez), 178–83
Segregation study, 293–96
     See also Desegregation; Residential
          segregation
Self-consciousness, 176
Self-esteem, socialization of girls in school
     and, 288–92
Self-help, 7
Self-interest, 76
Self-worth, beauty myths and, 354
Selves, 8
Sen, Amartya, "How to Judge Globalism," 175
"Sense and Nonsense about Surveys"
     (Schuman), 21–26
"September 11, 2001: Mass Murder and Its
     Roots in the Symbolism of American
     Consumer Culture" (Ritzer), 33–39
September 11 terrorist attacks, 390
     trust in government and, 391
Servants of Globalization (Parrenas), 163
Service industry, 305–12
     careers vs. jobs, 306–7
     experiment in low-wage workforce, 317–30
     gender, race, and stratification in, 307–11
     growth of occupations in, 315
     service work defined, 306
     statistics, 305–6
"Service Society and the Changing
     Experience of Work, The"
     (MacDonald and Sirianni), 305–12
Service-based economy, 137
Setting, 76–77
Sex, 197, 198
     hooking up (college campuses), 225–29
Sex before marriage, 388
Sexism, 7
Sexual abuse, 150
Sexual orientation, power elite and, 338
Sexuality, 197, 198–99
"Sexuality in the Workplace: Organizational
     Control, Sexual Harassment, and the
     Pursuit of Pleasure" (Williams, Giuffre,
     and Dellinger), 224
Shapiro, Thomas, 398
Shapiro, Thomas M., "Great Divides," 136–43
Shortchanging Girls, Shortchanging America
     (American Association of University
     Women), 290–92
Shrines, 29
"Silent Voices: 2000 Presidential Election
     and the Minority Vote in Florida,
     The" (Hines), 347
Sincerity, 75

Single mothers, 13–21
     poor, 256–57
     welfare reform and, 148–53
Single-parents, African American, 81, 83
Sissies, 230–31
Sit-ins, 399, 400
Slavery, 181, 398
Slevin, Kathleen and Toni Calasanti)
     "Gender and Aging," 70–73
Smalls, De Vonya, 16
Smithsonian World, 41
Smoking, advertising and, 48
Social capital, 379–85
Social changes, 388
Social connectedness, 390–91
Social construction of gender, 197–98
"Social Construction of Gender, The"
     (Andersen), 197–202
Social control, 124
     medicine and, 121–22
"Social Cost of America's Race to
     Incarcerate, The" (Mauer), 131
Social fronts, 75–77
Social inequality. *See* Social stratification
Social isolation, 82
Social movement studies, Black freedom
     struggle's impact on, 400–2
Social networks. *See* Network variety
"Social Networks: The Value of Variety"
     (Erickson), 111–15
Social Security, institutionalization of
     retirement and, 70–72
Social stratification, 136–43
     achieved status, 137–38
     ascribed status, 137–38
     in service industry, 307–11
     in the U.S., 138–39
          class, 139–40
          education, 141–42
          gender, 141
          race and ethnicity, 140–41
"Sociological Imagination, The" (Mills), 1–5
Sociological perspective, 1
Soft power, 396
Souls of Black Folk (Du Bois), 176–77
Southern state power, 399
Sport, 47
Sports
     globalization of, 170–74
     socialization and, 74
Standard of living, 136
Stanford Custody Project, 254–55
Statistics, 21–26
"Stay-at-home mom," 241
Steering, 147

Stepfamilies, 253–54, 258–59
Stereotypes
    Asian Americans (model minority), 182–83
    heterosexual men and, 231
    Latino, 182
    poor Black neighborhoods, 384
    toy stores, 197
Stereotyping, 15
Stigma, cliques and, 100–1
Strata (layers or hierarchy), 137
Stratification. *See* Social stratification
Street codes, African Americans and, 78–83
Street justice, 78
Street people, vs. decent people, 79–81, 83
Street violence, street codes and, 78–80
Structure, of societies, 2
Student health, 365
Studies
    African American mother-daughter
        relationships, 208–18
        background, 209–10
        community context, 214–15
        concrete learnings and critical
            understandings, 212–13
        double consciousness, 213–14
        participants, 211
        the study, 210–16
    Black freedom movement's impact on,
        400–2
    color blindness, 190–95
    fraternities and rape culture, 102–11
        gender relations, 106–8
        method, 104–5
        settings, 105–6
        treatment of women, 108
    pluralistic ignorance and hooking up,
        227–28
    security industry and networking, 112–13
    segregation, 293–96
        data and methods, 294
        findings, 294–95
        research questions, 293–94
    welfare reform, 149–53
    See also Research; Surveys
"Studying the Quagmire of Welfare
    Reform" (Hays), 148–53
Subconscious, 7
Suicide, 10–12
    patterns of, 11–12
    rates for males and females, 11
"Sunbelt" welfare office, 149
Sunshine modernist view, of global care
    chains, 167–68
Super Bowl advertising, 43–44
Supermodel of the Year, 40

Supersized, 36
Supplemental Security Income, 71–72
Supratribal, 59
"Suburban Transformation of the
    Globalizing American City, The"
    (Muller), 386
Surplus labor, 165
"Surplus" love, 164–65
Survey data, 24
Survey-based experiment, 24–25
Survey research, 21–26
Surveys
    comparative data, 25–26
    Current Population Survey, 313
    generational differences among Navajo
        women, 58–69
        conceptual framework, 59–60
        education, 60–61
        employment, 60–61
        family, 60
        historical periods, 61–65
        interview content, 60
        languages spoken, 61
        methods, 60–61
        negotiating Indianness, 61–65
        participants, 60
    of low-wage workers, 325
    omissions, 23–24
    questions, 24
    questions wording problem, 24–25
    sample size, 22–23
    sampling, 22
    Shortchanging Girls, Shortchanging America
        (American Association of University
        Women), 290
    trust (General Social Survey), 390–91
    of upper-middle class college-bound
        students, 51–58
    See also Research; Studies
Symbolism, of September 11 terrorist
    attacks, 33–34, 38–39
Systems, vs. individuals, 12–13
Szasz, Thomas, 122–23

Takaki, Ron, *Iron Cages,* 181
Tax deductions, 144
Teen mothers, 13–21
    researching, 16–21
Teeth rituals, 30
Televisions, number of world-wide, 171
Terminally ill, 363
Terrorism, 174
    symbolism of September 11 terrorist
        attacks, 33–34, 38–39
Thanksgiving family ritual, 245–52

Third World
  "debt-for-nature" swaps, 378
  immigrants from, 184–89
Thomas, Clarence, 340
Time Bind, The (Hochschild), 167
TNCs. *See* Transnational corporations
Tom-girls, 74
Tommy Hilfiger, 38
"Toward a 24–Hour Economy" (Presser),
  313–17
"Toward a Theory of Disability and
  Gender" (Gerschick), 94
Toxic Avengers of El Puente, 377
Toys, socialization and, 74
Transnational corporations (TNCs),
  155–61, 172
Tribal identity, 59
Tribalism, 394
Troubles, 1, 3–4
Truer self, 76
Truly Disadvantaged, The (Wilson), 370
Trust survey (General Social Survey), 390–91
Turner, Ted, 171
Twenty-four hour economy, work and,
  313–17

Underemployment, 144–45, 308–9
Unemployment, 4, 5, 144–45
Unemployment insurance, 145
United Nations, 171
United States
  day care in, 263–64, 267–69
  Jews in, 281–86
  religion in, 277–81
Upward mobility, 337
Urban population, 133
Urban underclass, 366–73
Urban violence, 78–83
Urine test, 320

Values, welfare mothers and, 149
"Vanishing American Jew, The" *(Look)*, 281
Varieties, 3
Vaughan, Diane, "Long Goodbye, The,"
  234–37
Vengeance, 78
Victoria's Secret, 43
Videology, 396
Vietnamese Americans, 185–86
Violence
  code of the street and, 78–80, 82
  manhood and, 231
Virgin/Whore paradox, 40
Virtual relationships. *See* Internet dating
Visa, 37
Voting participation, Latino, 342–46

Voting Rights Act, 345, 347, 399
Vouchers, 300–1

Wachner, Linda, 340
Wage laborers, 133
Waistland (waistsize), 40
Waitressing, 320–30
Wal-Mart, 34, 37–38
Walker, Alice, 211
Warnaco Group, 340
Wars, 4–5, 38
  War on Terrorism, 39
Water Temple, 29
Wattle and daub construction, 29
Wealth, 275, 336, 398
  importance of for families, 143–48
"Wealth Matters" (Conley), 143–48
"Weaving Work and Motherhood"
  (Garey), 239–45
Weber, Max, 140, 361, 399
  Protestant Ethic and the Spirit of Capitalism,
  The, 272–76
Weddings, 251
Weight loss, beauty and, 355–56
Weight Watchers Magazine, 47
Weight-loss industry, 356
Welfare disincentives, 370–71
Welfare mothers, 16–19, 148–53
  See also Welfare reform
Welfare offices, study of after welfare
  reform, 149–53
Welfare recipients, statistics, 150
Welfare reform, 318–19, 330
  cuts in welfare roll due to, 151–52
  effects on recipients, 152
  fallout from, 152
  new requirements, 150
  Personal Responsibility Act (1996), 148–49
  recipients' attitudes about, 151
  sanctions, 150–51
  supportive services, 151
West Harlem, 375
West Side Story, 48
Wheel of Fortune, 42
"When Agencies Sleep" *(The Christian*
  Science Monitor), 116
White Americanism, 176
White ethnic groups, Jews, 281–86
White House, 33
White Supremacy, 178–79
White terrorist organizations, 399
White, Vanna, 42
Whites
  advertising and, 47
  African American code for, 81
  Black-White model, 178–83

color-blindness and, 189–95
middle class, 142
Navajos and, 59, 62, 64
suicide rates, 11
wealth and, 143–48
"Why America Loves Reality TV" (Reiss), 50
Williams, Christine L., Patti A. Giuffre, and Kirsten Dellinger, "Sexuality in the Workplace: Organizational Control, Sexual Harassment, and the Pursuit of Pleasure," 224
Williams, Tennessee, 48
Wilson, Claudia, 16
Wilson, William Julius, 15, 371, 372
    Truly Disadvantaged, The, 370
Woman, 199
Women
    advertising and, 47
    African American mother-daughter relationships, 208–18
    African Americans, 142
    ageism and, 70–73
    Barbie Doll culture, 39–43
    beauty myths and realities, 353–60
        See also Beauty
    changing role of Middle Eastern, 218–23
    corporate "mothers" in workforce, 167
    custody and, 255–56
    Filipina immigrants, 165–66
    fraternities and date rape, 102–11
    generational differences among Navajo women, 58–69
    globalization of domestic workers, 162–70
    labor in Marxism, 134–35
    leaders, 222
    networking and, 114
    orientation model of work and family, 239–45
    power elite and, 337–42
    remarriages, 253–54
    restructuring and, 160
    social stratification and, 141–42
    suicide rates, 11
    unequal pay, 203–4
    in U.S. labor force, 166–67
    welfare reform and, 148–53
    work and, 137
    working women world-wide, 169
Women's movements, 136, 389, 400
    in Middle East, 220–21
Woodward, Kenneth L., "Peaceful Faith, A Fanatic Few," 287
Word of God, 277–78
Work
    mothering and, 239–45
    See also Employment
Work ethic, Protestantism and, 272–76
Work schedules, 313–17
Work-oriented, vs. family-oriented, 240–45
Workforce
    changing, 6
    low-wage workers in, 317–30
    welfare mothers and, 150
Working class, proletarians, 134–35
"Working mothers," 239–45
    oppositional images and, 241–43
Working poor, 152
Working women
    African-American, 211, 216
    See also Women
Workplace, age and, 73
Workweek
    nonstandard, 313–17
        effects on families, 315–16
        origins and causes, 314–16
        prevalence, 313–14
    standard, 313
World Bank, 157, 164
World Trade Center, 33

Xenophobia, 180

Yeats, William Butler, 393

Zero-sum game, motherhood and employment not a, 243–44
Zweigenhaft, Richard and G. William Domhoff, "Diversity in the Power Elite," 337–42